LIQUID METAL SYSTEMS

Material Behavior and Physical Chemistry in Liquid Metal Systems 2

LIQUID METAL SYSTEMS

Material Behavior and Physical Chemistry in Liquid Metal Systems 2

Edited by
H. U. BORGSTEDT

Kernforschungszentrum Karlsruhe
Karlsruhe, Germany

Assistant Editor
GUNTER FREES

Kernforschungszentrum Karlsruhe
Karlsruhe, Germany

SPRINGER SCIENCE+BUSINESS MEDIA, LLC

Library of Congress Cataloging-in-Publication Data

Liquid metal systems : material behavior and physical chemistry in
  liquid metal systems 2 / edited by H.U. Borgstedt ; assistant
  editor, Gunter Frees.
       p.   cm.
    "Proceedings of the Second International Seminar on Liquid Metal
  Systems: Material Behavior and Physical Chemistry in Liquid Metal
  Systems, held March 16-18, 1993, in Karlsruhe, Germany"--T.p. verso.
    Includes bibliographical references and index.
    ISBN 978-1-4613-5814-5   ISBN 978-1-4615-1977-5 (eBook)
    DOI 10.1007/978-1-4615-1977-5
    1. Liquid metals--Congresses. 2. Chemistry, Metallurgic-
  -Congresses.   I. Borgstedt, H. U. (Hans Ulrich), 1930-   .
  II. Frees, Gunter.   III. International Seminar on Liquid Metal
  Systems: Material Behavior and Physical Chemistry in Liquid Metal
  Systems (2nd : 1993 : Karlsruhe, Germany)
  TN689.2.L56 1995
  669'.94--dc20   95-9417
                                                            CIP

Proceedings of the Second International Seminar on Liquid Metal Systems: Material Behavior
and Physical Chemistry in Liquid Metal Systems, held March 16–18, 1993,
in Karlsruhe, Germany

ISBN 978-1-4613-5814-5

© 1995 Springer Science+Business Media New York
Originally published by Plenum Press New York in 1995
Softcover reprint of the hardcover 1st edition 1995

10 9 8 7 6 5 4 3 2 1

PREFACE

Liquid metal technology has been the subject of an impetuous development in the recent decades, mainly due to the application of liquid metals in nuclear techniques. The technological development has been supported by studies of the basic physical-chemical properties of liquid metals: One major concern is the material behaviour in contact with the liquid metals, corrosion and the possible deterioration of metallic and ceramic materials which are in use as constructional or functional materials in such systems.

Since the corrosion is in many cases not only a simple dissolution process, the chemical background of such processes had to be studied. Such studies included the determination of solubilities of metals and non-metals in liquid metals, the measurement of thermodynamic data of dissolved materials and of chemical equilibria. Several formerly unknown chemical compounds are formed in liquid metals and are only stable in this environment.

The research and development devoted to the fission reactor techniques were more or less completed in several countries, further work is in progress in some countries in which the interest in fast breeder reactors arose recently. Even the worldwide program on fusion reactor technology is related to liquid metals, and several laboratories are now contributing to this new technology.

The finalization of a research programme at the Nuclear Centre of Karlsruhe in Germany presented an occasion to organise a meeting in the field of *Material Behaviour and Physical Chemistry in Liquid Metal Systems*. This seminar was the second in a series of such meetings, it followed the previous one which was held in March 1981. The *LIQUID METAL SYSTEMS II* in March 1993 was held under completely different circumstances. In this seminar, several attendants from Eastern countries could participate, and also the researchers from Japan used the occasion to meet the colleagues from East and West. The presentations at the seminar gave an illustration on the worldwide state of knowledge within this field of research, and there is no further conference dedicated to the work on the application of liquid metals.

The topics of the seminar were *Activity Transport in Liquid Metal Systems, Corrosion by Liquid Metals, Influence of Liquid Metals on the Mechanical Properties of Materials, Purification of Liquid Metals and Purity Measurement, Chemical Reactions in Liquid Metals, Physical Chemistry of Solutions in Liquid Metals, Experiments in Relation to New Applications of Liquid Metals*, and *Technical Experiments with Liquid Metals*. The contributions are ordered into these headings.

Several colleagues assisted during the preparation of the seminar, namely Messrs. V. Archipov, M.G. Barker, W.F. Brehm, C. Guminski, Sh. Kano, C. Latgé, C.K. Mathews, H.D. Röhrig and A.W. Thorley as members of the programme committee. Messrs. Guminski and Zagorulko improved some of the contributions from Russia, and Mr. Frees served as the assistant editor. The contributions from different countries were completely retyped and all drawings were redrawn in order to get a uniform appearance of the proceedings. The editor received important help also from Mrs. Z. Peric and H. Glasbrenner.

The collection of the contributions to *Material Behaviour and Physical Chemistry in Liquid Metal Systems - LIQUID METAL SYSTEMS II* may help to conserve the knowledge collected during the four decades between the Second World War and the early nineties in the different areas of liquid metal research in order to preserve them for the time, when energy needs of the human communities demand more intensive uses of nuclear energy from advanced fission reactors and even fusion reactors.

Karlsruhe   H.U. Borgstedt

CONTENTS

LIQUID METAL SYSTEMS

Material Behavior and Physical Chemistry in Liquid Metal Systems 2

TRANSPORT OF RADIOACTIVE MATERIAL IN LIQUID SODIUM

W. F. Brehm

Westinghouse Hanford Company
Richland, Washington, USA

1. INTRODUCTION

This paper describes the development of measurement and control techniques for radioactive materials in liquid sodium systems, as applied to the Fast Flux Test Facility (FFTF). The FFTF is a three-loop, sodium cooled fast reactor located on the U.S. Department of Energy (USDOE) Hanford Site in Washington state. It achieved criticality in 1980, operated at its full rated power of 400 MWt until 1986, operated at 291 MWt from 1986 until 1992, and is now (March 1993) in standby condition. Three primary loops transfer heat from the core to intermediate heat exchangers (IHX), which in turn transfer the heat to the secondary sodium system. The secondary system consists of three independent loops, each connected to an air-dump heat exchanger. Sodium purification is achieved by one cold trap on the primary system and one cold trap on each of the secondary loops. The cold traps are normally on line, but have been taken off line on occasion for repairs. A trap to remove radioactive cesium from primary sodium has been operated since 1987.

The inlet temperature to the fuel bundles is 360 °C. During operation at 400 MWt, the average metal temperature at the top of the fuel pins is about 540 C; during operation at 291 MWt, the average temperature at that location is about 480 °C. At these operating temperatures, the effect of sodium on the structural integrity of fuel cladding, core materials, and reactor piping is negligible. The cold traps operate at about 110 C, corresponding to an oxygen level of less than 0.5 wppm [18]. This low level has been periodically verified by vanadium wire measurements.

Fuel cladding and most in-core materials at FFTF are Type 316 stainless steel. Primary sodium system piping and components are Type 304 and Type 316 stainless steel. Beginning in 1986, a fuel element test program used fuel cladding and duct material made from 12% chromium ferritic stainless steel.

The reactor has been used for a variety of purposes: as a fuels and materials test vehicle for the liquid metal fast breeder reactor (LMFBR) program, testing of metal fuel, safety studies, and testing materials for fusion reactor and space reactor applications. For each of these applications, the potential for radioactive material transport arising from the test program was assessed as part of the experiment or program review.

Radioactive corrosion products are removed from the in-core surfaces by flowing sodium and distributed throughout the rest of the primary system. Fission products are present only if fuel cladding is breached. They circulate throughout the system or are deposited near their release site, depending on their chemical interaction with sodium. Tritium diffuses through

Liquid Metal Systems, Edited by H.U. Borgstedt
and G. Frees, Plenum Press, New York, 1995

fuel and control rod cladding and circulates throughout the sodium systems by circulation and diffusion through component structures.

2. EXPERIMENTAL AND MEASUREMENT METHODS

Experiments directed toward understanding the movement of radioactive species in sodium started in the 1950's. Development of solid-state devices for radioactive material measurements, and of improved methods for sodium sampling and chemical analysis enabled experiments of greater sophistication. These experiments resulted in an understanding of the release and migration of the radioactive species, and led to the development of trapping devices for cesium and manganese.

3. ANALYTICAL MODELLING STUDIES

As soon as a conceptual layout of FFTF piping was available, work began on a predictive calculational model for estimating radioactive material buildup of corrosion products in the primary piping. The model was refined many times between 1968 and 1984. Locations for monitoring nuclide buildup in the FFTF heat transport cells were established early in the design phase of the reactor. The locations chosen were determined by the results of early experimental work.

4. SUMMARY

At FFTF, there has been little operational difficulty resulting from radioactive material transport. The only long-lived nuclides encountered in significant quantities are activation product ^{22}Na, corrosion product ^{54}Mn, fission products ^{134}Cs and ^{137}Cs, and ^{3}H (tritium). We believe that the factors important to radioactive material transport and ways to minimize and control it are well understood as a result of more than twenty years of worldwide laboratory and in-reactor experience. A significant contributing factor to this favorable result is the effort expended at FFTF during the design and by the operations program to provide some means to characterize radioactive material transport and attempt to minimize it. This effort has generated considerable practical knowledge regarding sodium and cover-gas sampling, radioactivity level measurements, and reactor design. The FFTF experience has taught us three important lessons:
- Adequate sampling capability and the ability to characterize and mitigate radioactive material transport must be provided in the design; these are very difficult to add later.
- Cover-gas spaces and cells containing cover-gas piping need to be designed to mitigate the effects of radioactive material transport beyond the effects of fission-gas incursion from breached fuel pins.
- The operational philosophy should emphasize minimizing radioactive material transport.

Corrosion product transport has resulted in radiation fields of several hundred mR/h in the primary system equipment cells, with very little corrosion product radioactivity in the sodium itself. These values are reasonably close to the predicted values, and the reasons for the discrepancy are well understood. "Nuclide traps" for removing radioactive manganese from sodium were tested, first in Experimental Breeder Reactor-II (EBR-II) and in FFTF, and performed successfully. Nuclide traps have not been incorporated into the core design, however. Fission-product transport has been limited almost exclusively to the two cesium isotopes, with cesium being released into the sodium between 1984 and 1989 during breached-cladding events. The cesium accumulated in the primary cold trap and also in the cesium trap after the cesium trap was placed in operation in 1987. Some cesium, transported into cover gas spaces, remains deposited there. Tritium release from FFTF into the environment has been minimal, far below regulatory requirements. Most of the tritium precipitates in the primary cold trap, with some migration into the secondary system and secondary cold traps. The amount of tritium migration through the IHXs into the secondary system agrees well with predictions. The amount of ^{22}Na in the sodium is about equal to predicted values.

5. DISCUSSION

5.1 Corrosion Product Transport and Control

The process of characterizing radioactive corrosion product transport has been extensively published [7]. Release and transport of radioactive corrosion products around a reactor primary sodium circuit is inevitable if operation goals are to be met.

Problems from ^{60}Co release can be minimized by specifying low cobalt content in materials exposed to the neutron flux. At FFTF, the cobalt content of core materials was restricted to 500 wppm, and actual fuel cladding was supplied at about 200 wppm cobalt. [Cobalt is an impurity in nickel]. Niobium and tantalum nuclides can also be avoided by restricting use of steels containing these elements and prohibiting tantalum- and niobium-containing alloys from contact with primary system sodium.

Conversely, the fact that cobalt is not readily soluble in sodium nor readily transported around a sodium circuit makes it possible to use cobalt-base bearing alloys in centrifugal pumps. The absence of large amounts of ^{60}Co deposition in FFTF or any other sodium-cooled reactor verifies that statement. However, ^{60}Co has been observed in the water from cleaning fuel assemblies at the Interim Examination and Maintenance Cell at FFTF. Its presence shows that some ^{60}Co is being produced in FFTF, but not transported. Cobalt-58, produced by (n, p) reactions with nickel and not by nuclear reactions with cobalt, has not been seen in any significant quantity in the primary circuit of FFTF or in any sodium-cooled reactor.

The amount of radioactive material release is not significantly changed by increasing the oxygen level in the sodium from 0.5 to 2.5 wppm nor by decreasing the oxygen level below 0.01 wppm (e. g., by hot trapping). Therefore, although purification measures that produce lower oxygen levels than those attained by cold trapping are not required for radioactive material transport control, they would not be deleterious if required for other reasons.

Most nations have developed models to predict buildup of radioactive species in the primary circuit [7]. These models show varying degrees of success when their predictions are compared to actual measurements. The FFTF predictions have been the subject of ongoing calculation and measurement for more than twenty years [3,8,12,14]. Figure 1 shows the buildup of corrosion product radioactivity in FFTF through 1988. Radiation levels since then have continued a slow steady downward trend. This trend is to be expected, as the reduced power level and temperature of the reactor means that less radioactive material is being transported to the piping than is decaying. Had the reactor power level and temperature remained constant, we would expect the radiation levels to remain constant. Testing fuel elements with ferritic steel cladding apparently does not significantly increase ^{54}Mn release. We know that the release of ^{54}Mn from ferritic steel by diffusion is significantly greater than such release from austenitic steel at the same temperature, because the diffusion coefficient is higher in the less closely packed ferritic structure [5]. The reduced temperature and power level and the relatively small number of fuel elements being tested probably offset the higher potential for ^{54}Mn release.

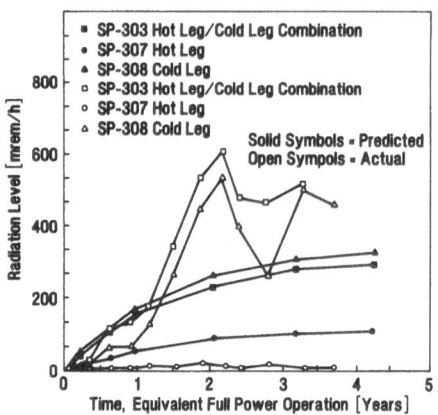

Fig. 1. FFTF Corrosion Product Reaction Level Buildup

The radiation level and gamma spectra measurements in the primary heat transport system cells are valuable, not only to calibrate a model, but more important, to provide information regarding radiation levels that must be taken into account if work in the vicinity of the primary system is necessary. For example, the measurements were used to develop procedures for entering a cell to repair instrumentation.

The discrepancy between calculated and measured radiation levels can be attributed to these factors:

- longer irradiation times in practice than in the model, meaning that corroding material has higher concentration of radionuclides;
- an incorrect temperature employed in the calculations (in fact, an unnecessary temperature correction; if we had not adjusted the model, the calculations would have been more accurate);
- lack of any ^{60}Co transport, which constituted a significant fraction of the predicted hot leg radiation level, to the primary circuit piping;
- lack of significant ^{54}Mn deposition in the hot leg as predicted by the model. For deposition predictions, the FFTF model relied almost exclusively on empirical assessments from results in our test loops. Incorporation of some information from mass transfer theory perhaps would have eliminated some of the discrepancy between calculated and measured values.

The collimated tubes through which the gamma spectra were obtained in the FFTF radionuclide buildup measurements were incorporated into the original reactor design. The radiation level measurements with the ion chamber were obtained through periscope holes also incorporated into initial design. This arrangement allowed calibration of instruments and calculation of conversion factors. It would have been very difficult, if not impossible, to attempt such measurements without the surveillance holes or to modify the reactor once nuclear operation began.

Deposited radioactivity diffuses into the material on which it is deposited. The diffusion penetration has been analyzed successfully by means of grain-boundary diffusion theory and measurement techniques developed for the study of radioactive corrosion product transport [13]. Typical penetration is 25-50 μm over several thousand hours. Because of the penetration, removing adherent sodium will remove only part of the deposited radioactivity. A decontamination process based on weak organic acids proved useful at removing deposited radioactivity from hot leg locations of experimental sodium systems [13]. A modification of that process showed promise at removing ^{54}Mn in nickel-manganese rich nodules characteristic of cold leg deposits. Thus far, a decontamination process has not been required for FFTF components.

The strong tendency of nickel and manganese to accumulate together in sodium systems [6] led investigators to develop devices to trap ^{54}Mn before it could reach the primary circuit [16]. Rolled sheet, fabricated from low-cobalt nickel, was used in a location above the fuel pins to trap radioactive manganese. Extensive hydraulic testing of the device and loop testing of models of the device were conducted before reactor irradiation. Six EBR-II tests showed the traps to function satisfactorily from both a hydraulic and safety standpoint and to be very effective for removing ^{54}Mn from sodium. An FFTF trap test, removed from the reactor after uneventful completion of its scheduled three cycle irradiation, has not been examined in detail. The EBR-II tests, supplemented by work done in test loops, showed that "full-flow" trapping was not required for effective removal of radioactive manganese. This result in turn means that not all fuel assemblies in a reactor need to be fitted with traps. A major advantage of incorporating the trap into the fuel assembly is that the trap can be removed with equipment already a part of the design, at intervals likely to maintain unsaturated nickel surfaces for trapping.

As part of the chemistry evaluation program at FFTF, overflow cup samples were withdrawn at regular intervals [4]. The only corrosion product ever seen was ^{54}Mn, and then in the nanocuries per gram range. We observed a significant difference, up to a factor of ten, between the ^{54}Mn levels in sodium collected in tantalum cups and those in sodium collected in stainless steel cups. This result, observed repeatedly, was not entirely unexpected. In early scoping work for trap material selection, tantalum was one of the materials shown to retard ^{54}Mn deposition. There was no significant difference between cesium isotope concentrations in sodium collected in the two different cup materials.

Because FFTF sodium was produced in a plant specifically built for FFTF sodium and because there was a rigorous cleanliness and material control program in effect during FFTF construction and startup, other nuclides such as ^{65}Zn have not been observed at FFTF. An occasional observation of ^{86}Rb, ^{124}Sb, or ^{110}Ag has been made, but not at high concentrations or at any regular basis. The ^{22}Na level in sodium is about equal to that calculated [3]. In pipes filled with sodium, radiation levels from ^{22}Na range from 100 to 500 mR/h, depending on location within the heat-transport system cell. The ^{22}Na radiation level from adherent sodium if the pipes were drained has been estimated at less than 10 mR/h.

5.2 Fission Product Transport and Control

For many years we have known that cesium isotopes are the most likely source of fission-product contamination because they tend to be readily released from fuel, they are soluble in sodium, and they tend not to form oxides in sodium or cover gas. [Cesium oxide is much less thermodynamically stable than other alkali metal, alkaline earth, or rare earth oxides.] We did not recognize the tendency for cesium to evaporate from the surface of the primary sodium pool and migrate to cover-gas spaces and cover-gas system piping. Several papers at the 1987 IAEA Specialists' meeting [7] noted the high radiation levels in cover-gas spaces from cesium deposition there. We have also known for some time that cesium tends to deposit reversibly on low temperature stainless steel surfaces from physical adsorption and irreversibly on carbonaceous surfaces from formation of a cesium-carbon compound. The practical significance of this knowledge is that cesium will deposit in cold traps in the primary sodium circuit, creating a problem for component disposal. Cesium can be made to deposit preferentially in traps filled with carbonaceous materials. Aggressive run-beyond-cladding-breach programs at EBR II and the BOR 60 reactor in Russia led to the placement of trapping devices in these reactors [10,17]. A series of cladding breaches and the discovery of high radiation levels in FFTF cover-gas piping cells led to the installation of a cesium trap in FFTF, in a closed-loop cold trap circuit cross-connected to the sodium characterization line outside containment [1].

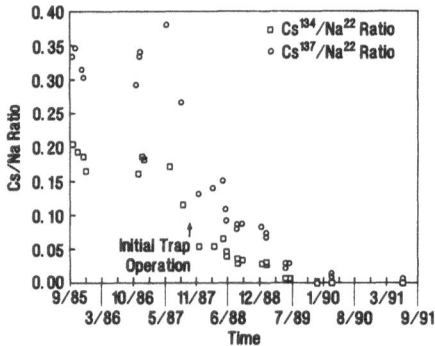

Fig. 2. Cesium Activity in FFTF Primary Sodium

The carbonaceous material used in the trap is reticulated vitreous carbon (RVC). The trap worked very well; Figure 2 shows a 20-fold decrease in cesium concentration in sodium after the trap was placed on line. However, the trap cannot be operated when the reactor is at power or when the main motors on the primary pumps are operating, because dissolved and entrained argon from the reactor cover gas impedes flow through the trap filter (integral to the trap and located downstream from the RVC elements), which unfortunately is probably too small (25 μm nominal pore size). A series of experiments conducted in 1989 attempted to remedy this situation; no remedy was found [2]. When the reactor was in refueling mode, at low temperature (200 °C) and with the primary pump main motors turned off, the gas blockage cleared and the trap resumed effective removal of cesium from the primary sodium. We estimate that the cesium trap, with a flow rate of about 1200 kg/hr, removes about 1% of the cesium inventory from the 450,000 kg of FFTF primary sodium per day. From the time before cesium trap operation until the cesium level had been substantially reduced, cesium isotopes were observed in the gamma spectra obtained when the piping system was monitored for radionuclide buildup.

An important point to be made here (other than the obvious one that a filter and flow configuration to avoid gas blockage should be selected) is that future reactors should incorporate designs for cesium traps into the original reactor layout. Adding the cesium trap to FFTF piping required considerable time and expense because of the need to work on radioactive, sodium-containing systems.

We can also emphasize the need to place systems for removal of cesium in cover-gas lines. Its low melting point (28 °C) and lack of oxidation in the presence of sodium aerosol in cover gas cause cesium to pass through conventional vapor traps and sodium aerosol filters and into cover-gas lines. A compressor replacement job at FFTF, with the compressor located 30 m from the reactor vessel, was made much more difficult because of the need to treat the job as one with a high potential for smearable contamination. Conventional high-efficiency particulate air filters have proven effective against cesium transport in gas lines.

The conservative fuel management scheme at FFTF, with a low shutdown limit for the delayed neutron signal and a policy of removing fuel assemblies with cladding breaches as soon as they are located, has precluded any introduction of nonvolatile fission products into the primary circuit or the sodium. Uranium and plutonium in the sodium samples have always been near or below the detection limit. Cup samples were always obtained after breached-pin events.

5.3 Tritium Transport and Control

Tritium produced in reactor fuel and control rods diffuses readily through the stainless steel cladding and enters the sodium. It is removed in primary cold traps, diffuses through the intermediate heat exchanger, enters the secondary sodium, and precipitates in the secondary cold traps. The oxide coating on the atmosphere side of the stainless steel piping and components provides a barrier to rapid release through the reactor piping. This barrier, combined with the other active sinks in the sodium system, minimizes tritium release from a sodium-cooled reactor. The tritium content of FFTF sodium, both primary and secondary, and of cover gas has been monitored extensively since reactor startup. Other monitors located outside the reactor building have shown the tritium releases to the environment are always well below regulatory limits. These results agree with published analyses of tritium behavior in sodium-cooled reactors [9,11,15].

The tritium content of the primary sodium in FFTF is now 100 to 200 nCi/gram. The secondary sodium contains about 30 nCi/gram. The ratio of the tritium level in the primary to that in the secondary is higher in FFTF than in EBR II, because the higher temperature in FFTF produces a higher diffusion coefficient of the tritium in stainless steel in the intermediate heat exchangers. The tritium content in primary cover gas is about 0.1 nCi/cm^3. It was measured more frequently during 1990-1991 when tritium-producing experiments, in support of a fusion energy program, were being irradiated. The experiments included a tritium removal and handling system that collected the gas for analysis and disposal. The system was very effective, as tritium levels did not increase markedly during that time.

Between December 1990 and February 1991 [4], higher than normal tritium, hydrocarbon, and hydrogen levels were noted in FFTF secondary loop 3 cover-gas samples. Also noted was an increased differential pressure across the cold trap. The gas sampling frequency was therefore increased during this time. We hypothesize that the cause of these phenomena was oil inleakage in December 1990 during an outage, followed by slow thermal decomposition during reactor startup and operation in the following months. The tritium increase in the cover gas is hypothesized to be the result of tritium atoms in the sodium exchanging with hydrogen atoms in the (assumed) oil, followed by thermal decomposition of the oil and subsequent entry of HT, T_2, and tritiated hydrocarbon molecules into the gas. The abnormal cover-gas behavior and higher differential pressure are the only indications of possible oil inleakage. The abnormal conditions returned to normal in March 1991.

In regard to tritium behavior in FFTF, problems were neither anticipated nor experienced. The apparent effectiveness of the cold traps at removing tritium suggests that they should be operated continuously even when not otherwise required.

6. CONCLUSIONS

The FFTF experience with radioactive material transport in the primary sodium circuit has been one of successful control. Emphasis on cleanliness of both reactor components and sodium, a strong sampling program, and conservative fuel management all contributed to the success. The addition of the cesium trap proved effective and would have been less expensive and time consuming if we had been able to install it before reactor startup.

ACKNOWLEDGMENTS

The author thanks J. B. Waldo, B. E. Sartoris, and the staff of Westinghouse Hanford Company Publications Services, Graphics Services, and Information/Word Processing for review and editing assistance. This work was sponsored by the United States Department of Energy under Contract Number DE-AC06-87RL10930.

REFERENCES

1. R.A Bechtold, and C. E. Grenard, *Proceedings of Fourth International Conference on Liquid Metal Engineering and Technology*, p. 609-1, Societe Francaise d'Energie Nucleaire, Paris, 1988.

2. W.F. Brehm, C. E. Grenard, and W. J. Schuck, *Transactions ANS*, 61, (1990) 333.

3. W.F. Brehm, and R. L. Simons, *Nuclear Technology*, 95, (1991) 148.

4. W.F. Brehm, W. J. Schuck, C. E. Grenard, R. A. Burk, and D. F. Hicks, USDOE report WHC-SA-1047-FP,and *International Conference on Fast Reactors and Related Fuel Cycles*, vol 1, p. 10.1-1, Kyoto, Japan, 1991.

5. W.F. Brehm, *Diffusion Processes in Nuclear Materials*, ed. R. P. Agarwala, p. 323; North-Holland, Elsevier Science Publishers B. V., Amsterdam, Netherlands, 1992.

6. R.P. Colburn, *International Atomic Energy Agency, International Working Group on Fast Reactors, Specialists Meeting on Sodium Removal and Decontamination Summary Report*, report IWGFR/23, 1978 p. 157.

7. IAEA, *International Atomic Energy Agency, International Working Group on Fast Reactors, Specialists Meeting on Fission and Corrosion Products in Primary Circuits of LMFBRs, Summary Report*, IWGFR/7, International Atomic Energy Agency, Vienna, Austria, 1975; IAEA, *Fission and Corrosion Products Behavior in Primary Circuits of LMFBRs, Proceedings of an IAEA Specialists Meeting, International Working Group on Fast Reactors*, report KfK-4279 (IWGFR/64), Kernforschungzentrum Karlsruhe, Germany, 1987.

8. T.J. Kabele, W. F. Brehm, and D. R. Marr, USDOE report HEDL-TME-72-71, Westinghouse Hanford Company, Richland, Washington, 1972.

9. T.J. Kabele, USDOE report HEDL-TME-74-6, Westinghouse Hanford Company, Richland, Washington, 1974.

10. N.V. Krasnoyarov, V. I. Polyakov, A. M. Sobolev, and E. K. Yashkin, *Proceedings of Third International Conference on Liquid Metal Engineering and Technology* Vol. 3, pp. 185-190, British Nuclear Energy Society, London 1984.

11. R. Kumar, USDOE report ANL-8089, Argonne National Laboratory, Argonne, Illinois, 1974.

12. W.L. Kuhn, in *Proceedings of International Conference on Liquid Metal Technology in Energy Production*, USDOE report CONF-760503-P1, p. 280, United States Energy Research and Development Administration (now   USDOE, Washington, D.C., 1976.

13. J.M. Lutton, H. P. Maffei, R. P. Colburn, W. F. Brehm, R. P. Anantatmula, R. A. Bechtold, M. B. Hall, and E. F. Hill, *Proceedings of Third International Conference on Liquid Metal Engineering and Technology* Vol. 2, p. 133, British Nuclear Energy Society, London 1984.

14. H.P. Maffei, W. F. Brehm, F. S. Moore, W. P. Stinson, W. L. Bunch, R.P. Anantatmula, and J. C. McGuire, *Conference on Liquid Metal Engineering and Technology* Vol. 1, p. 493, British Nuclear Energy Society, London, 1984.

15. J.C. McGuire, and T. Renner, *Atomic Energy Review*, 16, No. 4, (1978) 657.

16. J.C. McGuire, and W. F. Brehm, *Nuclear Technology*, 48, (1980). 101.

17. W.H. Olson, and W. E. Ruther, *Nuclear Technology*, 46, (1979). 318.

18. D.L. Smith, and R. H. Lee, USDOE report ANL-7891, Argonne National Laboratory, Argonne, Illinois, 1972.

TRANSPORT OF RADIOACTIVE CORROSION PRODUCTS IN PRIMARY SYSTEMS OF A SODIUM COOLED FAST REACTOR

K. Iizawa, K. Chatani, K. Ito, S. Suzuki,
M. Akutsu and K. Kinjo

Reactor Technology Section, Experimental Reactor Division,
O-arai Engineering Center
Power Reactor and Nuclear Fuel Development Corporation
JAPAN

1. INTRODUCTION

In order to elucidate the behaviour of radioactive corrosion products (CPs) in sodium cooling systems of fast reactor the radioactive deposition and distribution within the primary system have been measured and evaluated using a model for CP transport at JOYO, the Japanese experimental fast reactor. The main purpose of the studies has been to try and establish the calculational model for predicting the release and deposition and distribution of CPs in fast reactor sodium cooling systems. This paper reports the findings from these measurements and evaluations, and highlight the feature in CP transport behaviour which have to be taken into consideration to update the model.

2. EXPERIMENTAL

2.1 Radioactive CP Measurements in Primary Circuits of JOYO

In April 1977 the sodium cooled experimental fast reactor JOYO achieved the initial critically with the initial stages in Mk-I core (the fast breeder core). After 27.90 GWd operation, the core was changed to the current Mk-II core (the fast flux irradiation bed with stainless steel reflectors) in November 1982. The nominal power of the JOYO Mk-II core is 100 MWt. Each duty cycle consists of 32 to 70 days full power operation and 18 to 41 days shut down for re-fuelling as shown in Fig. 1.

By the half of 1991 the 23 duty cycles with Mk-II core had been completed with achievement of the cumulative reactor output of 150.6 GWd. Various irradiation tests to develop FBR fuels and core and structural materials were conducted during these periods, and radioactive CP deposition and distributions on the primary circuits were measured [1] in every 7 times of annual inspections since the 3rd one at the end of Mk-I core.

There are two (A and B) sodium main circuits and one overflow /purification circuit involving a cold trap in the JOYO primary system. Approximately 120 tons sodium is circulated in them with constant total flow rate (about 2200 t/h). Reactor inlet and outlet temperature are respectively 370 °C at full power. The primary sodium coolant has been purified by a cold trap at nominal temperature of 120 °C or 130 °C, except few periods with the cold trap operation at higher temperatures caused from its plugging by impurities in

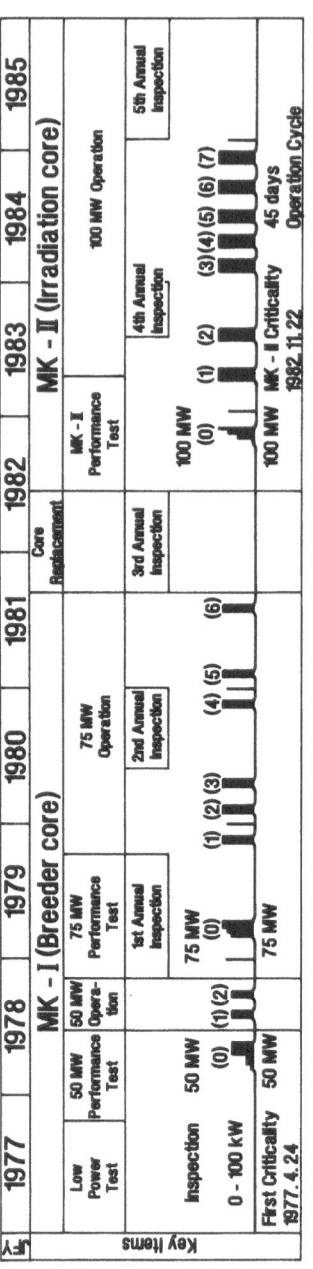
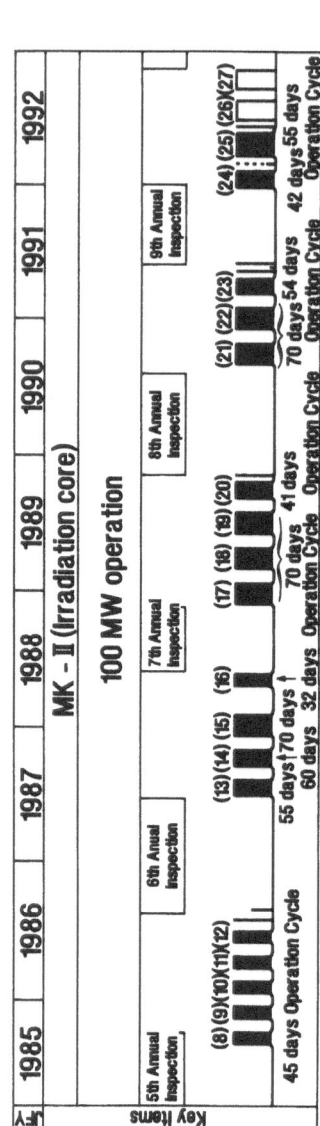

Fig. 1. Operation history of JOYO

10

sodium. In order to facilitate the identification of different regions of primary main piping, they have been designated as follows:

Hot-leg (HL) is located from the outlet of reactor vessel to inlet of IHX,

Cold-leg (1) (CL/1) from the outlet of IHX to inlet of main pump and

Cold-leg (2) (CL/2) from the outlet of pump to inlet of reactor vessel.

The sodium temperatures are respectively 500, 370 and 370 °C at the HL, CL/1 and CL/2 regions.

The sodium velocities are respectively 1.9, 2.3 and 4.8 m/s along nominal piping diameters of 20, 18 and 12 inch.

During annual plant inspections, to prevent hazard of radiation from long-lived ^{22}Na, all primary sodium in the loops except the reactor vessel is drained into a sodium tank. Under these conditions the special dose rate distribution is dominated by the radioactive CPs which have deposited on inner surfaces of primary piping and components. In every annual inspection, radioactive CP deposits on inner surfaces of primary main piping were measured at the 14 locations shown in Fig. 2, using a Ge solid state detector system. The detector system employed in these CP measurements was calibrated by the use of a piping mock-up with two planar type standard gamma sources, ^{54}Mn and ^{60}Co, so that absolute amounts of CP deposits could be obtained from the counting rate.

Gamma dose rate distributions near the primary main piping, IHX and pump were measured by using calcium sulphate (CaSO$_4$) thermo-luminescent dosimeters (TLDs). The distribution around the piping was measured in detail for 93 locations, at one meter intervals along the loop (A) from the outlet to inlet of the reactor vessel. For each location TLDs were placed every 90 degrees around the thermal insulator cover. The distributions around IHX (A) and (B) or pump (A) and (B) were measured in more detail for respective about 600 or 700 points, at 200 or 100 mm intervals in vertical. For each location TLDs were placed every radial 45 degrees around the thermal insulator cover.

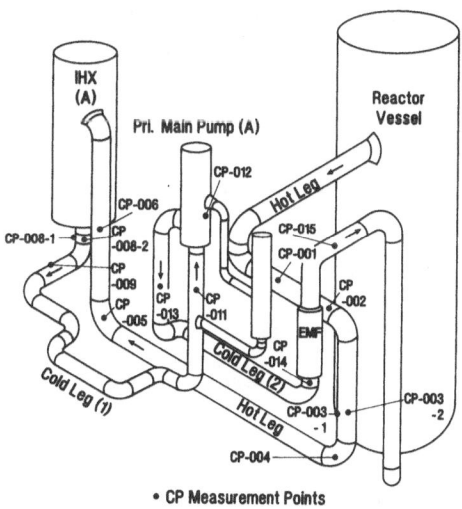

Fig. 2. Measurement points for CP deposits

2.2 Radioactive CP Measurements in Sodium Removal and Liquid Waste Treatment Facility

Some findings of interest about the radioactive CPs deposited on the surfaces of fuel subassemblies or reflectors and pump intervals have been obtained from the radioactivity measurements for the liquid waste in the JOYO facility, produced during sodium removal from them using steam and demineralised water. The liquid samples containing CPs taken from the liquid waste tank for neutralisation were measured by the use of Ge solid state detector. These amounts at a subassembly were evaluated from the least square analysis for the measurement results, involving 10 batches of sodium cleaning for 40 driver fuel subassemblies and 10 outer radial reflectors.

3. RESULTS

3.1 Radioactive CP Measurements and Evaluation in Primary Sodium Circuits

In the CP deposits measurement in the primary sodium circuits only ^{54}Mn and ^{60}Co could be significantly detected. Other radionuclides expected in the deposits i.e. ^{58}Co, ^{51}Cr, ^{59}Fe and ^{22}Na, could be found in a very small amounts or not at all. It can be concluded that the gamma dose rate around the primary circuit piping and components after draining the sodium is determined by these two CP radionuclides during the maintenance operation of plant. The amounts of radioactive CP deposits averaged within each leg are tabulated in Table 1, along with cumulative reactor output. ^{54}Mn is the most dominant radionuclide in the out-of-reactor primary sodium system and it is about 12 to 4 times greater than ^{60}Co for the HL, the magnification decreasing along the reactor operation, while about 16 times greater within the CLs throughout the operation.

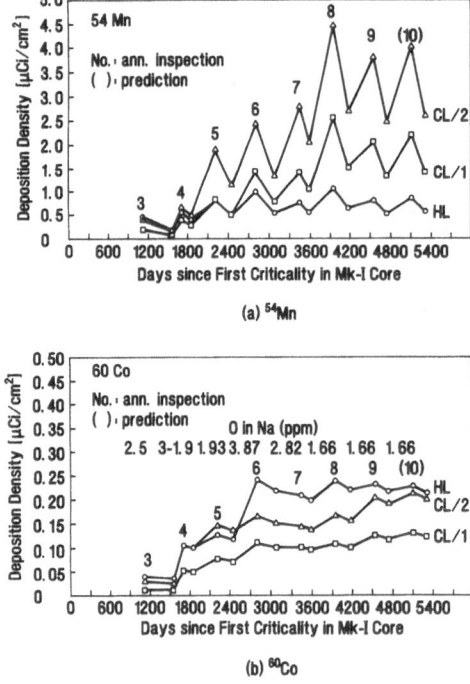

Fig. 3. CP build-up in primary main piping (A)

Table 1. Comparison of calculated vs. measured valves for ^{54}Mn an ^{60}Co on inner surface of JOYO primary circuit piping (A)

Nuclide	Region	3rd annual inspection (1981 12/23 0:00)			4th annual inspection (1983 12/3 0:00)			5th annual inspection (1985 5/1 0:00)			6th annual inspection (1986 12/6 0:00)			7th annual inspection (1988 9/7 0:00)			8th annual inspection (1990 1/23 0:00)			9th annual inspection (1991 9/11 0:00)		
		Cal.	Meas.	C/E	Cal.	Meas.	C/E	Cal.	Meas.	C/E	Cal.	Meas.	C/E	Cal.	Meas.	C/E	Cal.	Meas.	C/E	Cal.	Meas.	C/E
^{54}Mn	HL	0,474	0,459	1,03	0,457	0,542	0,84	1,02	0,775	1,32	0,979	0,987	0,99	0,966	0,748	1,29	1,34	1,064	1,26	1,02	0,809	1,26
	CL 1	0,189	0,183	1,03	0,712	0,370	1,88	2,04	0,818	2,49	2,32	1,428	1,62	2,15	1,409	1,53	2,84	2,546	1,11	2,13	2,039	1,04
	CL 2	0,326	0,380	0,86	1,26	0,657	1,92	3,60	1,895	1,90	4,10	2,44	1,68	3,79	2,786	1,36	5,01	4,458	1,12	3,77	3,77	1,00
^{60}Co	HL	0,0432	0,0380	1,14	0,0491	0,103	0,48	0,139	0,125	1,11	0,222	0,239	0,93	0,229	0,206	1,11	0,246	0,236	1,04	0,232	0,231	1,00
	CL 1	0,0153	0,0101	1,51	0,0205	0,0507	0,40	0,0731	0,0745	0,98	0,151	0,108	1,06	0,121	0,0983	1,23	0,133	0,106	1,25	0,128	0,122	1,05
	CL 2	0,0249	0,0273	0,91	0,0342	0,102	0,34	0,126	0,164	1,23	0,201	0,142	1,42	0,212	0,142	1,42	0,232	0,165	1,41	0,222	0,200	1,11
^{58}Co	HL	0,0007	-	-	0,294	-	-	0,239	-	-	0,297	-	-	0,134	0,0507	2,64	0,148	0,082	1,80	0,0866	-	-
	CL 1	0,0003	-	-	0,168	-	-	0,136	-	-	0,149	-	-	0,0753			0,0858	0,064	1,34	0,0504	-	-
	CL 2	0,0004	-	-	0,291	-	-	0,237	-	-	0,260	-	-	0,0131	0,0347	3,78	0,149	-	-	0,0874	-	-
Cummul. Output (GWd)		27,90			39,50			61,67			83,85			105,3			130,4			150,6		

13

The measured results are shown for ^{54}Mn and ^{60}Co build-up within the HL, CL/1 or CL/2 as a function of elapse days since the initial critically of Mk-I core in Fig. 3, where the decrease during every annual inspection was estimated from the radioactive decay. The ^{54}Mn build-up aspects are different between the HL and CLs as shown in the figure.

It can be seen that the build-up within the HL region had approached to the saturation level in an early operation stage, however, for the CL regions the continuation of build-up has been observed over the operating periods, exceeding the activation saturation for the radionuclide in core materials.

This is made clear by evaluating the deposition rate on each leg from the measured results as shown in Fig. 4, where the rate was assumed to be constant within one operating period between any annual inspection and the next, involving several duty cycles. The deposition rate has been nearly constant for the HL over the periods, while it has been accelerated within the CLs, especially being significant for the CL/2.

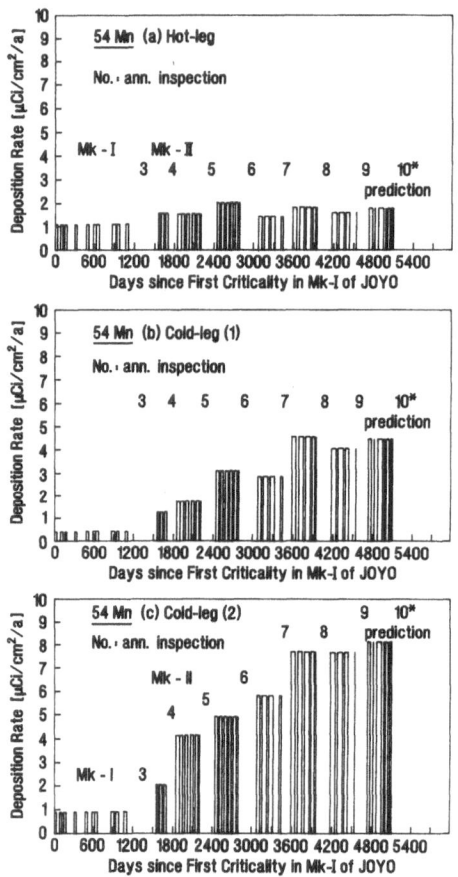

Fig. 4. ^{54}Mn deposition rate on primary main piping (A) wall

As shown in Fig. 3 the ^{60}Co build-up has gone on slowly over the whole region in the primary circuits, excepting the periods between the 3rd and 4th annual inspections or between the 5th and 6th annual inspections. The deposition rate, as shown in Fig. 5, has varied from period to period. This variation appears to be resulted from that of the operating conditions such as the controlling oxygen impurity levels in sodium, the content of cobalt impurity in core materials and re-fuelling pattern. The higher value for the above periods could be assigned to the higher oxygen impurity levels because of > 3.6 ppm for the former and 3.9 ppm for the latter, and below 2.5 or 1.9 to 1.7 ppm for the other periods with Mk-I or Mk-II core, respectively. It can be concluded that the reactor operating conditions affect more on the deposition rate of cobalt than manganese. The predictions for CP build-up or deposition rate at the next 10th annual inspection are also shown in the Figs. 3, 4 and 5, estimated by using the experimental expressions for deposition rate determined from the least square fitting to the measured data.

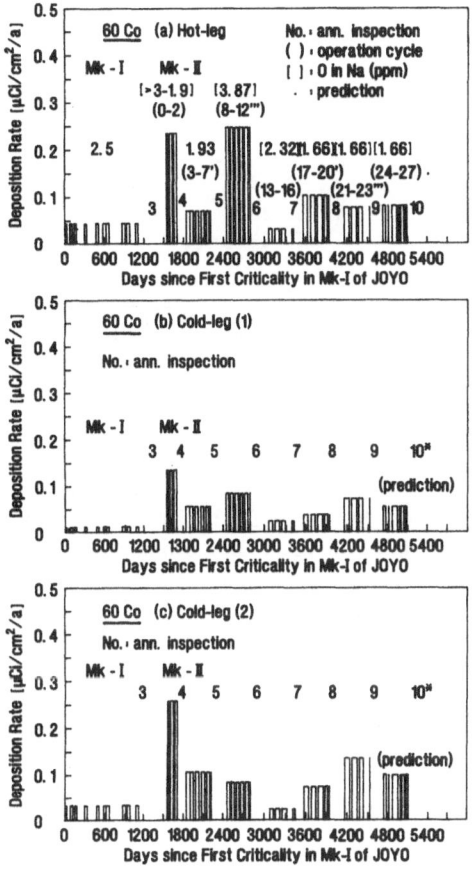

Fig. 5. ^{60}Co deposition rate on primary main piping (A) wall

The measured deposition and distribution for ^{54}Mn or ^{60}Co in the primary circuit piping (A) at each annual inspection are shown in Fig. 6. Although ^{54}Mn had transported and deposited appreciably more to the HL region in the initial operating stage i.e. Mk-I phase, the transport and deposition on the CL regions has been distinctly overcome along operating times in the Mk-II phase, resulting from the deposition rates at each leg, so that ^{54}Mn has preferentially transported and deposited to the CL regions. ^{60}Co has appreciably more transported and deposited to the HL and the increasing rate has broad speaking been similar over the whole regions in the primary main piping, so that an similar distribution pattern has been being maintained throughout the operating periods.

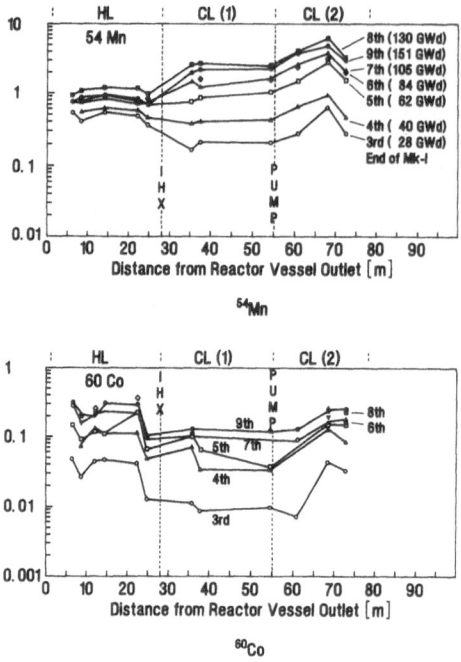

Fig. 6. Deposition and distribution of CP deposits along primary main piping (A)

The CP deposition levels are higher in the CL/2 than CL/1 for both isotopes. This appears a sodium velocity within the CL/1 and CL/2 i.e. 2.3 and 4.8 m/s, respectively, at the same temperature. The respective deposition densities for ^{54}Mn and ^{60}Co in the CL/2 are 1.8 and 1.5 times to those in CL/1, leading to the respective exponents of 1.6 and 1.1 in the expression for the velocity effect on CP deposition rate (Reynolds No.).

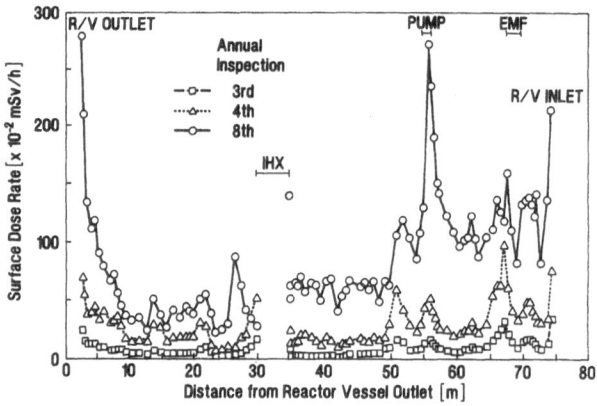

Fig. 7. Dose rate distribution along primary main piping (A)

Fig. 7 shows the gamma dose rate distribution near the primary main piping (A) at the 3rd, 4th and 8th annual inspections, where that at the 8th has given the highest levels by so far. As shown in this figure the dose rate levels along the primary circuit piping increase in order of the HL, CL/1 and CL/2, corresponding to the distribution of ^{54}Mn deposits, a major radioisotopes in the circuit. The average levels were respectively 0.65, 0.78 and 1.28 mSv/h for each region at the 8th annual inspection. The peaks in measured levels at the inlet or outlet piping of the pump and IHX can be attributed to influences from the radioactive CP deposits in their respective components. The increase in the inlet or outlet region of the reactor vessel is resulted from streaming of gamma-rays from the core.

The highest dose rate distribution around the IHX (A), observed at the 8th annual inspection and given in a way of stereo-mapping in Fig. 8, appears broad speaking symmetric one around the axis. The radial average values were respectively 0.92, 1.02 and 2.48 mSV/h for the upper, middle and lower zones. It appears that the higher levels measured at the lower part could be resulted from the highly turbulent flow at the lower structure in the IHX. The build-up aspects of those rate levels for the three zones are similar each other, corresponding well to those for the ^{54}Mn deposits in the primary circuit piping. The dose rate distribution around the main pump (A) is also shown in the figure.

The average levels are respectively the values of 1.78, 2.59 and 1.82 mSv/h for the upper, middle and lower zones in the pump, higher for the pump than IHX. It was found that the ratio of ^{60}Co to ^{54}Mn in the relative counting rates was higher in the main pump (0.0915) than the inlet and outlet piping i.e. CL/1 and CL/2 (0.0677 and 0.0529), using the Ge detector system.

3.2 Radioactive CP Evaluation in Liquid Waste from Cleaning

The evaluated results for radioactive CP amounts released from a driver fuel subassembly or an outer radial reflector by sodium cleaning are shown in Fig. 9. The released amounts from an outer radial reflector is four times as much as that for a driver fuel subassembly . ^{60}Co is the most dominant radioisotope for the both cases, respectively followed by ^{51}Cr, ^{58}Co, ^{54}Mn, ^{65}Zr-^{65}Nb, ^{182}Ta and the other radioisotopes (^{181}Hf etc.) for the driver fuel subassemblies, and by the other radioisotope, ^{65}Zr- ^{65}Nb, ^{182}Ta, ^{51}Cr, ^{54}Mn and ^{58}Co for the outer radial reflectors. The compositions are obviously different from those observed in the out-of-reactor primary circuit. In contrast ^{54}Mn was the most dominant radionuclide in the liquid waste from the pump internals cleaning [1], similar to the deposits in the sodium loop. It should be attended that the thermal neutron induced radioisotopes such as ^{60}Co, ^{181}Hf, ^{65}Zr-^{65}Nb and ^{182}Ta are more dominant in the outer radial reflector than in the driver fuel subassembly. The ^{65}Zr-^{65}Nb and ^{182}Ta are respectively produced by the activation of a small amount of Zr (^{64}Zr), a controlled element in core materials, and a trace impurity of Ta (^{181}Ta) in core materials.

4. CALCULATIONAL MODEL AND COMPUTER CODE FOR PREDICTING RADIOACTIVE CP TRANSPORT BEHAVIOUR IN PRIMARY SODIUM SYSTEMS

In the previous attempt [2] to express mathematically the description for CP transport behaviour in flowing sodium system, the transfer between steel and sodium was assumed to be mainly due to the solution and precipitation of atomic or molecular species. The model to calculate release and deposition of CP, "solution-precipitation model", requires the solution of diffusion equation in steel walls, with the boundary condition at moving sodium-steel interfaces. This model was originally advanced by M.V. Polley [3] and W.L. Kuhn [4], and was improved in some points by K. Iizawa [2].

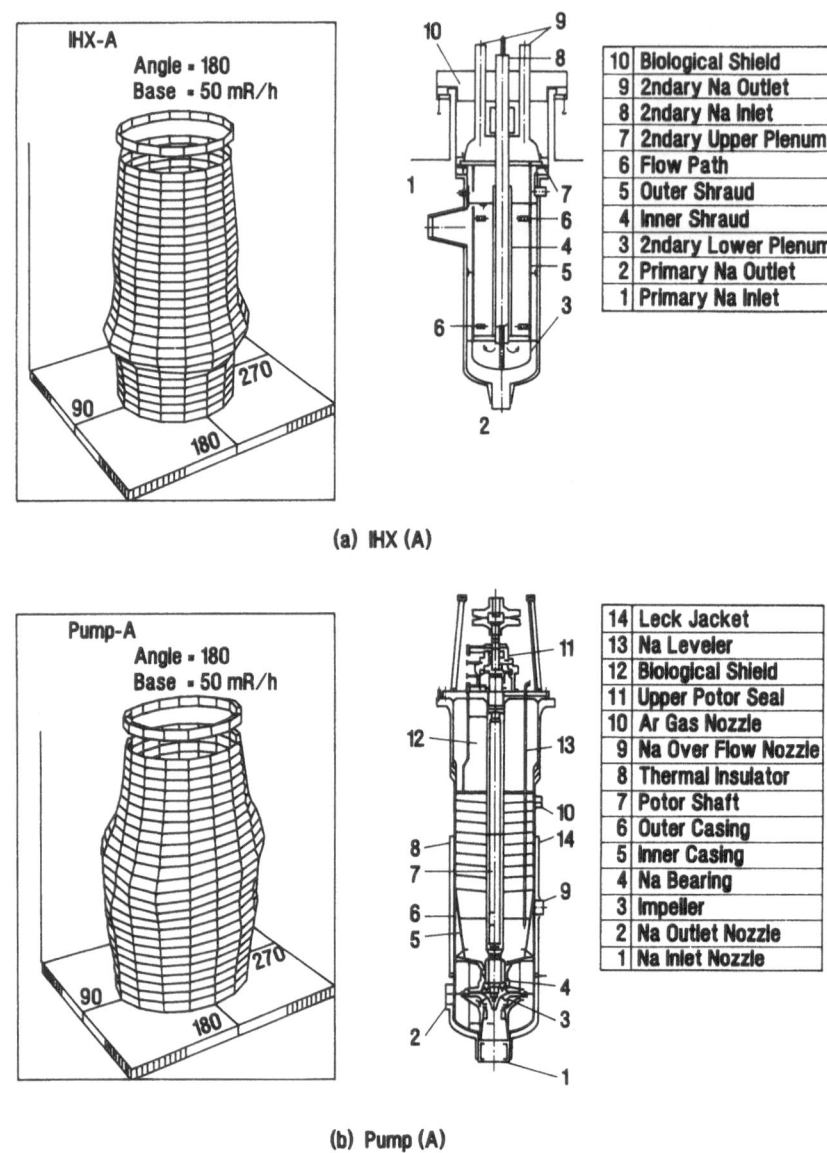

10	Biological Shield
9	2ndary Na Outlet
8	2ndary Na Inlet
7	2ndary Upper Plenum
6	Flow Path
5	Outer Shroud
4	Inner Shroud
3	2ndary Lower Plenum
2	Primary Na Outlet
1	Primary Na Inlet

(a) IHX (A)

14	Leck Jacket
13	Na Leveler
12	Biological Shield
11	Upper Potor Seal
10	Ar Gas Nozzle
9	Na Over Flow Nozzle
8	Thermal Insulator
7	Potor Shaft
6	Outer Casing
5	Inner Casing
4	Na Bearing
3	Impeller
2	Na Outlet Nozzle
1	Na Inlet Nozzle

(b) Pump (A)

Fig. 8. Dose rate distribution around IHX and main pump

The model parameters required to describe the model are as follows: the diffusion coefficient of radioisotope in steel (D), the overall mass transfer coefficient ($K' = k' k_d' / (k' + k_d')$), the chemical partition parameter ($\beta = k'_d / k_s$ or $\beta' = \beta/ \Theta'$) where k_s or k_d' is respectively the solution or precipitation kinetic constant and Θ' is the dimensionless oxygen concentration in sodium i.e. the actual oxygen concentration divided by the reference oxygen concentration, and the magnitude of the interfacial velocity (u_c for bulk corrosion or u_d for bulk deposition). The analytical solutions of these diffusion equations, thus the resulting expressions for the concentration of radioisotope in the steel and bulk sodium, and the mass flux, can be approximately obtained by taking account of the boundary condition and mass balance, described in the previous paper [2] in detail.

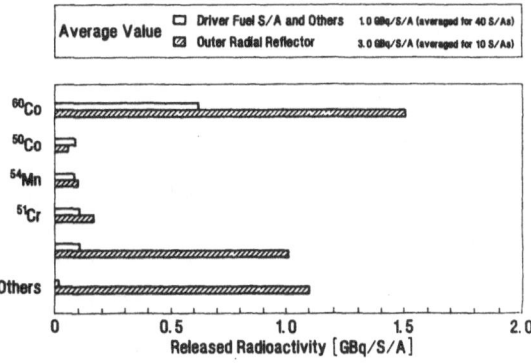

Fig.9. Released CP amounts during sodium cleaning for core subassembly

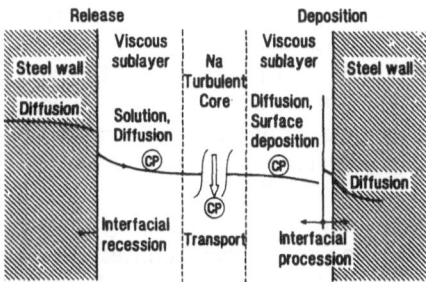

Fig. 10. "solution-precipitation model" for CP behaviour analysis

Fig. 10 shows schematically the solution-precipitation model. Following this model, CP transfer in sodium system is characterised by four steps:

(a) solid state diffusion in the steel and surface loss or gain by bulk corrosion or deposition, respectively,

(b) mass transfer (reverse or inverse reaction) at the sodium-steel interface,

(c) diffusional mass transfer across the sodium viscous boundary sublayer,

(d) transport with the circulating sodium, followed by depositions at the surface.

The results of measurements have been compared with the calculations to validate a radioactive CP transport behaviour model for primary sodium systems of a fast reactor, using the computer code "PSYCHE" [2] (The first application of this code for JOYO was reported in the reference [1]. This code consists of two sub-programs for source term and radiation field calculations, linked together. Source term calculations are done on the basis of the solution-precipitation model. The radiation levels are calculated using the QAD-CG code modified by evaluating the JOYO data [5], named as "JOANDARC".

The model parameters, D, K', β or β' and u (surface recession rate, u_c, or surface procession rate, u_d) needed to predict fast reactor radioactive CP transfer have been determined by fitting the model to results of sodium loop experiments [2] and measurements for the primary circuits of JOYO in the previous [1] or this work. Each of the model parameters is classified to two types: manganese and cobalt types. The manganese type involves strongly leached species into sodium such as manganese or nickel isotopes, while cobalt type involves weakly leached species such as cobalt, iron or chromium.

6. DISCUSSION

[54]Mn is most highly activated in the fuel cladding made of type 316 stainless steel in sodium corrosion regions of a fast core, followed by [58]Co and ^{60}Co, e.g. [54]Mn : [58]Co : [60]Co= 1 : 0.28 : 0.02 for the JOYO Mk-II core just after the 23rd duty operation with Mk-II core (the 9th annual inspection). In addition sodium loop experiments [6], [7], [2] have shown that CP species are either selectively leached or retained in the steel, i.e. [54]Mn release is often super-stoichiometric, while ^{60}Co is often sub-stoichiometric, and that [54]Mn migrates throughout the following sodium system. Selective leaching can be interpreted as follows: [54]Mn is released into sodium from the surfaces as well as through bulk corrosion (Surface loss) by diffusing towards the surface from the interior of steel, whereas the release accompanying bulk corrosion is more of importance for ^{60}Co because of its tendency to be retained in steel. This appear to result in the prevalence of [54]Mn in the out-of-reactor primary sodium system.

The measured CP deposition and distributions along the primary circuit piping of JOYO, shown in Fig. 6, lead the following estimations about the transport and deposition mechanism of CPs in the following sodium system. Firstly, it appears that [54]Mn deposits on the wall surfaces of HL region with a higher sodium temperature by diffusion into the interior of steel, while mainly on the wall surfaces of CL regions by formation of alloying precipitates containing [54]Mn. The acceleration of the deposition rate found in the CL regions could be resulted from the formation and growth of micro - structures in iron, nickel or chromium, easily alloying manganese, on the wall surfaces in sodium environments [8]. Secondly, ^{60}Co appears to precipitate mainly on the wall surfaces in a whole region of primary sodium systems by the formation of alloying particles containing [60]Co. It has been proposed that cobalt isotopes could form various ferritic cobalt particles in sodium [9], so that the micro-structures rich in iron and etc. could be formed on the wall surfaces. Thirdly, it appears that the different mechanism for deposition on the HL region between [54]Mn and ^{60}Co could be interpreted by the relatively higher solubility of manganese in sodium and the very lower that of cobalt, and the temperature dependency of their deposition sites formation in the sodium environments.

It is estimated that [54]Mn could be mainly transported as soluble species in the sodium circuit, whereas ^{60}Co could be something transported in particle forms such as ferritic cobalt particles. N. Yokota found the ferritic cobalt particles with such the composition as Fe/61 at%, Co/ 36 at% and Cr/ 3 at% on the test specimens when the stellite No.6 test specimens were exposed in the sodium experimental loop [10]. These particles containing cobalt isotopes (both radioactive and stable) could occasionally transfer from the sodium main stream to stagnant regions, such as the upper zone above bearing housing of the main pump and the outer reflector regions in the core, and gradually accumulated in there as operating the primary sodium cooling system. In the case of outer reflector these particles carry strongly ^{60}Co through thermal neutron capture by the concentrated [59]Co in them in addition to the concentration of radioactive isotopes through mass transfer in sodium system. This estimation is consistent with some features observed for the dose rate build-up in the pump and the production of liquid waste with high concentrations of ^{60}Co particles.

The release deposition inventories of [54]Mn increase with increasing diffusion coefficient, D, so that [54]Mn release and deposition is strongly affected by the parameter, where the influence of interfacial procession rate, u_d, on the inventories is little found. This parameter controls rather the deposition and distribution than the inventories in primary systems Therefore it can be seen that the observed acceleration phenomena of [54]Mn deposition rates on the CLs do not presumably yield the increase of release from core, but the re-distribution from core to out-of-reactor primary circuits. While the release inventories of ^{60}Co are strongly affected by chemical partition parameter, β', but not by D or u_d, and thus also so for the deposition inventories. The smaller effect of D on ^{60}Co release in this model appears to indicate wall surface loss rather than diffusion from the interior of wall as the main release mechanism.

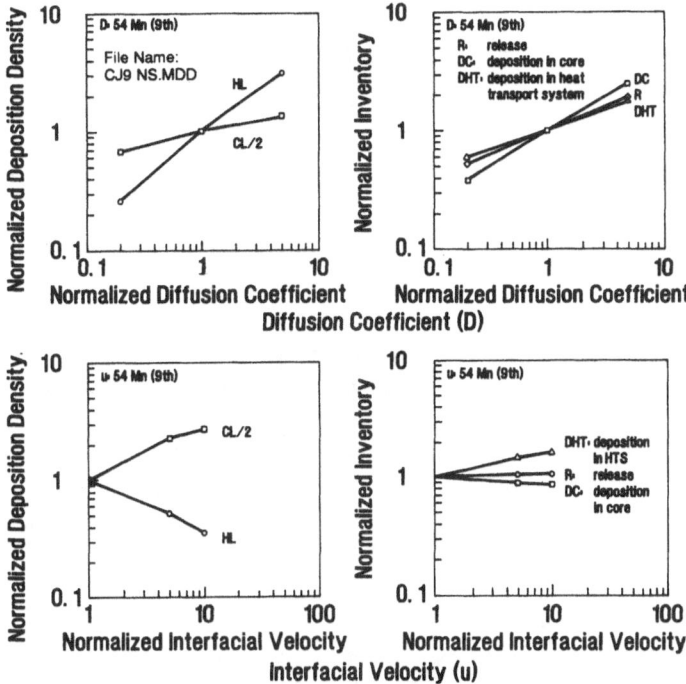

Fig.11. Sensitivity of model parameters (D and u_d)for ^{54}Mn deposition density and inventory calculation

In order to adjust the model parameters, D, u_d and β' their sensitivities to ^{54}Mn, ^{60}Co or ^{58}Co deposition or release are tested about the depositions densities on the primary piping and the radioactivity inventories of release or deposition in the core and out-of-reactor primary circuits. The test values of each parameter were referred to that given in the previous work [2].The other parameters, K', β for ^{54}Mn and u_c, are the same as given in the previous work. The sensitivity test results for D, u_d and β' are respectively shown in Fig. 11 and 12.

^{54}Mn deposition and distribution in the primary circuits is strongly affected by D and u_d. Its levels increase more in HL than CL along increasing D, more markedly in the early stage of plant, while the increase of u_d yields increasing the amounts on the CL piping, lower temperature regions of IHX and pump, associating with the decrease of the amounts in the higher temperature regions such as HL and outlet regions in core. While the controlled parameter for ^{60}Co deposition and distribution is β', and the influence of D or u_d can be only slightly found, which indicates the compensative aspects between the higher and lower temperature regions. Its deposition and distribution levels increase with decreasing β'.

The model parameter re-adjustment in this work required the operating time (practically Mk-I and Mk-II phase) dependency of diffusion coefficient, D, and the new surface procession rate, u_d, for ^{54}Mn, and the new u_d for ^{60}Co or ^{58}Co, respectively, where u_d should appear rather a virtual quantity than an actual one in the character.

The measured and calculated results for ^{54}Mn and ^{60}Co deposition distribution in the primary main piping of JOYO are given in Table 1 and those at the 3rd and 9th annual inspection are compared in Fig. 13 as an example.

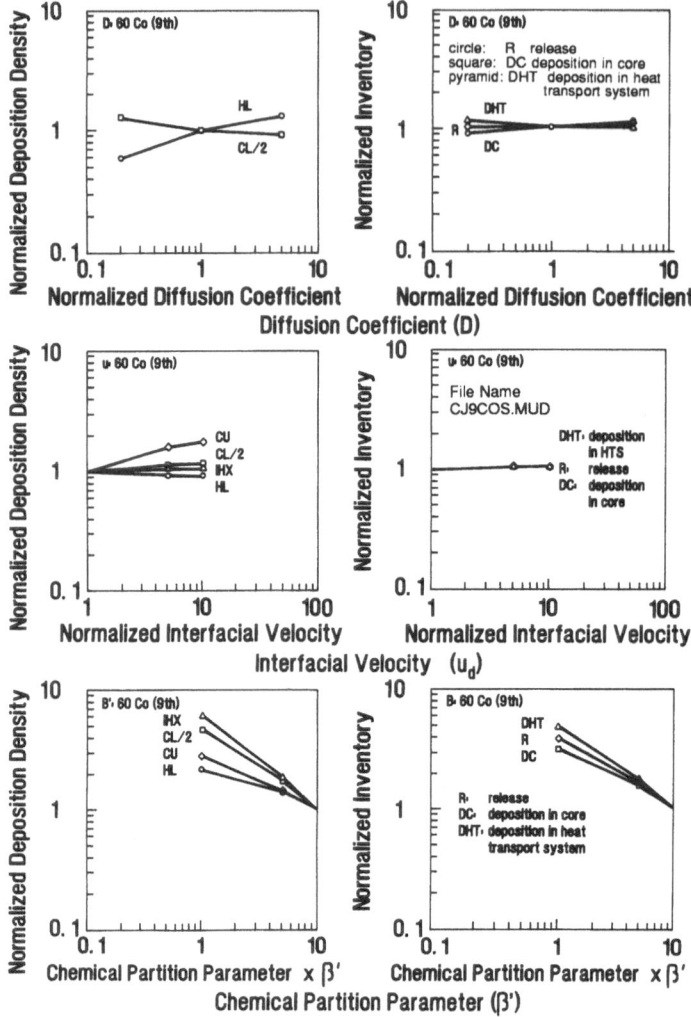

Fig. 12. Sensitivity of model parameters (D, u_d and β') for ^{60}Co deposition density and inventory calculation

The calculated values, as shown in the figure, are in good agreement with the measured. The calculated to measured values (C/Es) are respectively 1.36 and 1.03 for ^{54}Mn and ^{60}Co from the Table 1, averaging over all the operating periods. It can be seen that results show reasonable agreement on the whole. The measured and calculated results are show for ^{54}Mn and ^{60}Co build-up within the HL, CL/1 or CL/2 in Fig. 14, including the prediction for future operation.

These results seem reasonable, expect that the calculated results for ^{54}Mn in the CL/1 or CL/2 are higher than that measured from the 4th to the 6th annual inspection. This implies that the interfacial procession rate, u_d, for ^{54}Mn in Mk-II phase, which was adjusted carrying weight on the results from the 7th to 9th annual inspection, is too high. That is, the most adjustable parameter, u_d, appears it should have gradually increased, being consistent with the acceleration phenomena observed for the ^{54}Mn deposition rates on CLs.

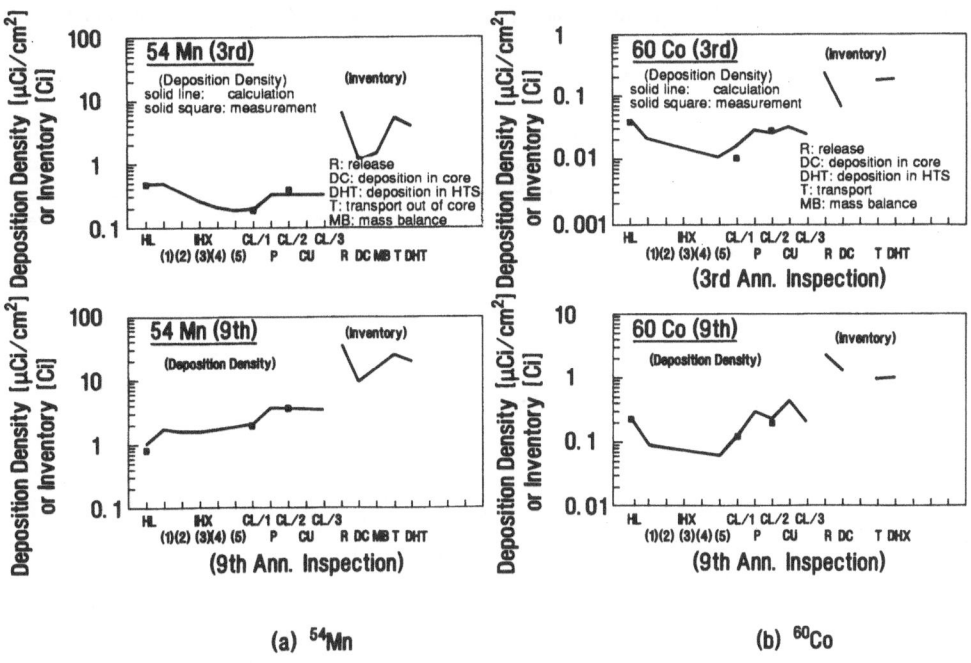

Fig. 13. Comparison of calculated vs. measured CP deposits in primary main system (A)

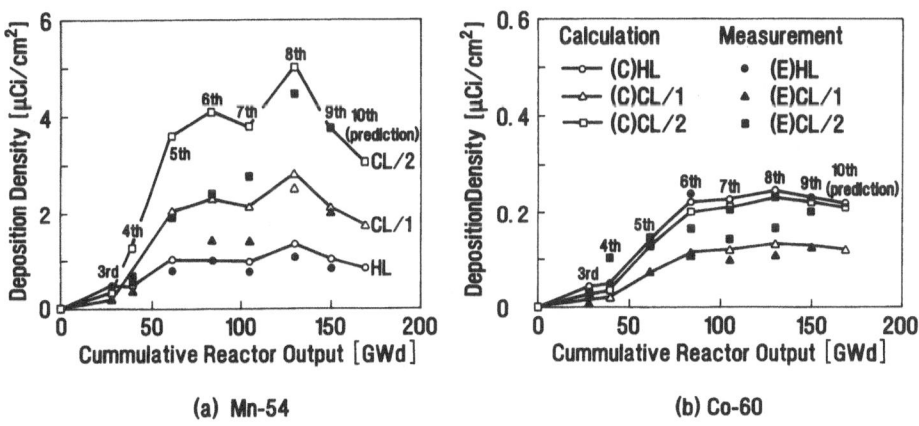

Fig. 14. Comparison of calculated vs. measured CP build-up in primary main piping (A)

The measured and calculated results for radiation levels exterior to the primary main piping just after the 23rd duty cycle (the 9th annual inspection) are compared in Fig. 15.

It can be seen that the measured values are successfully reproduced for the HL, CL/1 or CL/2 within an average factor of 1.48, 1.70 or 1.45 by the calculation, respectively. The contribution of ^{54}Mn, ^{60}Co and ^{58}Co to the radiation levels are respectively estimated to be about 75 %, 25 % and 5 % from the calculation. The measured and calculated radiation build-up is compared for the primary main piping in Fig. 16, and for the upper and lower regions in IHX or pump in Fig. 17, respectively.

This reproducibility for the lower regions IHX or pump is reasonable. However, the calculated curve for the upper region in IHX is higher than the measured and lower for there in pump. The tendency in IHX appears to indicate that the interfacial procession velocity should be smaller in there, higher temperature region in IHX, while that in the pump could presumably come from the accumulation of ^{60}Co rich particles in there, because the build-up variation in radiation levels seems to be if anything similar to that for ^{60}Co deposits observed in the primary main piping.

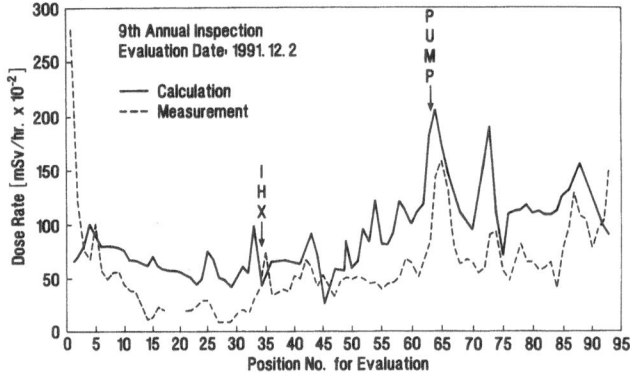

Fig. 15. Comparison of calculated vs. measured dose rate distribution along primary main piping (A)

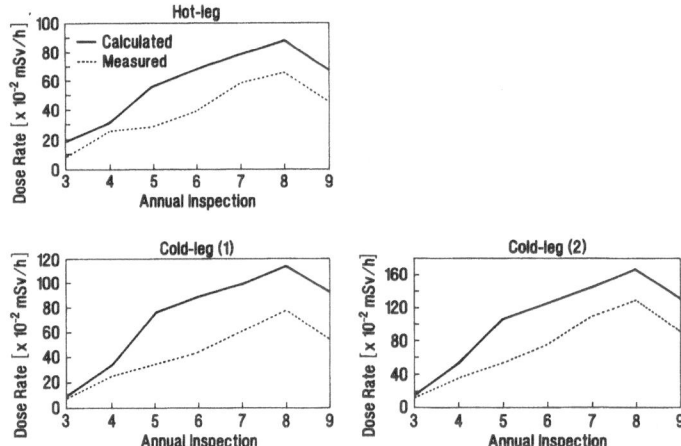

Fig. 16. Comparison of calculated vs. measured dose rate build-up around primary main piping (A)

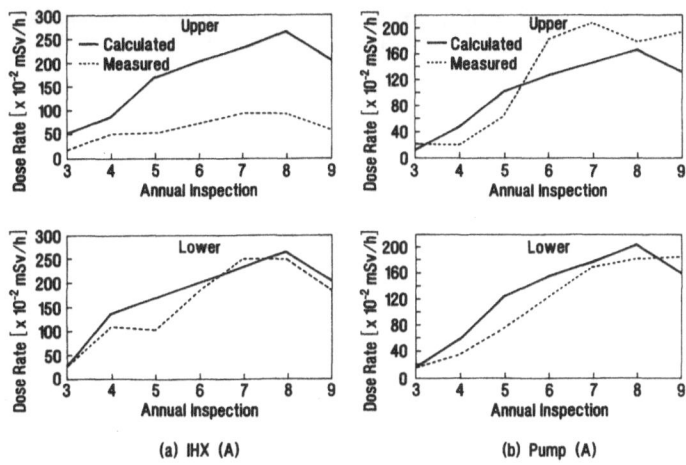

(a) IHX (A)

(b) Pump (A)

Fig.17. Comparison of calculated vs. measured dose rate buildup around IHX (A) and main piping (A)

5. CONCLUSIONS

The transport behaviour of [54]Mn in primary sodium systems, including both out-of-reactor primary piping and components and core, could be well reproduced by the "solution-precipitation model", taking account of the acceleration phenomena of the deposition rates within CL regions. However, ^{60}Co appears to transfer something as particle forms in sodium, tending to accumulate in sodium stagnant regions such as the overflow zone in main pump, the radial outer reflector zone in core and etc., despite that model being reasonable for its transport behaviour in the primary main piping with higher sodium velocities. It can be seen that the calculated results using "PSYCHE" and "JOANDARC" show reasonable agreement with the measured values in JOYO within an average factor of 1.36 or 1.03 for [54]Mn or ^{60}Co deposits in the primary main piping, and within a factor of about 1.5 for the radiation levels around there.

Acknowledgements

The authors gratefully acknowledge the contributions of Mr. Tomohiko Masui, Mr. Takuya Saikawa, Inspection and Development CO., Ltd. and Mr. Jun Takeuchi, Mr. Hidehiko Teruyama, Hitachi Division Engineering Co,. Ltd.

References

1. K. Iizawa, S. Suzuki, M. Tamura, S. Seki, T. Hikichi, , Proc. Specialists' Meeting on Fission and Corrosion Products Behaviour in Primary Circuits of LMFBRs, Karlsruhe, KfK 4279 IWGFR/64 (1987) p.227.
2. K. Iizawa, T. Kikuchi, I. Nihei, J. Horie, ibid., p. 191.
3. M.V. Polley, G. Skyrme, J. Nucl. Mater., 75 (1978) 226.
4. W.L. Kuhn, USAEC Report HEDL-TME 76-10, September 1977.
5. S. Suzuki, K. Iizawa, N. Ohtani, T. Kobayashi, J. Horie, H. Handa, Proc.Topical Conference on Theory and Practices Radiation Protection and Shielding, Knoxville, Tennessee, 1987.
6. N. Sekiguchi, K. Iizawa, H. Atsumo, Proc. Specialists' Meeting on Fission and Corrosion Product Behaviour in Primary Circuits of LMFBRs, Dimitrovgrad (1965) p. 82.
7. W. Brehm, ibid. (1975) p. 186.
8. N. Yokota, S. Shimoyashiki, J. Japan Inst. Metals, 53, No.2 (1989) 175.

9. W.F. Brehm, R.P. Colburn, H.P. Maffei, W.P. Stinson, W.L. Bunch, R.A. Bechtold, Proc. Specialists' Meeting on Fission and Corrosion Products Behaviour in Primary Circuits of LMFBRs, Karlsruhe, KfK 4279, IWGFR/64 (1987) p. 75.
10. N. Yokota, J. Japan Inst. Metals, 55, No.2 (1991) 229.

Transport Phenomena of Iodine and Noble Gas Mixed Bubbles Through Liquid Sodium

S. Miyahara, K. Shimoyama

Safety Engineering Division, O-arai Engineering Center
Power Reactor and Nuclear Fuel Development Corporation
4002 Narita, O-arai machi, Ibaraki 311-13, Japan

1. INTRODUCTION

In the fuel pin failure accident in a liquid metal fast breeder reactor, volatile fission products play an important role in the assessment of radiological consequences. Especially the radioisotopes of elemental iodine are important because of their high volatility and of the low permissible dose to human thyroid. Although the coolant, liquid sodium, has a high retention capability of elemental iodine due to the chemical affinity between alkali metals and halogens, it may be directly released from the fuel to the cover gas, if the iodine might be carried by means of bubbles of xenon or krypton through the sodium pool. So far, the only few experimental results [1] were available concerning the decontamination factor (DF: the ratio of the initial mass in a bubble to the released mass into the cover gas) of iodine in this phenomenon. Thus, an experimental study was performed to investigate the transport phenomena of iodine by means of iodine and noble gas mixed bubbles rising through the liquid sodium pool.

2. ANALYTICAL ASSUMPTION

The iodine mass transport from a bubble into sodium can be described by the iodine vapor diffusion model as shown in Fig. 1, if we assume that the iodine vapor reeacts with liquid sodium instantaneously at the gas-liquid interface and the product NaJ is removed into the sodium due to the sodium flow. According to this assumption, the iodine mass transfer rate, dM/dt (mol/s) is expressed as the following equation:

$$- dM/dt = JA = kCA = kMA/V \tag{1}$$

where M is the iodine mass within the bubble (mol), t the time (s), J the iodine mass flux ($mol \cdot m^{-2} \cdot s^{-1}$), A the surface area of the bubble (m^2), k the mass transfer coefficient ($m \cdot s^{-1}$), C the iodine concentration within the bubble ($mol \cdot m^{-3}$), and V the volume of the bubble (m^3).

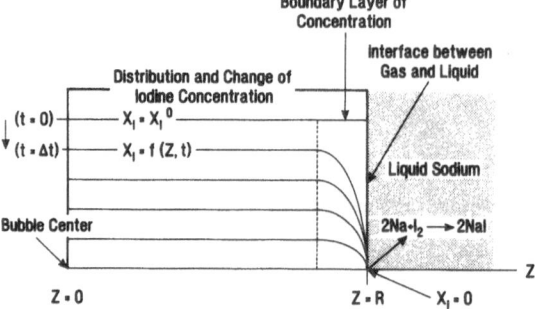

Fig. 1. Analytical model for iodine vapour diffusion from a xenon-iodine mixed gas bubble into liquid sodium

Therefore, we tried to obtain the iodine mass transfer rate from our experimental results under various initial conditions of the bubble volume, the iodine concentration, the sodium temperature, and the depth of the sodium pool (related to the rising time of the bubble). The mass transfer coefficient was then deduced using equation (1).

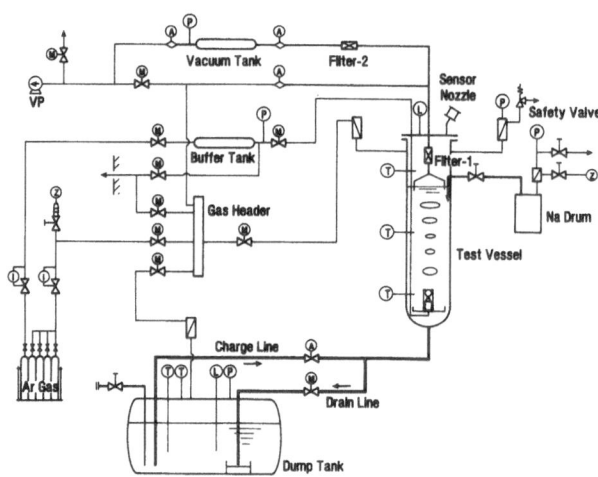

Fig. 2: Arrangement of the test rig for volatile fission product bubble behaviour in sodium

3. EXPERIMENTAL EQUIPMENT

Figure 2 shows the flow sheet of the test rig. This sodium loop mainly consists of a stainless steel test vessel with an axial length of 3.0 m and an inner diameter of 0.3 m, a sodium storage tank, an argon gas supply system, and a cover gas sampling system.

In the vessel, a quartz glass ball which contained mixed gas of iodine and xenon and its crushing device were installed at the bottom, and an inverted funnel to collect rising bubbles was located at the surface of the sodium pool, as is shown in Fig. 3. Chen-type void sensors were also installed at axial and radial positions in the vessel to measure the rising time and the velocity of bubbles through the sodium pool.

In the tests, the bubble was generated at the bottom of the sodium pool by means of crushing the quartz ball. The iodine containing bubble which was released to the sodium surface was collected together with sodium vapour in the inverted funnel which was connected with the sampling system and was aspirated by a vacuum tank through aerosol filters installed in the system. After that, the collected amounts of iodine and sodium were dissolved in distilled water; the iodine contents were then measured by means of ion chromatography. The accuracy of these measurements was $\approx 10\%$.

The conditions and parameters of the tests are listed in Table 1. The temperature of the quartz ball and internal gases was the same as the sodium temperature in all experiments, and the initial gas pressure was adjusted to atmospheric pressure added to the sodium head under the heated-up condition. The size of the quartz ball and the iodine concentration varied from 5 to 12 cm in diameter and 1 to 50 mol %, respectively, in each of the experiments.

4. RESULTS AND DISCUSSION

4.1 Iodine Mass Transfer Rate

The iodine mass transfer rate from the bubble into sodium was calculated from the test results concerning the trapped iodine mass in sodium and of the rising time of the bubble through the sodium. The trapped iodine was determined here from the difference of the initial mass of iodine in the quartz ball and the mass collected by the cover gas sampling system. The rising time of the bubble was obtained from the signals of the void sensors.

Table 1. Test conditions

Initial bubble volume ($\cdot 10^{-4}$)	0.6 to 8.5 m^3
Initial iodine concentration	1 to 50 mol %
Temperature of the sodium pool	400 to 600 °C
Depth of the sodium pool	1.0 to 2.0 m

Figures 4,5, and 6 show the relation between the obtained iodine mass transfer rate and the initial bubble volume, sodium temperature, and rising time of the bubble, respectively. In these figures, the results are also indicated for different initial iodine concentrations as a parameter. It is obvious from these results that an adequate correlation is observed between the iodine mass transfer rate and the initial bubble volume, and also the initial iodine concentration. However, the effects of the sodium temperature and the rising time of the bubble upon the iodine mass transfer rate were not clearly observed.

4.2 Mass Transfer Coefficient

In order to deduce the mass transfer coefficient from the relation between the iodine mass flux and the iodine concentration by means of Eq. (1), the iodine mass flux should be previously calculated from the results of the iodine mass transfer rate. To determine the surface area of the

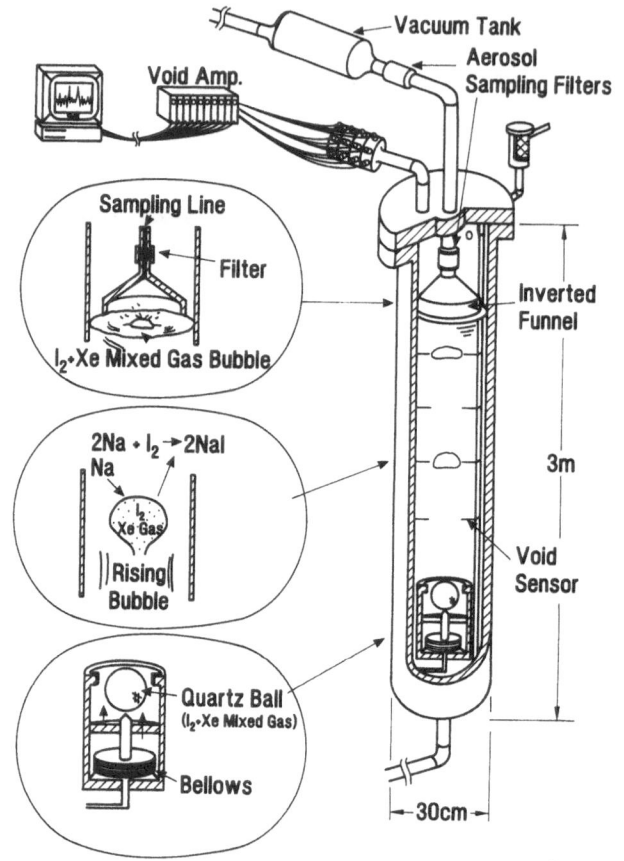

Fig. 3. Test vessel for volatile fission product bubble behaviour in sodium

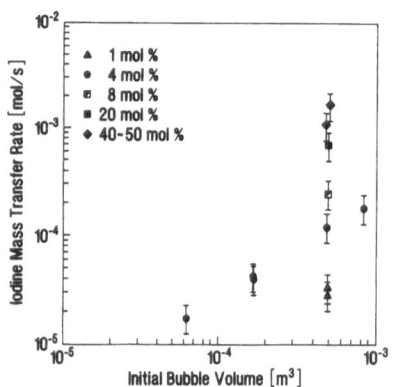

Fig. 4. Relation between iodine mass transfer rate and initial bubble volume (test results)
Test conditions:
initial iodine concentration: 1 to 50 mol %
sodium pool temperature: 500 °C
rising time through sodium pool: 2.4-3.7 s
depth of sodium pool: 1.5 m

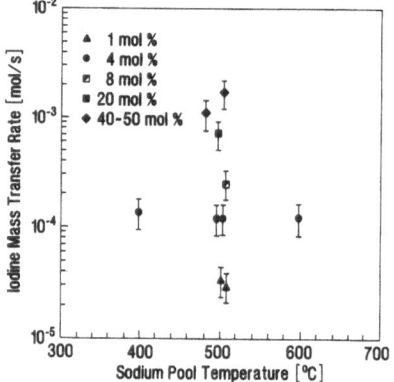

Fig. 5. Relation between iodine mass transfer rate and sodium pool temperature (test results)
Test conditions:
initial iodine concentration: 1 to 50 mol %
initial bubble volume: $5 \cdot 10^{-4}$ m^3
rising time through sodium pool 2.4-3.7 s
depth of sodium pool: 1.5 m

bubble in the calculation, we introduced an assumption that the shape of the bubble in the experiments is a single spherical-cap as shown in Fig. 7 [2], since the actual surface area of the bubble could not directly be measured in the liquid sodium pool. The curvilinear radius R of the bubble in Fig. 7 is estimated as [3].

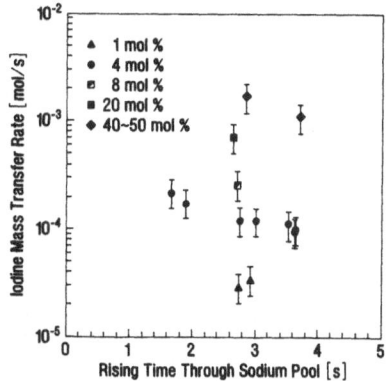

Fig. 6. Relation between iodine mass transfer rate and rising time through the sodium pool (test results)
Test conditions:
initial iodine concentration: 1 to 50 mol %
sodium pool temperature: 500 °C
initial bubble volume: $5.0 \cdot 10^{-4}$ m^3
depth of sodium pool: 1.0 to 2.0 m

4.2 Mass Transfer Coefficient

In order to deduce the mass transfer coefficient from the relation between the iodine mass flux and the iodine concentration by means of Eq. (1), the iodine mass flux should be previously calculated from the results of the iodine mass transfer rate. To determine the surface area of the bubble in the calculation, we introduced an assumption that the shape of the bubble in the experiments is a single spherical-cap as shown in Fig. 7 [2], since the actual surface area of the bubble could not directly be measured in the liquid sodium pool. The curvilinear radius R of the bubble in Fig. 7 is estimated as [3].

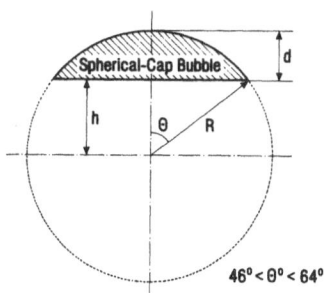

Fig. 7. Spherical-cap bubble shape

$$R = (V/\pi(2/3 - \cos \Theta + 1/3 (\cos \Theta)^3))^{1/3} \qquad (2)$$

Thus, the surface area of the bubble A can approximately be expressed as:

$$A = \pi(R \sin \Theta)^2 + \pi^2 R^2 \sin\cdot\Theta(\Theta/180) \qquad (3)$$

The angle Θ is in the range between 46 and 64 degrees [4]. The average value 53 degrees was chosen for this calculation.

Figure 8 shows the relation between iodine mass flux and initial iodine concentration. It is clear that the iodine mass flux is proportional to the initial iodine concentration, it does not depend on the temperature of sodium and the depth of the pool. From these results, the following correlation was obtained for the $5\cdot10^{-4}$ m^3 bubble data by the use of the least-square technique,

$$J = 3.39\cdot10^{-3}\cdot C_0 \qquad (4)$$

and thus, the mass transfer coefficient k = $3.39\cdot10^{-3}$ (m·s^{-1}) was deduced as an average value. The fitting line is also presented in Fig. 8. Although the fitting is only based on the $5\cdot10^{-4}$ m^3 bubble data, the line represents all of the plotted data within the accuracy of ≈ 45 %.

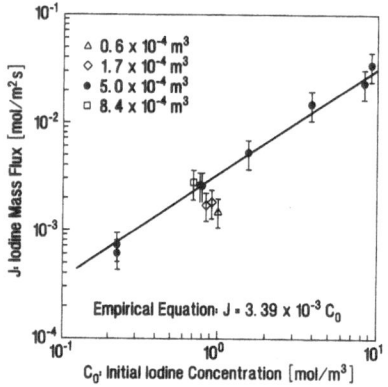

Fig. 8. Relation between iodine mass flux and initial iodine concentration
Test conditions:
initial iodine concentration: 1 to 50 mol %
sodium pool temperature: 500 °C
initial bubble volume ($\cdot10^{-4}$): 0.6 to 8.4 m^3
rising time through sodium pool: 2.4 to 3.7 s
depth of sodium pool: 1.5 m

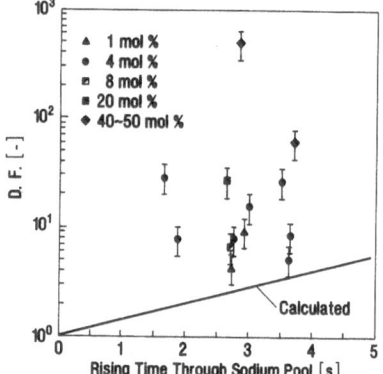

Fig. 9. Comparison of decontamination factors between experimental results and calculated ones
Test conditions:
initial iodine concentration: 1 to 50 mol %
sodium pool temperature: 500 °C
initial bubble volume: $5.0\cdot10^{-4}$ m^3
rising time through sodium pool: 1.7 to 3.7 s
depth of sodium pool: 1.0 tp 2.0 m
Calculational conditions:
volume of bubble: $5\cdot10^{-4}$ m^3
surface area of bubble: $5\cdot10^{-2}$ m^2

4.3 Decontamination Factor

The decontamination factor (DF) is described as Eq. (5) by resolving Eq. (1), if the bubble volume and its surface area would not change during the rising process:

$$DF = M_0/M = 1/\exp(-kAt/V) = 1/\exp(-3.39\cdot10^{-3}At/V) \qquad (5)$$

where
M_0 = iodine mass in initial bubble (quartz ball), and M = iodine mass released into the cover gas. Figure 9 compares experimental DFs with the calculated ones. The calculated DFs were obtained by means of Eq. (5) with the above k value under the condition of $5\cdot10^{-4}$ m^3 in volume and the surface area of $5\cdot10^{-2}$ m^2 which is equivalent to the single spherical-cap bubble. Although the experimental DFs scattered within two orders in the figure, the calculated DFs indicate the

conservative lower limit of the experimental ones. This discrepancy might be attributed to the fact that the iodine mass transfer rate was calculated as the average value of two positions. Therefore, further experimental study was necessary to investigate the transport phenomena of iodine from the xenon-iodine mixed gas bubbles.

5. CONCLUSION

(1) The mass transfer rate of iodine from the bubble into the sodium correlated adequately with both the bubble volume and the iodine concentration.
(2) An empirical equation to evaluate the conservative decontamination factor (DF) of iodine in the sodium was obtained by deducing an average mass transfer coefficient from the relation between the iodine mass flux and the iodine concentration.

References

[1] C.T. Nelson, R.P. Johnson, E.U. Vaughan, C.A. Guderjahn, H.A. Morewitz, Proc. Intern. Meeting on Fast Reactor Safety and Related Physics, Chicago 1976, p. 1937.
[2] R. Clift, J.R. Grace, M.E. Weber, "Bubbles, Drops, and Particles", Academic Press, New York 1978, p. 26.
[3] D.S. Azbel, S.L. Lee, T.S. Lee, "Two-Phase Momentum, Heat and Mass Transfer in Chemical Process and Energy Engineering Systems, Hemisphere Publ. Comp. 1978 Vol. 1, p. 159.
[4] R.M. Davies, G. Taylor, Proc. Royal Soc. A 200 (1950) 375.

ACTIVATION, CORROSION AND CONTAMINATION IN FAST BREEDER REACTORS - VALIDATION OF MODELS WITH EXPERIMENTAL DATA

F. Masse and G. Rouviere

DER/STML - CEA-Cadarache
F-13108 Saint Paul lez Durance,
France

1. INTRODUCTION

In the field of Fast Breeder Reactors, it is essential to be able to evaluate the radioactivity levels of the various circuits, and in particular those of the large primary circuit components which may be removed, cleaned and decontaminated for maintenance or repairs.

Knowledge of their contamination levels allows both the dose rates corresponding to the activity deposited on each component to be predicted and the composition and activities of the cleaning and decontamination wastes to be evaluated ; the latter is also vital information for the waste processing plants.

The main radioactivity sources are :
- activation of the sodium and its impurities ;
- activation of core materials which, through corrosion, dissolution enter the sodium;
- active fission products mainly emitted when there is clad failure, but which can also be due to a small amount of external pollution from sub-assembly fissile material ;
- tritium released by ternary fission, by activation of the boron in the control rods and by activation of certain sodium impurities, such as lithium 6.

Experience [1] has shown that if there is no cladding failure, the main source of radioactivity is due to the deposits of activated corrosion products, manganese 54 in particular. The other sources of contamination are not taken into account in this paper. In order to evaluate contamination due to activated corrosion products, a certain number of phenomena such as activation, corrosion and deposition have to be taken into account. The constituent steel elements of the core, fuel cladding in particular, are in fact activated under the effect of the neutron flux.

Due to the corrosion phenomenon, these active metallic elements enter the primary sodium. The activated corrosion products tend to deposit in the hot areas for cobalt (on the upper parts of the core) and in the cold areas for manganese, iron and chrome (first on the intermediate heat exchangers, then on the primary pumps).

Evaluating the contamination level consists in calculating the cumulated mass of the corrosion products and their respective activities per isotope, particularly as regards core characteristics, the neutron flux and thermal hydraulics, and according to operating conditions (duration, oxygen content).

In order to carry out these estimates, a computer calculation code is absolutely necessary. The French computer calculation code previously used, CORONA, was not efficient enough to give an accurate evaluation of the contamination levels. A new computer calculation code, ANACCONDA, was therefore developed. It allows the use of more recent corrosion models perfected on specifically designed experimental facilities. They are P. BAQUE's model (FRANCE) for low oxygen contents ([O] < 5 ppm), and A. THORLEY's model (U.K), for high oxygen contents ([O] > 5 ppm). Core modelling has been considerably improved in this code.

2. RADIOACTIVE CONTAMINATION

The three main phenomena involved in the contamination process are :
   activation,
   corrosion,
   deposition.

2.1 Activation

A fast breeder reactor is the site of nuclear reactions generating high energy neutrons. These neutrons are able not only to maintain a nuclear chain reaction, but also to bring about supplementary reactions in stable nuclides constituting the core and reactor structures.

The production of radioactive species from stable radionuclides is called activation. This mainly depends on neutron energy as well as on neutron density which varies tremendously according to distance in relation to the core.

For example, in Superphénix, the neutron density in the core centre is about 10^{15} n/cm^3 ; it is only 10^7 n/cm^3 in the lateral neutron shielding. This is why the activation phenomena are mostly situated in the vicinity of the active core.

The main activation reactions as regards contamination by activated corrosion products are as follows :

$$^{54}Fe + {}^1_0n \rightarrow {}^{51}Cr + {}^4_2\alpha$$
$$^{54}Fe + {}^1_0n \rightarrow {}^{54}Mn + {}^1_1p$$
$$^{58}Fe + {}^1_0n \rightarrow {}^{59}Fe + {}^0_0\gamma$$
$$^{58}Ni + {}^1_0n \rightarrow {}^{58}Co + {}^1_1p$$
$$^{59}Co + {}^1_0n \rightarrow {}^{58}Co + 2{}^1_0n$$
$$^{59}Co + {}^1_0n \rightarrow {}^{60}Co + {}^0_0\gamma$$
$$^{60}Ni + {}^1_0n \rightarrow {}^{60}Co + {}^1_1p$$

2.2 Corrosion

The corrosion process of austenitic steels is a complex one. It consists of a dissolution of the surface in contact with the sodium and a diffusion of the steel elements. It can be divided into four successive stages (Figure 1) in time, whose existence and duration depend on the temperature.

-Stage 1. Surface Cleaning.
This corresponds to the dissolution of the oxides and surface inclusions during the first few hours. Mass transfer during this stage is negligible.

-Stage 2. Austenite Dissolution.
At a temperature of 570°C or more, the following stage is the dissolution of the austenite layer in contact with the sodium and the diffusion of the steel elements towards the sodium. If the temperature is lower than 590°C, the following stages do not occur.

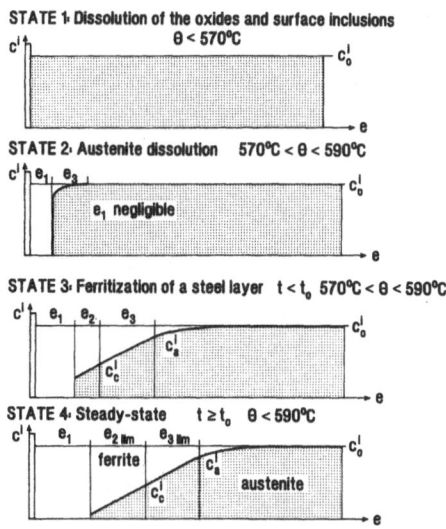

Fig. 1. Schematic of the four stages of the sodium corrosion process of stainless steel

-Stage 3. Formation of a ferrite layer.

During a longer period of time and at a temperature of 590°C or more, the diffusion of elements from the external austenite layer leads to a ferritization of the steel. This layer then dissolves and the steel elements diffuse to the surface.

-Stage 4. Steady - State.

The ferrite layer reaches a limit value ; its dissolution and the diffusion of elements to the sodium is equivalent to stoichiometric corrosion of the basic austenite. The contents reached at the sodium-steel interface are of the order of 1 to 2% for nickel, 5 to 7% for chrome and less than 0,5% for manganese.

The overall mass loss rate decreases in time during stages 2 and 3 until it reaches a constant value during stage 4. Mass losses during stages 2 and 3 are enriched in alloy elements. During stage 4, the corrosion rate is constant, the total amount of dissolved metallic elements corresponds to an alloy which is identical to the initial alloy. This is what is commonly called stoichiometric corrosion. In addition to the parameter of time corresponding to the above stages, the parameters affecting corrosion kinetics are as follows:

2.2.1 The Oxygen Content in the Sodium

Alloy element contents in the outer area depend on the sodium oxygen content : those of chrome, nickel and manganese increase with the oxygen content, that of iron decreases.

Moreover, the corrosion process changes when the oxygen content exceeds a threshold of low value (of the order of 5 ppm). Dissolution is then replaced by chemical reactions giving rise to complexes. Depending on the corrosion model being considered, the exponent of the oxygen term [O] ranges from 0,8 to 1,5. In BAQUE's model, the exponent of the oxygen parameter is 1.

2.2.2 The Sodium Velocity

In the literature, there is a critical sodium flow rate beyond which the corrosion rate does not increase. This phenomenon is explained by the disappearance of the laminar layer : the dissolved elements enter the sodium directly at a maximum, constant rate. The limit value may depend on the oxygen content.

2.2.3 The Temperature of the Steel

Temperature distribution is complex and difficult to model. This parameter however has a considerable effect on corrosion. It is also on this point that the various models in the literature differ the most.

2.2.4 The Downstream Effect

In fact, the downstream effect mainly appears in the isothermal areas, and, in extreme cases, leads to saturation of the sodium in dissolved elements, thus limiting the corrosion process. In the present case, the downstream effect can be disregarded in view of the positive or only very slightly negative temperature gradient along the fuel cladding.

Stages 2 and 3 also involve :

2.2.5 The Metallic Structure

Strain hardening in particular can affect the diffusion of the alloyed elements into the sodium.

2.2.6 The Chemical Composition of the Steel

A re-equilibrium of the crystalline network in the austenite area depleted alloyed elements results in a change of phase (passage from the gamma to a alpha phase) when the nickel and carbon contents are sufficiently low -due to chrome carbide dissolution. Molybdenum tends to slow down this process.

2.3 Deposition

The greater the temperature, the greater the solubility of the corrosion products. ^{54}Mn, ^{51}Cr and ^{59}Fe deposit according to a law as a function of temperature, with the exponent $(13200/\Theta)$ with Θ in K. The radio cobalt deposit mainly in the hot areas, that is to say 90 % in the core. The 10 % fraction that deposits on the components is considered not to be dependent on the temperature.

The sodium velocity affects the deposit linearly. The inventory of corroded matter is deposited in its entirety during cold shut-downs of the reactor.

3. ANACCONDA CODE

3.1 General

Evaluating contamination in handleable components of LMFBRs is based not only on experience but also on modelling tools that have sufficiently high performance characteristics to give a real representation of all the different physical and chemical phenomena at work in the processes of activation, corrosion and mass transfer responsible for radioactive contamination.

Any solution to this problem must involve devising a computer code.

The CORONA code gave a first approach to the problem, but its possibilities are limited. A new, more efficient code was therefore developed ; this was ANACCONDA (ANalysis of

Activation, Corrosion, COntamination of Na and Deposition of Activity).

This code is more flexible in use and allows several corrosion models to be used (CORONA, BAQUE and THORLEY) as well as giving a better representation of the thermal characteristics of a LMFBR core.

The code is also designed to meet requirements concerning the evaluation of activities deposited on the components more efficiently.

3.2 Modelling

All the phenomena mentioned above are modelled in the ANACCONDA computer calculation code.

3.2.1 Activation

The reactor core is the principal site of activation reactions. Reactions on the vessel and the components can be disregarded in this context. Our hypothesis considers ten distinct neutron areas (Figure 2).

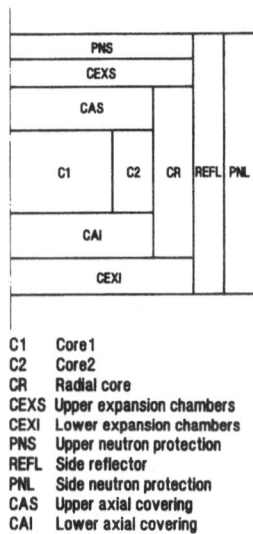

C1   Core1
C2   Core2
CR   Radial core
CEXS Upper expansion chambers
CEXI Lower expansion chambers
PNS   Upper neutron protection
REFL Side reflector
PNL   Side neutron protection
CAS   Upper axial covering
CAI   Lower axial covering

Fig. 2. The neutron areas of the hypothetical core used for the code to calculate the activation

- the core C1
- the core C2
- the radial core CR
- the upper expansion chambers CEXS
- the lower expansion chambers CEXI

- the upper neutron shielding PNS
- the side reflector REFL
- the lateral neutron shielding PNL
- the upper axial blanket CAS
- the lower axial blanket CAI.

This representation covers most situations and is enough to represent the main variations in intensity and in the neutron flux spectrum. The neutron flux considered constant within these areas. The code is designed to calculate activation according to fifteen activation reactions.

At present, only the main activation reactions presented in paragraph 2.1 have been introduced into the code.

In CORONA, the activation calculation is an approximation since the decrease in the radioactivity of the products formed during the operating period is not taken into consideration. Only the decrease in the products formed during the preceding periods is accounted for. In ANACCONDA, the activation calculation is not approximated.

3.2.2 Corrosion

3.2.2.1 Description of corrosion models

One of the first empirical relationships to describe the corrosion behaviour of stainless steels in sodium was produced by General Electric in USA.

The equation used in the previous French code CORONA was mainly taken from this G.E. work. For the reactor situation, the equation ignores the downstream effect and the coefficient has been adjusted to accommodate direct measurements taken on Rapsodie.

The CORONA code equation is of the form :

$$R = a \cdot v^{0,4} \cdot [O]^{1,16} \cdot \exp(-13200/T) \, \exp(0,12/(0,082 + t))$$

where :
  R = rate of metal loss in kg m^{-2}.s^{-1}
  a = coefficient
  v = Sodium velocity m/s
  $[O]$ = oxygen concentration in ppm (G.E)
  T = temperature in K
  t = time in years.

In order to provide a better understanding of the behaviour of radio nuclides in sodium systems, a number of organisations have undertaken research programmes in support of the subject of this study.

Studies on corrosion were carried out under different temperature, oxygen level and sodium velocity conditions.

BAQUE's model is therefore used in the ANACCONDA code for low oxygen levels (<5 ppm) and THORLEY's model for high oxygen levels (>5 ppm).

BAQUE's model equation is :

$$R = a \cdot v^{0,435} \cdot [O] \cdot \exp(-150,5/(T-544))$$

where :
  R = rate of metal loss in kg m^{-2} for 1 year
  a = coefficient
  v = Sodium velocity m/s
  $[O]$ = oxygen content in ppm (Eichelberger)
  T = temperature in °C.

THORLEY's model equation is :

$$S = \exp(9,48 + 1,5 \ln [O] - 9000/T - 0,18 \ln(L/D))$$

Where :
  S = rate of metal loss in µm/a
  $[O]$ = oxygen content in ppm (THORLEY)
  T = temperature in K
  L/D = downstream ratio (not taken into account)

$$\log S = a + 1,106 \log [O] - 3913/T$$

Where :
  S = rate of metal loss in mg cm^{-2}.a^{-1}
  a = coefficient
  $[O]$ = oxygen content in ppm (Eichelberger)
  T = temperature in K

3.2.2.2 Comparison of results

A comparison of these two models gave rise to a parametric study. Metal loss rates were calculated and compared at several oxygen levels. The results are presented in table 1 and figures 3 to 10.

On the one hand, there is no significant difference between the two series of results for temperatures of 600°C or more. On the other hand, the results are quite different for lower temperatures because, with BAQUE's model, there is no significant corrosion at temperatures lower than 550°C.

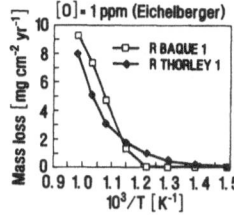

Fig. 3. Mass losses at [O]= 1ppm based on the models of Baque and Thorley

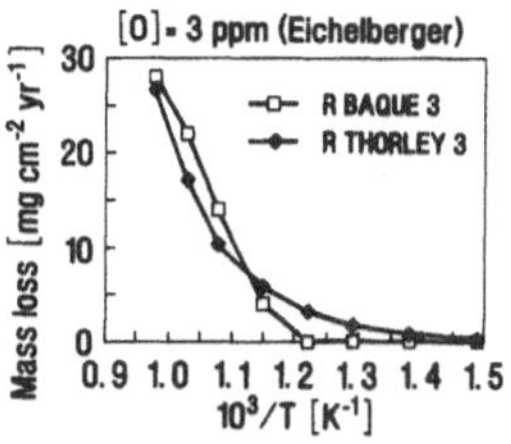

Fig. 4. Mass losses at [O]=3 ppm based on the models of Baque and Thorley

3.2.2.3 Modelling

The main parameters affecting the corrosion rate are as follows :
    sodium velocity
    oxygen content
    steel temperature.

Modelling the first two parameters (V_{Na} and [O]) is relatively easy. Sodium velocity varies little between the inlet and the outlet of the assembly. The velocity per area can be considered constant and calculated in terms of the Na flow rate, the cross section and the density of sodium at the temperature under consideration.

The oxygen content can also be considered constant. It is generally considered that the oxygen content corresponds to that of sodium at saturation point at the temperature of the cold point of the purification system, i.e. the cold traps.

The trickiest parameter to model is the temperature. In CORONA, the temperature profile is linear for the entire core. This does not reflect the real situation. Moreover, the temperature taken into account is that of sodium. The temperature that should be used, particularly with BAQUE and THORLEY's models, is that of the cladding steel.

Once the specific activities per neutron area and per time step are known, the activity per isotope of the total corroded mass can be calculated for a given time-period.

There can be remarkable variations in temperature inside the different neutron areas. It is therefore vital to calculate representative surface/temperature histograms for each neutron area. The relevant metal surface is evaluated for each 5°C temperature variation. The corrosion model is then applied for the mean temperature in the sub-area (T + 2.5°C) for each time step (10 days).

Once the specific activities per neutron area and per time step are known, the activity per isotope of the total corroded mass can be calculated for a given time-period.

In ANACCONDA, the temperature profiles are linked to the operating power of the reactor. The corrosion model used for oxygen concentrations lower than 5 ppm is usually BAQUE's. For concentrations higher than 5 ppm, THORLEY's model is used. Different choices of model can be imposed.

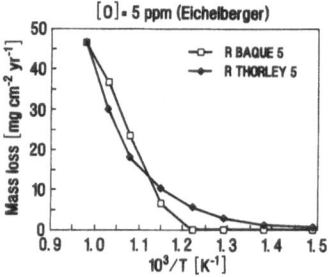

Fig. 5. Mass losses at [O]=5 ppm based on the models of Baque and Thorley

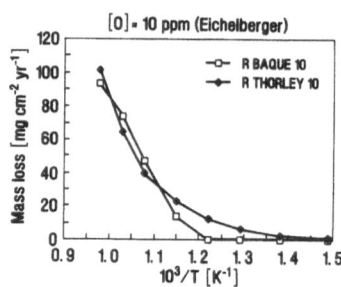

Fig. 6. Mass losses at [O]=10 ppm based on the models of Baque and Thorley

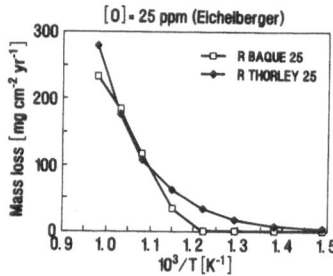

Fig. 7. Mass losses at [O]=25 ppm based on the models of Baque and Thorley

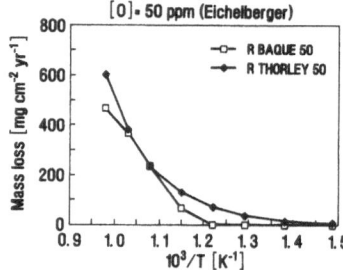

Fig. 8. Mass losses at [O]=50 ppm on the models of Baque and Thorley

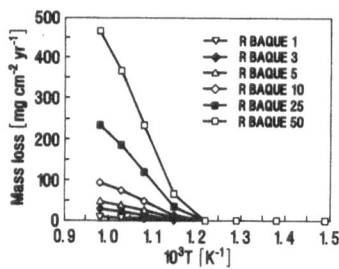

Fig .9. Mass losses at different O concentrations in Na based on the model of Baque

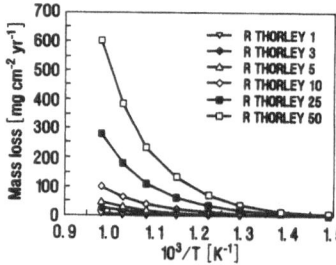

Fig. 10. Mass losses at different O concentrations in Na based on the model of Thorley

Table 1. Comparison of results obtained using Baque's and Thorley's models: Metal loss rates in (mg cm^{-2}a)

Temp. °C	Baq..	Th.	Baq.	Th.	Baq.	Th.	Baq.	Th.	Baq.	Th
	Oxygen Content in (wppm)									
	1	3,5	3	7,7	5	11,3	10	18,8	25	37
400		0		0		1		1		3
450		0		1		1		3		7
500		0		2		3		6		16
550	0	1	0	3	0	6	0	12	6.10^{-9}	33
600	1	2	4	6	7	10	13	22	33	61
650	5	3	14	10	23	18	47	39	117	107
700	7	5	22	17	37	30	73	64	184	177
750	9	8	28	27	47	47	93	101	232	279

Baq. using Baque's model,   Th. using Thorley's model

3.2.2.4 Data Required for the Calculation

The small amount of data required for ANACCONDA is significantly greater than that required for CORONA because of the greater complexity of the model.
1 - Neutron flux file.
   per neutron area (10)
   per energy group (25).
2 - Cross section file.
   per neutron area
   per energy group
   per activation reaction.
3 - Material characteristics file.
   composition of cladding steel
   isotopic composition of cladding steel
   radioactive constant.
4 - Core thermal data file.
   standard profile of cladding temperature
   nominal, maximum and minimum cladding temperatures
   number of thermal sub-areas.
   core inlet temperature
   number of sub-assemblies per area.

5 - Geometric data file.
   for each area :
      height
      number of fuel pins per sub-assembly
      number of sub-assemblies per area
      external diameter of the pin.
   for the hexagonal tubes :
      perimeter
      height.

6 - Reactor operating file.
   number of operating phases
   percentage of nominal power per operating period
   operating time
   oxygen content
   primary pump speed
   temperature of sodium inlet into core.
7 - Initial activity file.

4. EXAMPLES OF APPLICATION

4.1 Experimental facilities

As shown in Table 2, there is a good agreement between the calculated metal loss rates and the experimental results.

4.2 Parametric study on Phénix and Superphénix

Mass loss rates were calculated for Phénix and Superphénix using the two models for different oxygen levels : 1, 3, 5 and 10 ppm.

The results are presented in Table 3 and 4.

The differences are due to the fact that even though release rates were low at temperatures below 600 °C, the corresponding surfaces were very large. There is a factor of 3 between Thorley's and Baque's results for Phénix, and a factor of 10 for Superphénix.

4.3 Comparison: Calculation Experiment on Phénix

As shown in Table 5, both the measurements taken during waste cleaning and decontamination and the calculations gave the same order of magnitude for released and deposed mass on heat exchanger. It should be noted that the experimental value corresponds to the sum of the corrosion products deposited then dissolved, plus a thickness of base metal dissolved by the decontamination treatment.

5. CONCLUSION

In order to improve the evaluation of radioactivity levels in LMFBRs, a new corrosion and contamination code, ANACCONDA, was developed.

The corrosion models introduced into ANACCONDA are BAQUE's model, qualified for low oxygen contents (< 5 ppm), and of THORLEY's model, which is more suitable for oxygen contents of over 5 ppm. The essential difference between the two models is that in BAQUE's model, corrosion at temperatures lower than 540°C is negligible.

Table 2. Measured and calculated mass losses on various CEA facilities

Facility	temperature (°C)	v_{Na} (m/s)	duration (h)	mass loss (mg/cm^2) measured	calculated CORBAC	THORLEY
TIGIBUS	550	2	5750	< 0,2	0	
	620	2	9700	0,9	1,1	
	635	2	8800	1,2-1,3	1,4	
	650	2,15	8050	1,7	1,7	
	700	2	2000	1,06	0,94	
	700	2	3000	1,33	1,25	
GRANGIBUS	650	2	2000	0,8	0,7	
	650	2	3000	1,0	0,9	
	650	2,4	11700	2,6	2,4	
	650	4	6000	1,76	1,8	2,1
	650	4	9000	2,5	2,5	3,2
OCTAVE	600	7	7100	0,75-0,8	0,75	1,45
	600	7	7200	0,9-1,0	0,75	1,45
	600	7	11700	1,45-1,6	1,1	2,4
MASSBROS	580	0,9	10200	0,1	0,1	
	600	0,9	10200	0,45	0,4	
	620	0,9	10200	0,8	0,8	
	650	0,9	10200	1,45	1,4	
	670	0,9	10200	2,0	1,7	
	580	0,9	13600	0,1	0,1	
	600	0,9	13600	0,3	0,45	
	620	0,9	13600	0,85	0,9	
	650	0,9	13600	1,7	1,6	
	670	0,9	13600	2,3	2,0	
	580	0,9	23800	0,2-0,6	0,13	
	600	0,9	23800	0,8	0,8	
	620	0,9	23800	1,3-1,8	1,6	
	650	0,9	23800	3	2,8	
	670	0,9	23800	4	3,6	

Thermal modelling of the core has been considerably improved in ANACCONDA, allowing a representative surface/temperature histogram to be obtained for each area.

The calculation results obtained were in agreement with the few experimental values available. It would now be appropriate to test this code on other LMFBRs. It would therefore be essential to obtain accurate and complete data files such as those described in this paper and, if possible, any available experimental data.

Table 3. Comparison of results obtained using Baque's and Thorley's model: Application on Phénix

PHENIX	RELEASED MASS (kg/year)	
OXYGEN (ppm)	BAQUE	THORLEY
1	3,4	8,2
3	10,1	27,7
5	16,8	48,7
10	33,8	105

Table 4. Comparison of results obtained using Baque's and Thorley's model: Application on Superphénix

SUPERPHÉNIX	RELEASED MASS (kg/year)	
OXYGEN (ppm)	BAQUE	THORLEY
1	3,4	37,5
3	10,1	126
5	16,9	222
10	33,8	478

Table 5: Comparison: calculation/ experience on Phénix

PHENIX EXPERIMENT	BAQUE MODEL	THORLEY MODEL
Mass released in cleaning and decontamination waste	Mass released by corrosion	
activated corrosion products and dissolved base metal	activated corrosion products	
10,3 kg	4,7 kg	11,5 kg
measured ^{54}Mn	calculated ^{54}Mn	
4 TBq	2-5 TBq	

References

1. R. Clerc, J. Guidez, P. Michaille, J. Misraki, Specialists Meeting on Fission and Corrosion Product Behaviour in Primary Systems of LMFRs, Karlsruhe, 1987, Rep. IWGFR/64, IAEA, Vienna, KfK Report 4279 (1987).
2. P. Baque, unpublished work.
3. A.W. Thorley, C. Tyzack, "Liquid Alkali Metals" BNES, London, 1973, 253-273.

RADIOACTIVE SODIUM CHEMISTRY LOOP FOR ACTIVITY TRANSPORT STUDIES AND METER TESTING

V. Ganesan, P. Muralidaran, K. Chandran, K.C. Srinivas,
T. Gnanasekaran, G. Periaswami and C.K. Mathews

Materials Chemistry Division, Chemical Group
Indira Gandhi Centre for Atomic Research
Kalpakkam-603 102, India

H.U. Borgstedt and G. Frees

Kernforschungszentrum Karlsruhe GmbH.
Institute of Materials Research III
D 76021 Karlsruhe, Germany

1. INTRODUCTION

Transport of radioactive nuclides from the reactor core to out-of-core components is a common phenomenon occurring in Liquid Metal Cooled Fast Breeder Reactors (LMFBRs)[1,2]. The major source of activity released during normal operation of the reactor is the activated corrosion products such as ^{54}Mn, ^{60}Co, ^{140}Ba etc. arising from the materials of construction of the core components. In the event of fuel pin failure, fission product nuclides such as ^{137}Cs, ^{131}I etc. and fuel materials are released into the coolant stream. As these nuclides possess high gamma energies, they pose serious problems during operation and maintenance of the reactor systems. In order to understand and study such activity release behaviour from the reactor core, it is imperative to generate data in loop systems with controlled chemical and process conditions. With this in view, Radioactive Sodium Chemistry Loop (RASCL) has been built at IGCAR, Kalpakkam.

Another important area of concern in any operating sodium circuit of LMFBR is the presence of non-metallic impurities such as oxygen, hydrogen and carbon in the coolant. Under normal circumstances, hydrogen finds its way into sodium by ingress of hydrocarbon oil occurring at the pump shaft seal and from the water side corrosion in steam generator units. However, failure of steam generator tubes can lead to steam ingress into sodium resulting in severe sodium water reactions. Thus, continuous monitoring of hydrogen levels in both primary and secondary sodium circuits of LMFBRs becomes essential.

This paper describes the salient features of the Radioactive Sodium Chemistry Loop built at the Radiochemistry Laboratory, Kalpakkam under the Indo-German collaboration. The paper also describes the calibration of an electrochemical meter for detection of hydrogen in ppm and sub-ppm levels in sodium. The response of such a meter for addition of sodium hydroxide at relatively low temperature of sodium has also been described.

Liquid Metal Systems, Edited by H.U. Borgstedt
and G. Frees, Plenum Press, New York, 1995

2. RADIOACTIVE SODIUM CHEMISTRY LOOP

2.1. Objectives

In order to carry out experiments related to activity transport due to radionuclides of interest under controlled conditions with respect to sodium purity and other process parameters such as temperature and sodium velocity, RASCL was designed and fabricated [3]. The major objectives of this loop are:

i) to study the release, transport and deposition behaviour of radionuclides as a function of sodium temperature, velocity, purity and downstream position,

ii) to study the release behaviour of various fission products from defected carbide fuel pins simulated to various burnups and to study the kinetics of oxide fuel sodium chemical interactions,

iii) to develop radionuclide traps for nuclides such as ^{54}Mn, ^{137}Cs etc. and

iv) to evaluate the long term performance of on-line monitors for oxygen, hydrogen and carbon in well characterised sodium

2.2. Salient features of the loop

The general outline of RASCL is given in Fig. 1. The total sodium hold-up of the loop is 150 kg. The main loop is of "figure of 8" configuration with a recuperative heat exchanger, bulk heater and test section. A flat linear induction pump with a capacity of 2 m^3h^{-1} at 4.5 bar pressure circulates sodium through the main and auxiliary loops, the sodium flow being measured by means of a permanent magnet flow meter. Typical fast reactor core conditions such as sodium velocity of 5 ms^{-1} and temperature of 923 K can be achieved in the test section to facilitate study of the release behaviour of activated corrosion products such as ^{54}Mn, ^{60}Co and ^{59}Fe by employing suitable specimens. Con-o-seal coupling made of stainless steel AISI type 347 body with stainless steel AISI type 321 gaskets is used for introduction of specimens into the test section. Chemical interactions between the fuel and the coolant can be studied by introducing defected fuel pins with simulated burnups. To facilitate gamma scanning of the various nuclides, the main loop is made planar.

The operation of RASCL can be carried out in different modes namely, circulation of sodium through

(a) main loop only
(b) auxiliary loop only
(c) cold trap and dump tank and
(d) entire loop.

The auxiliary loop consists of a cold trap for impurity control, a plugging indicator, on-line meters for oxygen, hydrogen and carbon and a foil equilibration port for characterising the sodium. In addition to these, a TNO type overflow sodium sampler is incorporated in the main loop bypass section for periodic sodium sampling. Thus the auxiliary loop aids in characterising the sodium during activity transport studies. The cold trap is designed such that the sodium contained in it is not drained during dumping operation. This prevents unwanted mixing up of impurities with the sodium in the dump tank. In addition to this, the crystalliser and the economiser of the cold trap were kept separated to facilitate incorporation of radio nuclide trap for ^{137}Cs. The foil equilibration and sampler sections have the facility to drain sodium selectively while the rest of the loop is in operation.

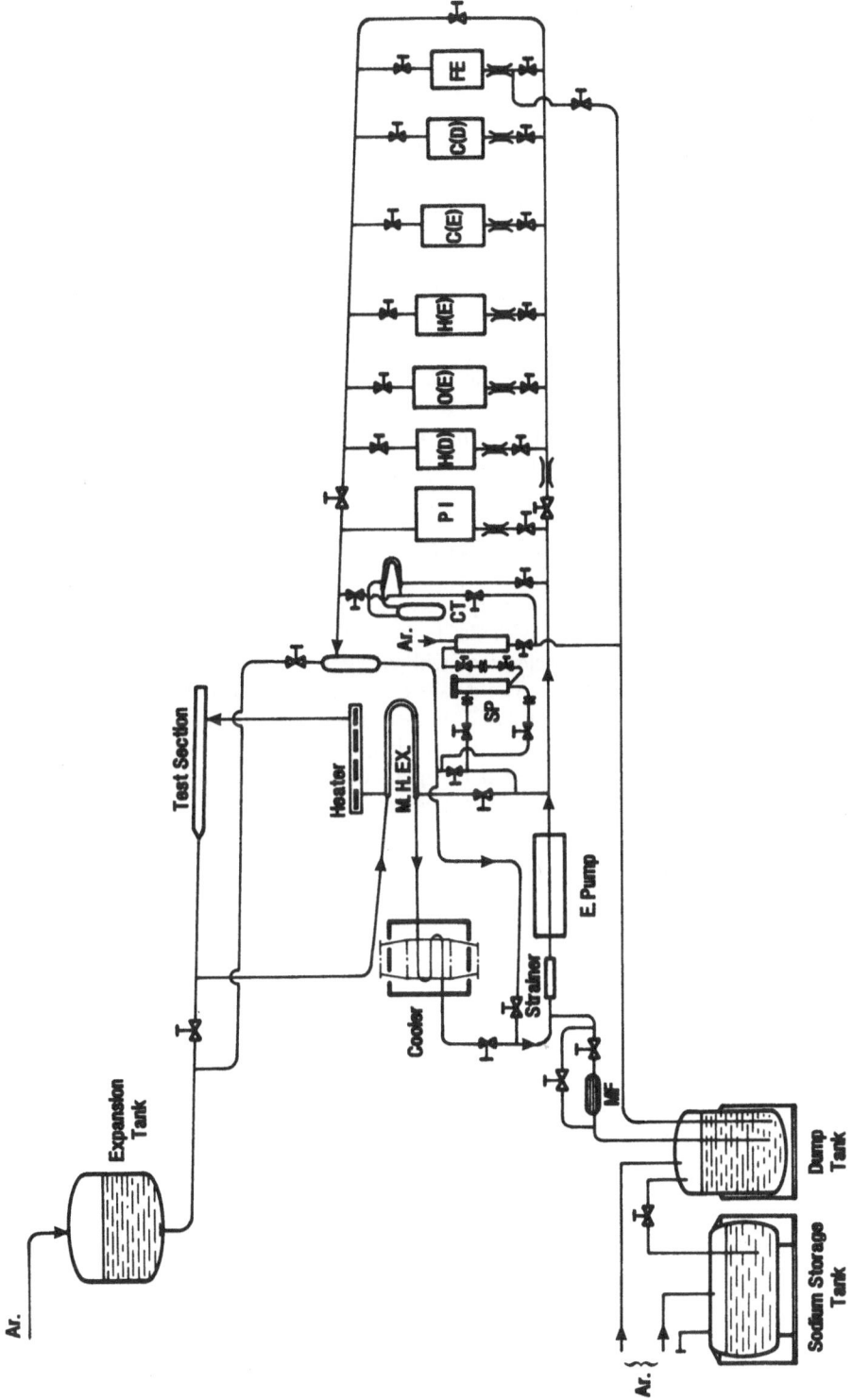

Fig. 1. Schematic diagram of the Radioactive Sodium Chemistry Loop

The control, surveillance and data logging of the various loop parameters are performed by a personal computer. The microprocessor module handles the prime task of maintaining temperature control over the various sections of the loop. Using suitable software the processor is made to scan the thermocouple outputs in the loop and apply the values derived from specially identified thermocouples to switch the corresponding heaters on/off. This arrangement effectively realises a bank of on/off temperature controllers in a compact hardware environment.

The processor also scans the leak and level detectors. In the event of a leak being detected, the dump routine is invoked to perform the dumping operation. The level signals are monitored in order to assess the presence of sodium in the loop. Provisions have been made to log the EMF signals from the on-line monitors using custom made high input impedance volt meters.

The personal computer forms the front end intelligence of the system. User friendly routines guide the operation of the loop. The PC receives user intelligible mnemonics to denote the various parameters and performs a translation into the respective channel numbers. In order to do this, the PC consults an internal database where field tag labels are set off against the corresponding channel numbers.

Graphical representation of the loop is being developed so that alarm messages can be displayed on the locations concerned for ready assimilation of its impact on operation.

3. CALIBRATION OF HYDROGEN METER

An electrochemical hydrogen meter is incorporated in the auxiliary section of RASCL. This meter is a hydrogen concentration cell with $CaCl_2$ 5% CaH_2 as hydride-ion conducting solid electrolyte. Li/LiH serves as reference electrode with well defined hydrogen partial pressure at the operating temperature of the meter. The details of construction of such meters have been described elsewhere [4,5]. This meter (RASCL H-meter) was operated at 723 K at a sodium flow rate of 50 l h^{-1} and the hydrogen content of the loop sodium was controlled by operating the cold trap suitably. After about 750 h of operation at constant meter and cold trap temperatures, the temperature coefficient was measured by reducing the meter temperature from 723 to 713 and then to 703 K.

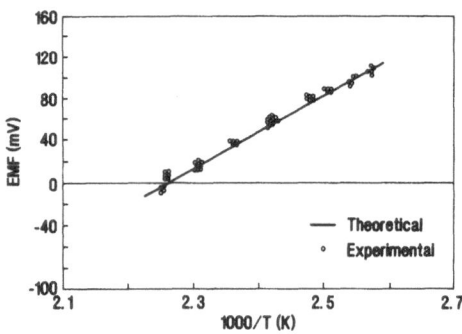

Fig. 2. Variation of EMF as a function of reciprocal cold trap temperature for RASCL H-meter

The temperature coefficient was found to be 1.2 mV K^{-1} which is in agreement with the theoretically calculated values. The response of the meter for changes in hydrogen concentration was tested by changing the cold trap temperature from 388 to 443 K in steps of 10 K. Fig. 2 shows the variation of meter output as a function of reciprocal cold trap temperature. The change in signal was found to be in excellent agreement with the theoretically calculated values [6]. In addition to this meter, four other hydrogen meter probes were calibrated in RASCL before incorporation in the secondary circuits of the Fast Breeder Test Reactor (FBTR) at Kalpakkam. The calibration of such meters are shown in Fig. 3. Though some of these meters did not show theoretical behaviour, still the temperature coefficients of the individual meters were reasonable and the response to changes in hydrogen concentrations was linear.

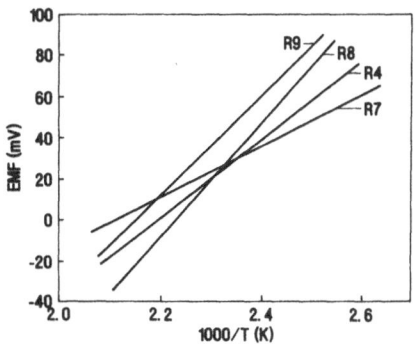

Fig. 3. Calibration of FBTR H-meters in RASCL

4. RESPONSE TO SODIUM HYDROXIDE ADDITION

The response of the electrochemical hydrogen meter for sodium hydroxide addition at relatively low temperature of 523 K in the cold leg section of the loop was studied and the output of the meter was followed as a function of time. Since direct addition of hydrogen in gas phase is not possible with the present design and since the product of sodium-water reaction in the steam generator is partly sodium hydroxide, the same was chosen for changing the hydrogen levels in sodium. About 5 g of sodium hydroxide in the form of pellets packed in fine stainless steel wire mesh was loaded into the sampler vessel and secured in its place in the sodium loop. The sampler was valved-in with sodium at 523 K flowing at the rate of about 100 l·h⁻¹. The EMF output of the meter decreased rapidly within a minute which was approximately the time taken for sodium to travel from the sampler section to the hydrogen meter section thereby showing the high sensitivity and very low response time of the meter for change in hydrogen levels. The decrease in the meter output and the corresponding increase in hydrogen levels in sodium are shown in Fig. 4.

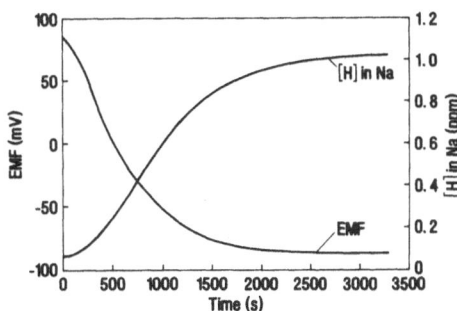

Fig. 4. Response of hydrogen meter to sodium hydroxide addition

After stopping the sodium hydroxide addition by isolating the sampler section and upon valving-in the cold trap, the meter output increased gradually and attained values corresponding to the hydrogen concentration dictated by the cold trap temperature as shown in Fig. 5.

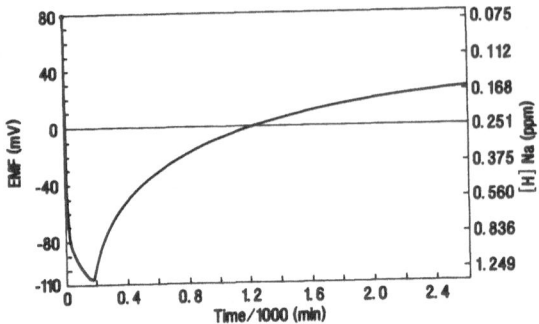

Fig. 5. Response of hydrogen meter to sodium hydroxide addition followed by cold trapping

5. STEAM LEAK EXPERIMENTS

During operation of LMFBRs, owing to the high differential pressure between the tube and shell side, any failure in the steam generator tubes may lead to violent sodium-steam reactions releasing enormous amounts of hydrogen and oxygen into the secondary sodium circuits. In many cases, the tube failure results initially in micro leaks of steam into sodium as stated earlier. If the hydrogen released in such instances can be detected instantaneously, enlargement of leaks leading to violent reaction can be prevented. This calls for study of the response of the electrochemical hydrogen meter for changes in hydrogen levels as low as 50 ppb in sodium. In order to study the usefulness of the hydrogen meter for detection of steam leaks into sodium at low temperatures as envisaged during reactor start-up and low power operation, a small stainless steel capsule with a fine orifice as shown in Fig. 6, has been designed and fabricated. The volume of the capsule is 5 ml and the orifice diameter is 1 mm.

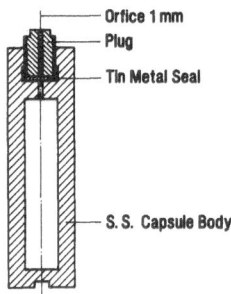

Fig. 6. Capsule for steam leak experiments

The steam produced in the capsule could be let out into sodium at any desired temperature by choosing a suitable metallic seal at the orifice. Since it was required to simulate steam leak conditions at relatively low temperature of 523 K as envisaged in FBTR secondary circuits, tin was chosen as the orifice seal. The performance of tin metal seal was tested in air by taking calculated quantities of water and heating the capsule slowly. The operation of the seal was tested several times and the seal was found to blow open at 507 K, giving out a fine jet of steam. It is proposed to incorporate this capsule in the sampler section of RASCL for subsequent testing in sodium.

6. CONCLUSIONS

The Radioactive Sodium Chemistry Loop for activity transport studies has been put into operation with all the sections commissioned. The electrochemical hydrogen meters based on $CaCl_2$-5% CaH_2 as hydride-ion conducting solid electrolyte were successfully tested and calibrated in this loop. The response of hydrogen meter to sodium hydroxide addition was studied and the response was found to be very fast allowing for the time taken for hydrogen to reach the meter chamber from the point of injection in the loop.

References

1. H. Feuerstein, A.J. Hooper and F.A. Johnson, Atomic Energy Review, 17 (1979) 697.
2. H.U. Borgstedt and C.K. Mathews, Applied Chemistry of the Alkali Metals, Plenum Press, New York, 1987.
3. V. Ganesan, P. Muralidaran, T. Gnanasekaran, G. Periaswami and C.K. Mathews, Proc. Fourth Int. Conf. on Liquid Metal Engineering and Technology, Avignon, France, 1988, vol.3, p. 634-1.
4. T. Gnanasekaran, V. Ganesan, G. Periaswami, C.K. Mathews and H.U. Borgstedt, J. Nucl. Mater., 171 (1990) 198.
5. T. Gnanasekaran, K.H. Mahendran, R. Shridaran, V. Ganesan, G. Periaswami and C.K. Mathews, Nucl. Technol., 90 (1990) 408.
6. T. Funada, K. Nihei, S. Yuhara and T. Nakasuji, Nucl. Technol. 45 (1979) 158.

POST-CORROSION AND METALLURGICAL ANALYSES OF SODIUM PIPING MATERIALS OPERATED FOR 100,000 HOURS

Eiichi Yoshida, Shoichi Kato and Yusaku Wada

Materials Development Section, O-arai Engineering Center,
Power Reactor and Nuclear Fuel Development Corporation
4002 Narita, O-arai-machi, Ibaraki-ken 311-13, Japan

1. INTRODUCTION

For liquid metal (sodium) cooled fast reactor systems, the effects of sodium environment on the corrosion and the mechanical strength properties of structural materials have to be evaluated to maintain the material integrity through out the design life. Generally, under the operation conditions of LMFBRs with the sodium purity control, the compatibility of structural materials to sodium is good from the engineering viewpoint that no large problem has been recognised in practical use [1]. However, for the integrity of the structural materials over a long service life of 30-40 years, it is important to testify based on the longer duration data.

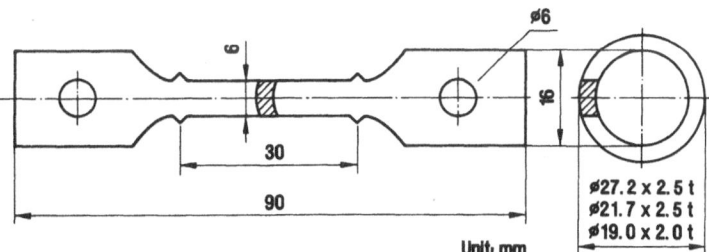

Fig. 1. Configuration of tensile and creep-rupture specimen

For the purpose of long time extrapolation, the detailed metallurgical analyses and mechanical strength tests were conducted on pieces of Type 304 stainless steel pipes cut out from two units of sodium testing loop (the Material Test Sodium Loop No. 1 installed in 1970, and No. 2 in 1972) which have recently achieved a long-time operation of about 100,000 hours. As for basic researches, test duration is usually less than 20,000 hours, and 100,000 hours is not practical. Therefore the precious data could be provided by this study,

Liquid Metal Systems, Edited by H.U. Borgstedt
and G. Frees, Plenum Press, New York, 1995

and it was expected to clarify the metallurgical change of material after the long-time exposure to sodium. As the material examinations on the sodium test apparatus with a long operation experience, the studies on intermediate heat exchanger by Oak Ridge National Laboratory in US. [2] and on the sodium loop by Mol Research Institute in Belgium [3] were well-known. However there are few researches on the materials operated in sodium over 100,000 hours.

In this paper, the change of the surface corrosion morphology, micro structure, chemical composition and high-temperature strength of the sodium piping materials after a long time exposure to sodium were analysed metallurgically using the test results on the above test loops with an about 100,000 hours operation history.

2. EXPERIMENTAL PROCEDURE

2.1 Test materials

The test materials were solution treated Type 304 stainless steel pipes. Chemical composition is shown in Table 1. The total number of material heats were 4; AM, AD, BM and BD, which were used in main circulation systems and test sections of the sodium testing loops. The carbon contents of 0.054-0.056% in BM and AD heats were slightly higher than those in the other heats. The contents of impurities like Al , N an O, especially 0.051% of Al, of BD heat were also high.

The histories of test specimens are summarised in Table 2. Two specimens, AM-1 and BM-1, were sampled from main circulation systems of loops 1 and 2. Another eight specimens, AD-1 to AD-4 and BD-1 to BD- 4, were sampled from corrosion and mass transfer test sections in the daughter systems of loops 1 and 2 called M-3 and M-12, respectively. Specimens AM-1 and BM-1 were operated for 105,000-108,000 hours in the temperature range of 420-440°C. Specimens AD-1 to AD-4 were operated for 53,500 hours in the temperature range of 420-600°C. Specimens BD-1 to BD-4 were operated for 82,000 hours in the temperature range of 420-650°C. In the case of specimens BD-1 to BD-4, the average temperature is shown in Table 2, because the operation temperature changed to 50°C higher after 47,000 h of operation. The oxygen concentration in the sodium during the operation period was less than 10 ppm.

Table 1. Chemical composition of tested materials (Type 304 stainless steel) in mass %

Heat	C	Si	Mn	P	S	Ni	Cr	Al	N	O
AM	0.046	0.28	1.69	0.022	0.014	9.38	18.40	0.003	0.024	0.019
AD	0.056	0.44	1.70	0.024	0.006	9.48	18.47	0.016	0.029	0.006
BM	0.054	0.34	1.67	0.029	0.016	9.65	18.79	0.018	0.031	0.006
BD	0.048	0.47	1.78	0.035	0.021	9.42	19.92	0.051	0.039	0.026

2.2 Methods of Material Testing

The change in the surface corrosion morphology, micro structure and composition of each test specimen sampled from the piping, was examined by the optical microscopy and the electronic microscopy(SEM, TEM). Tensile and creep-rupture tests in air were also conducted on the typical piping materials to survey the change of their mechanical properties. Figure 1 shows the configuration of tensile and creep test specimens. In the tensile test, the test specimen of which sodium affected surface was cut out to the depth of 0.5 mm (hereinafter "the thermal aged material") was also used in order to separately evaluate the effect of sodium environment.

Table 2. Operation histories of tested materials

Sodium loop	Section		Specimen	Operation temp. (°C)	Operation time (h)
Test loop 1	Main	(∅ 27.2 x 2.5t)	AM-1	440	108,000
Test loop 1	M-3	(∅ 19.0 x 2.0t)	AD-1	500	53,500
Test loop 1	M-3	(∅ 19.0 x 2.0t)	AD-2	550	53,500
Test loop 1	M-3	(∅ 19.0 x 2.0t)	AD-3	600	53,500
Test loop 1	M-3	(∅ 19.0 x 2.0t)	AD-4	420	53,500
Test loop 2	Main	(∅ 27.2 x 2.5t)	BM-1	420	105,000
Test loop 2	M-12	(∅ 21.7 x 2.5t)	BD-1	525*	82,000
Test loop 2	M-12	(∅ 21.7 x 2.5t)	BD-2	575*	82,000
Test loop 2	M-12	(∅ 21.7 x 2.5t)	BD-3	625*	82,000
Test loop 2	M-12	(∅ 21.7 x 2.5t)	BD-4	420*	82,000

 * Average temperature
** Material: Type 304 ss; Oxygen level: < 10 ppm; Sodium velocity: 1 ~ 0.3 m/s

3. TEST RESULTS

3.1 Corrosion Morphology and Composition Change

Figure 2 shows the typical SEM micrographs of main circulation piping materials(AM-1 and BM-1). The sodium exposed surface was covered all over with coral-like corrosion products (Fig. 2a) and particulate corrosion products of the thickness of less than 1 μm were observed over the surface (Fig. 2b). The coral-like corrosion products were rich in Fe content, while the particulate ones contained a greater amount of Fe or Mn-Ni compared with those of the base metal.

Figure 3 shows the typical corrosion morphology at each position (BD-1~4) of M-12 piping materials. On the surface of the pipes (BD-1~3) used at the high-temperature (up-stream) region, it was observed that alloy elements in the steel were dissolved out into the sodium. For the BD-4 pipe used at the low-temperature (down-stream) region, the deposit of coral-shaped or particulate corrosion products were shown. The deposit consisted mainly of Mn-Ni, because of the mechanism that Mn and Ni were dissolved out into sodium in the up-stream region and precipitated in the down-stream region. This phenomenon of dissolution and deposition was also shown by the analysis on the surface of pipes in Figure 4, and was also observed by previous examinations of the short-time corrosion test [4,5]. Similar results were observed in M-3 piping materials (AD-1~4).

Figure 5 shows the cross-sectional distribution of Ni, Cr, Mn and Si contents in M-12 piping materials. In the up-stream region (BD-1~3), the above contents decreased from the sodium exposed surface toward inner section due to their dissolution into the sodium, and the depth of sodium affected zone tended to increase with the rise of sodium temperature. Also, the increased carbon content, i.e., carburization, was recognised at the sodium exposed surface of pipes. The carburization becomes remarkable with the higher temperature, but its depth was as small as 10 μm at its maximum.

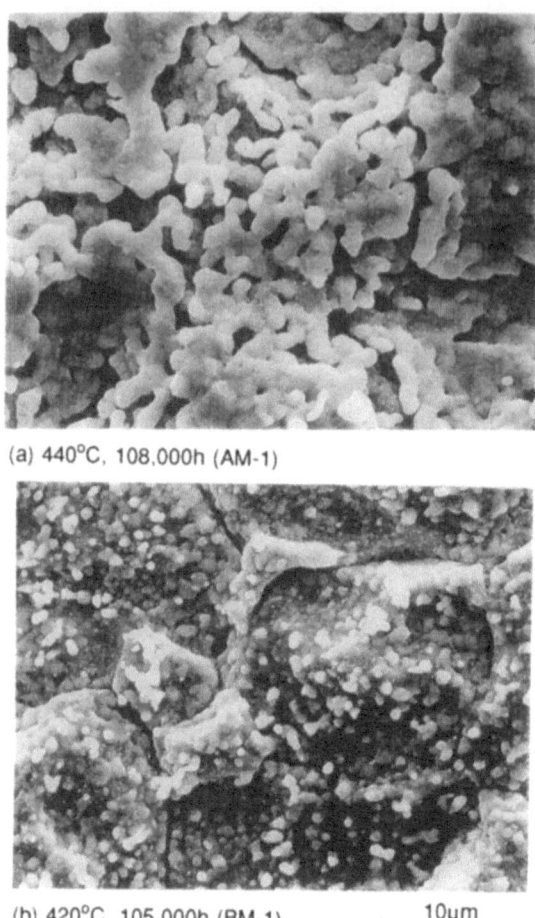

(a) 440°C, 108,000h (AM-1)

(b) 420°C, 105,000h (BM-1)   10μm

Fig. 2. SEM micrographs of Type 304 sodium piping surface after exposure to flowing sodium for 105,000~108,000 hours

3.2 Microstructural Changes

Figure 6 shows the typical cross-sectional micro structures after long-time sodium exposure. In the main circulation piping materials (AM-1, BM-1), although the precipitation of very small carbide particles and some carburization were observed near the surface. In the M-12 piping materials (BD heat), many sigma phases were observed on the grain boundary in the case of the temperature above 575°C. In the M-3 piping materials (AD heat), more precipitation of carbides $M_{23}C_6$ at higher operation temperature was observed on the grain boundary. The degraded layer was also observed at the sodium exposed surface.

Figure 7 shows the TEM observation results for the typical micro structures. At a low temperature (420°C), $M_{23}C_6$ was observed on the grain boundary, α-Fe was also observed adjoining the $M_{23}C_6$. In the piping materials with temperatures of 525°C and 550°C, $M_{23}C_6$ has grown and become coarse, with the precipitation of α-Fe being more remarkable. Considering that $M_{23}C_6$ consisted mainly of Cr and Ni, it seems that the increased precipitation of $M_{23}C_6$ reduced the nearby Cr and Ni contents and accelerated the precipitation of α-Fe. But at the high temperature above 550°C, the M$_{23}$C$_6$ on the grain boundary did not grow much but rather tended to reduce in number through disintegration.

Sodium flow

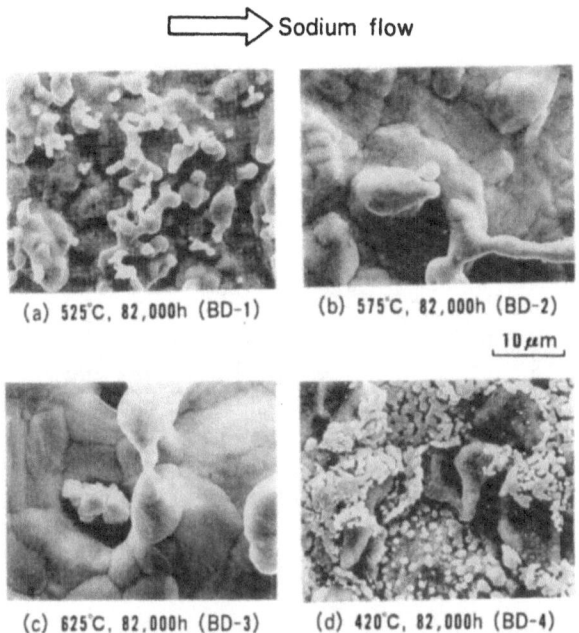

(a) 525°C, 82,000h (BD-1)　　(b) 575°C, 82,000h (BD-2)

10 μm

(c) 625°C, 82,000h (BD-3)　　(d) 420°C, 82,000h (BD-4)

Fig. 3. SEM micrographs of Type 304 steel sodium piping surface after exposure to flowing sodium for 82,000 hours

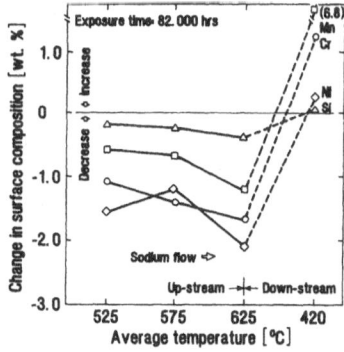

Fig. 4. Change in the surface composition of Type 304 steel sodium pipes after exposure to flowing Na for 82000 h

In the BD heat, sigma phases (Cr-Fe) were observed along the grain boundary, which became coarse as the temperature was higher. It seems that the disintegration of M$_{23}$C$_6$ at the high-temperature has been caused by the reduced carbon dispersion within the grain boundary or the absorption of Cr due to the precipitation of sigma phase on the grain boundary. Further, the thin-plate-like AlN which associated the sigma phase on the grain boundary was also observed in the case of the temperature at 625°C. This was only observed in the BD heat containing much Al and not in other heats with a small Al content. AlN has been also recognised by N. Shinya et al [6] in the materials rich in Al, showing accordance with the present study results.

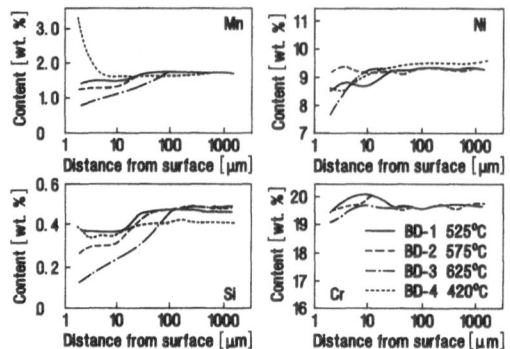

Fig. 5. Cross-sectional distribution of Mn, Ni, Si and Cr for Type 304 sodium pipes after exposure to flowing sodium for 82,000 hours

3.3 Mechanical Properties

Figure 8 shows the high-temperature tensile test results for BM and BD heats of piping materials, along with the design yield strength S_y and design tensile strength S_u of Type 304 stainless steel. Both 0.2% yield and ultimate tensile strengths of sodium piping materials obtained from this study have satisfied S_y and S_u values irrespective of the operation temperature and time. These values were not significantly different from those of thermal ageing materials under the same operation conditions, and no effects of sodium environment were recognised. But in BD heat, the trend was observed that the fracture elongation reduced on the whole. AM and AD heat piping materials satisfied the Sy and Su values and showed no significant difference in property changes from the thermal ageing materials.

Figures 9 and 10 show the creep-rupture test results for AD and BD heat piping materials, along with the in-air creep data [7,8] so far obtained on the as-received Type 304 stainless steel plates. The AD heat (AD-3) piping material showed the equal creep-rupture strength to that of as-received material, but indicated no reduction in the fracture elongation. The creep-rupture strength of BD heat piping materials was higher than average curve at 525°C (BD-1) but the trend was observed that their creep rupture strength lowered as the piping material is exposed to such higher operation temperatures as 575°C (BD-2) and 625°C (BD-3), especially in BD-3. The fracture elongation was also decreased following this creep-rupture strength reduction.

4. DISCUSSION

4.1 Influence of Sodium Environment after Long-Time Exposure

It is generally known that in the cooling system of a fast reactor plant which is provided with the heating section (up-stream) and heat exchange (down-stream) section, i. e., with temperature distributions, metal loss(corrosion) can occur at the high temperature region due to the dissolution of alloy elements from steel and that these dissolved alloy elements are deposited in supersaturation conditions at the low temperature region, increasing the weight there [5,9,10].

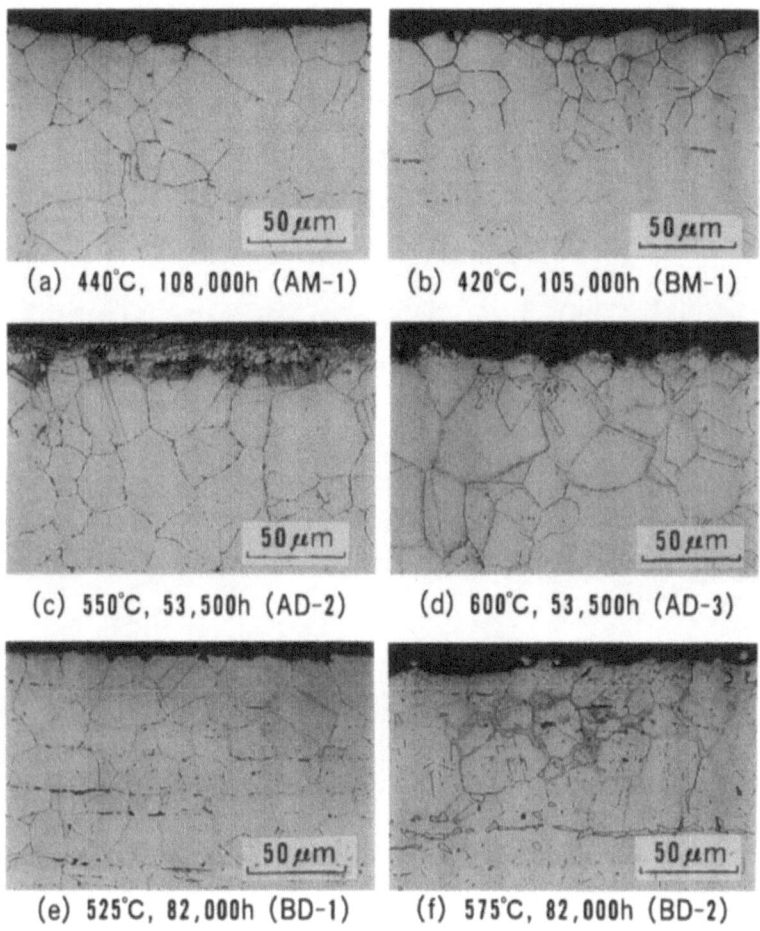

(a) 440°C, 108,000h (AM-1)   (b) 420°C, 105,000h (BM-1)

(c) 550°C, 53,500h (AD-2)   (d) 600°C, 53,500h (AD-3)

(e) 525°C, 82,000h (BD-1)   (f) 575°C, 82,000h (BD-2)

Fig. 6. Photographs of cross-sectional micro structure of Type 304 sodium pipes after sodium exposure

This study also observed the dissolution of Ni, Cr, Si, Mn on the surface of the piping materials used at the high temperature region, and recognised the deposit of these elements at the low temperature region. These results were same as the previous examination of relatively short-time corrosion tests and it was confirmed that the thermal gradient mass transfer phenomenon was stable for such a long-time sodium exposure as 100,000 hours. It could also be confirmed as a result of observation of the surface morphology and cross-sectional micro structure that the corrosion phenomena at the high temperature resulted in the general corrosion at the pipes operated for a long time, with no local corrosion such as pitting and intergranular corrosion. Based on the observation of these phenomena, attention is paid in design to the material thickness loss due to dissolution to establish the allowable material corrosion by sodium. These tests results lead to the judgement that the conventional method on corrosion prediction based on general corrosion mechanism are reasonable at long-time exposure.

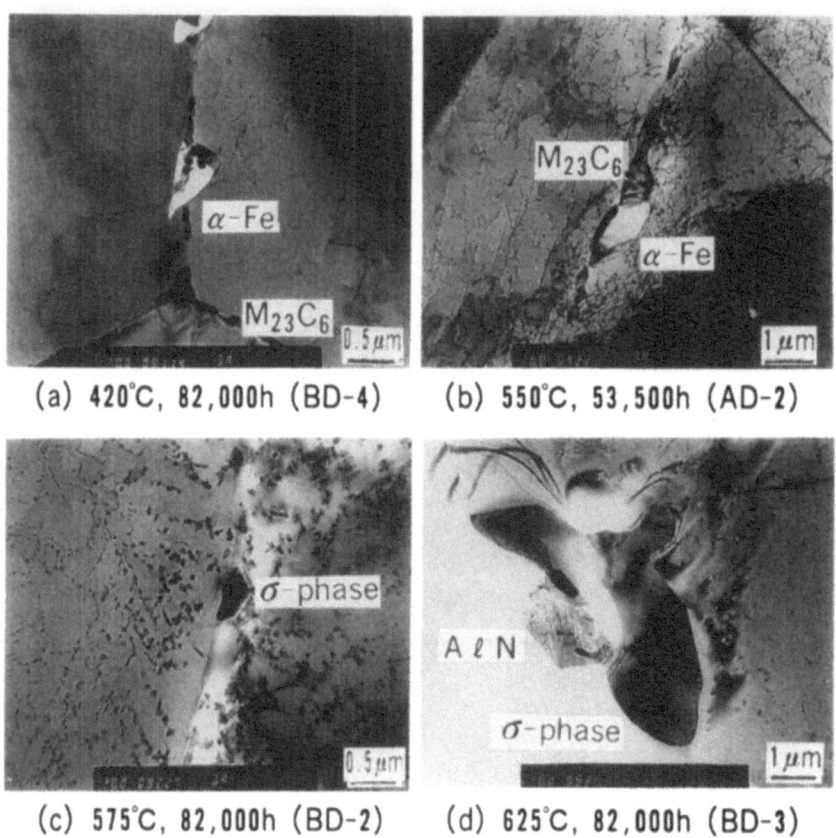

(a) 420°C, 82,000h (BD-4) (b) 550°C, 53,500h (AD-2)

(c) 575°C, 82,000h (BD-2) (d) 625°C, 82,000h (BD-3)

Fig. 7. Electron micro structures of Type 304 sodium pipes operated for 53,500-82,000 hours

Further, if austenitic stainless steel is exposed in the high-temperature sodium, a thin degraded layer is formed on the sodium exposed surface of material [11]. The formation is considered due to the dissolution of Ni, Cr, Si, Mn in steel into the sodium. The depth of surface degraded layer mainly depended on time and temperature, and the layer has been formed by ferrite phase which changed from the original austenite phase. Figure 11 shows the relations between the depth of surface degraded layer and sodium exposure time for Type 304 stainless steel. The data in this study show the values obtained through the observation by an optical microscope and the concentration distributions of alloy elements. The depth of surface degraded layer increases with a sodium exposure temperature and time, but it was clarified that the amount obtained from this study was conservative relative to the conventional standard equation on surface degraded layer for stainless steel.

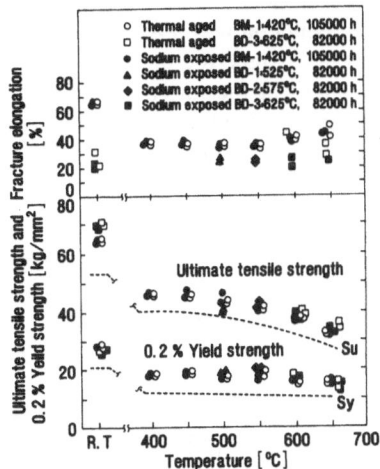

Fig.8. Tensile properties of Type 304 sodium piping materials

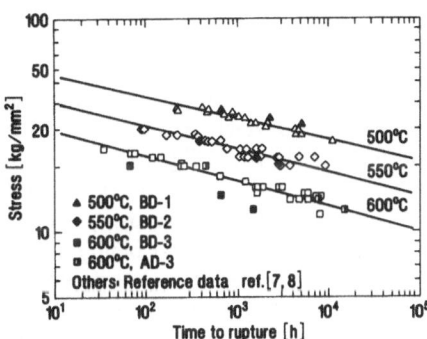

Fig. 9. Creep-rupture properties of Type 304 sodium piping materials in air

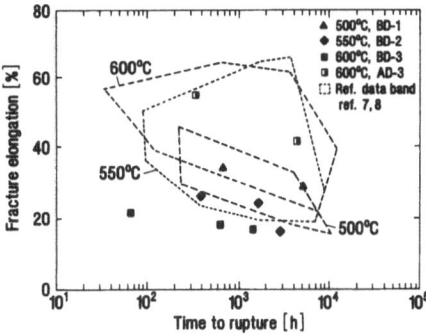

Fig. 10. Fracture elongation of Type 304 sodium piping materials in air

4.2 Influence of Microstructural Changes on Mechanical Properties

In BD-3 piping materials, some reduction was observed in the creep rupture strength and fracture elongation compared with the as-received material. The possible causes of reduced creep-rupture strength include changes in the micro structure due to the precipitation of carbide by long-time exposure to high temperature and the influence of sodium environment used.

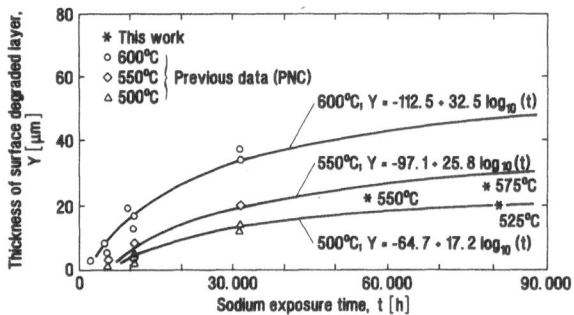

Fig. 11. Relation between surface degraded layer and sodium exposure time for Type 304 stainless steel

To study the causes of strength reduction, creep-rupture tests at 600°C were conducted under same stress conditions on the BD-1 piping material used at a relatively low temperature and BD-3 piping material of which sodium affected layer had been cut to a depth of about 0.5 mm. The results are shown in Figure 12 which indicates that the creep-rupture strength can slightly increase by removing the affected layer from the sodium exposed surface, but it is still below the average creep-rupture strength curve of the as-received materials. The creep-rupture strength of BD-1 piping material was same as average curve, showing no significant strength reduction. It is considered based on these results that the reduction of creep rupture strength at 600°C was due to changes in micro structure by thermal ageing during operation rather than the effects of sodium environment.

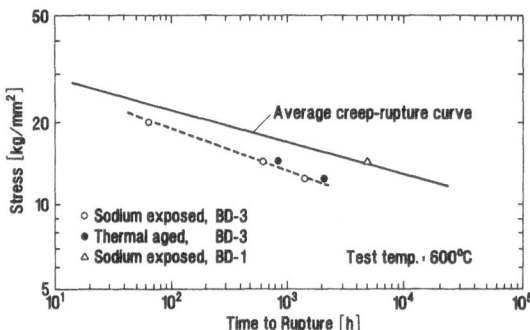

Fig. 12. Influence of operation history on the creep-rupture strength for Type 304 sodium piping materials at 600°C

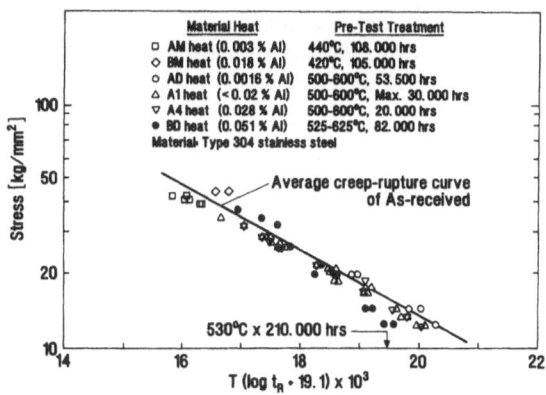

Fig. 13. Creep-rupture properties of Type 304 stainless steel exposed to flowing sodium for long-time

In other words, in the BD-3 piping material which showed the largest creep strength reduction of all BD heat materials, the precipitation of AlN was observed together with the sigma phase on grain boundary as shown in Figure 7. N. Shinya et al [6] suggest that the precipitation of AlN attached to the sigma phase on grain boundary may accelerate the growth of cavities and cracks at the boundary surface and cause a remarkable reduction of the creep rupture strength. The BD heat materials in this study contained a high AlN concentration (0.051%), which seems to have accelerated the AlN precipitation and resulted in the reduced creep-rupture strength.

Figure 13 shows the creep-rupture properties of the Type 304 stainless steels after exposure to sodium for 10,000-108,000 hours. In other high temperature materials with a lower Al content (<0.028%), no precipitate of AlN was observed at the high temperature region, nor was any significant strength reduction, and thus their material integrity could be maintained even in a long-time service life. This is also true of the conventional Types 304 and 316 stainless steel of which Al content has already been regulated.

5. CONCLUSIONS

(1) The thermal gradient mass transfer phenomena such as dissolution and deposition which were observed in relatively short-time corrosion tests in the past were still observed after long-time operation. With no intergranular and pitting corrosion observed, it was considered that conventional method on corrosion prediction based on general corrosion mechanism is reasonable for long-time operation.

(2) The formation of degraded layer and chemical compositional changes were observed near the surface of piping material used at the high temperature, but it was confirmed that the depth was as thin as maximum 30 μm and that the conventional standard equation could provide the sufficiently conservative evaluation even in the long-time operation.

(3) All the high-temperature tensile strengths of piping materials have satisfied the Sy and Su values, and the creep-rupture strength also showed the average strength level of all the as-received materials. However, in the piping materials with a high Al content, some reduction in the creep rupture strength and fracture elongation was recognised at high temperature. It was considered that it is not due to the effects of sodium environment but due to the microstuctural changes caused by the AlN precipitation on the grain boundary.

(4) It was concluded through the detailed analyses of sodium piping materials operated for about 100,000 hours that the extrapolation of the present evaluation method on sodium environmental effects for Type 304 stainless steel is satisfied with high reliability.

REFERENCES

1. Proc. of an Int. Atomic Energy Agency Specialists' Meeting, Int. Working Group on Fast Reactors, Report KfK 4935, IWGFR/84, Karlsruhe(1991).
2. G.M. Goodwin, J.H. DeVan, B.E.Foster, et al, Proc. of Int. Conf. on Liquid Metal Tech. in Energy Prod., (1976) p584.
3. F. Casteels, H. Tas, J. Dresselaers, et al, ibid, (1976) p 577.
4. A. Maruyama, S. Nomura, M. Kawai, et al, J. At. Energy Soc. Japan, 26 (1984) 327.
5. A.W. Thorley, C. Tyzack, et al, Proc. of Int. Conf. on Liquid Metal Tech. in Energy Prod., (1976) p 685.
6. N.Shinya, J.Kyono, et al, ISIJ, 69 (1983) 138.
7. A.Yoshitake, Y.Wada, et al, Proc. of Int. Conf. on Creep (1986) p 441.
8. Y.Wada, E.Yoshida, S.Kato, et al, Proc. of an Int. Atomic Energy Agency Specialists' Meeting Int. Working Group on Fast Reactors, Report KfK 4935, IWGFR/84, Karlsruhe(1991) p 17.
9. H.Atsumo, S.Yuhara, et al, Proc. of Int. Conf. on Liquid Metal Tech. in Energy Prod., (1976) p 849.
10. S.A.Shiels, et al, WARD-NA-3045-34 (1976).
11. H.U.Borgstedt and C.K.Mathews, Applied Chemistry of the Alkali Metals, Plenum Press, New York and London (1987).

CORROSION BEHAVIOUR OF STAINLESS STEEL AND ITS WELD IN HIGH TEMPERATURE SODIUM UNDER TENSILE STRESS

Chen Xueding, Xia Tiandong, Lu Wenjiang

Welding Institute, Technology University of Gansu,
Lanzhou 730050, China
Xu Yongli, Zhang Daode, Wang Jiaying, Qin Jingxiang
China Institute of Atomic Energy, P.O.Box 275(53), Beijing

1. INTRODUCTION

It is well known that the main corrosion types of LMFBR cladding in high temperature sodium are of uniform corrosion, intergranular corrosion, selection solution of composition elements and mass transfer effect [1-4]. On the other hand, the cladding of fuel pins suffers many kinds of stress, such as mechanical interaction between fuel and cladding due to the difference in thermal-expansion when power changes and reactor restart; increasing of internal pressure of cladding with fission products release, and so on. For the weld, the chemical-metallurgical process caused by heating and cooling rather quickly will occur in the welding seam, and the heating effect and residual stress will appear in it. The corrosion, creep, and tensile behaviour of weld in sodium have been studied [4-7]. Its corrosion behaviour is different obviously from those of the material itself. Therefore, the investigation of the corrosion behaviour of cladding materials, especially their weld, in high temperature sodium under tensile stress is rather important for the development of FBR fuel cladding material.

In this paper, the corrosion behaviour of domestic Cr13Ni17Mo3W3 used for cladding material is compared with those of AISI 316 S.S. and SUS 316L S.S.. The corrosion behaviour of SUS 316L weld, and the effect of the tungsten content on the corrosion behaviour of Cr13Ni17Mo3W3 are investigated.

2. EXPERIMENTAL METHODS

2.1 Specimens

AISI 316 SS., SUS 316L SS. and Cr13Ni17Mo3Wx SS. were used as the testing specimens. The φ4.0mm ok 63.41 material as the welding rod of the weld specimens. Their chemical compositions are listed in Table 1. The size of the specimens is shown in Fig.1. The metallurgical structure of the welds is 5.5 % delta - ferritic and austenitic.

Table 1. Chemical composition of the test specimens

Material	Content of elements (wt %)										
	C	Si	Mn	S	P	Cr	Ni	Mo	W	N	Fe
AISI 316	<0.08	<1.0	<2.0	-	<0.045	16/18	10/14	2/3	-	-	balance
SUS 316L	0.020	0.68	1.07	0.003	0.027	16.99	12.64	2.20	-	-	balance
Cr13Ni17Mo3W	0.10	0.75	1.20	<0.003	<0.05	13	17	2.5	0-3	<0.01	balance

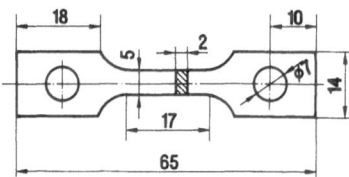

Fig.1. Testing specimen

2.2 Test Facility and Conditions

The test facility is shown in Fig.2. The test conditions are listed in Table 2.

Table 2. Testing conditions in the sodium test facility

Parameter	Quantity / Dimension
Temperature	550, 600 °C
Period of exposure	500, 1000 h
Tensile stress	0.117; 176 MPa
Oxygen level in sodium	≈ 50 ppm

3. RESULTS AND DISCUSSION

3.1 Effect of Tensile Stress on the Corrosion Resistance of Cr13Ni17Mo3Wx SS., AISI 316 S.S. and SUS 316L S.S. in High - Temperature Sodium

The test results shows that the surfaces of Cr13Ni17Mo3Wx, SUS 316L SS., and AISI 316 SS. specimens without stress were not attacked by sodium under testing conditions (see Fig.3 a. b. c.). If the tensile stress applied to the specimen is less than 117 MPa, no considerable changes were observed for tested specimens. However, under 176 MPa tensile stress, the changes in surface and matrix of these specimens appear after sodium exposure. For AISI 316 SS. specimens, both the grain boundary crack at exposure surface and the triangle grain boundary hole at matrix were observed. The depth of the largest crack is 50 μ m (Fig.4 a.). For SUS 316L SS., the grain boundary corrosion of the surface is 25 μm deep (Fig.4 b.). The situation of Cr13Ni17Mo3W3 SS. seems to be better. Slight widening-grain boundary at sodium exposure surface and slip-line at some grains of matrix were only observed (Fig.4 c.). However, the triangle grain boundary hole in matrix can also be seen for some specimens (Fig.4 d.).

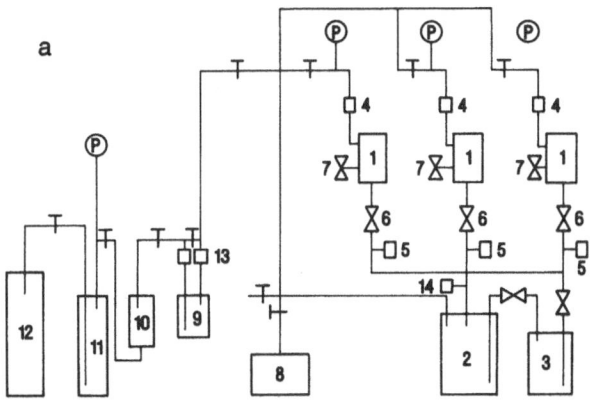

a

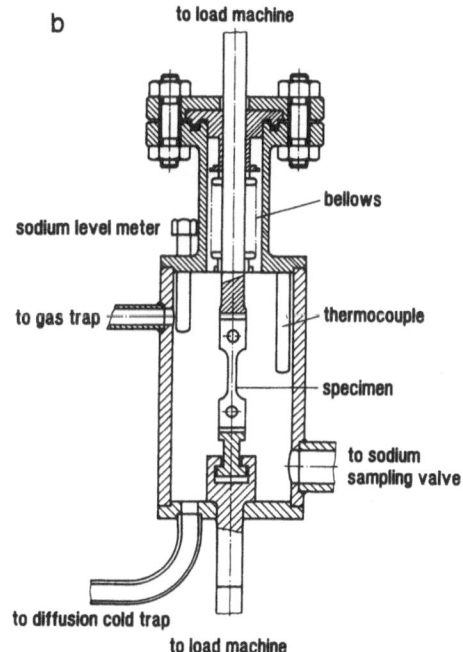

b

to load machine

sodium level meter

bellows

to gas trap

thermocouple

specimen

to sodium sampling valve

to diffusion cold trap

to load machine

Fig.2. Facility for testing a. schematic diagram of the facility b. testing container

1. testing container
4. gas trap
7. sodium sampling valve
10. molecular sieve
13. pressure balance container

2. sodium tank
5. diffusion cold trap
8. vacuum pump
11. gas storage tank
14. sodium inlet

3. filter
6. sodium valve
9. Na-K alloy
12. Ar gas bomb

It is obvious that the grain boundary of materials is damaged when the tensile stress is high enough. Sodium and impurities in sodium penetrate along the damaged grain boundary, attack and weaken them. The attacked and weakened grain boundary in turn enhances the effect of tensile stress. Therefore, the corrosion of materials under stress is more serious than that without stress.

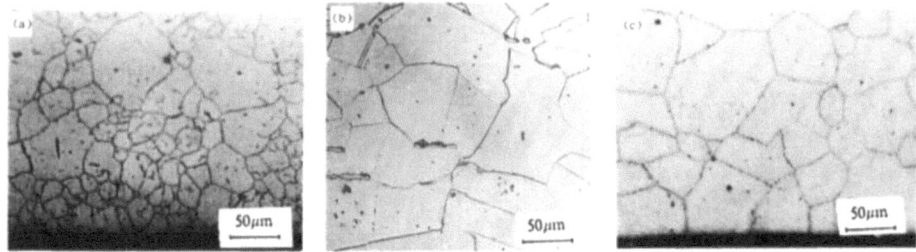

Fig.3. surface morphology of materials exposed in 600°C sodium without stress
a. AISI 316 SS.   b. SUS 316L SS. c. Cr13Ni17Mo3W3

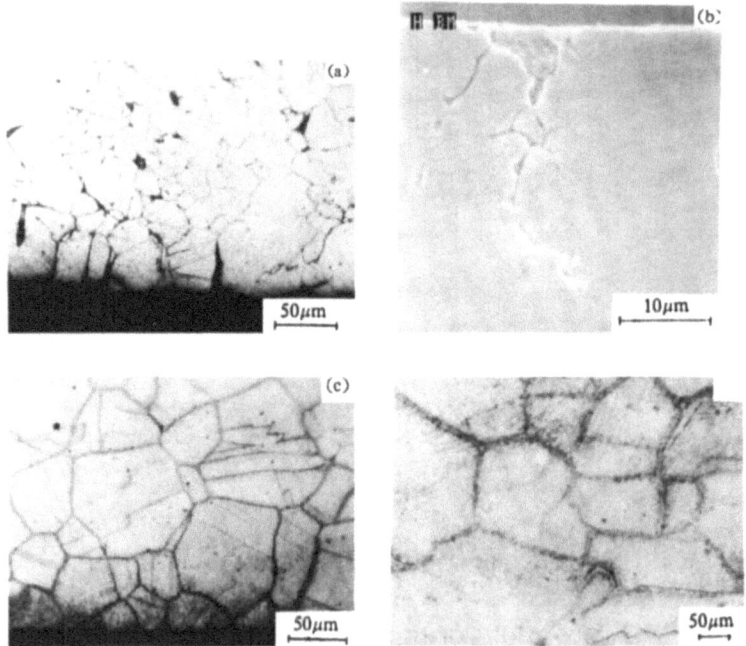

Fig.4. Surface Morphology of Materials under 176 MPa Tensile Stress and 600 °C sodium exposure
a. AISI 316 SS   b. SUS 316L SS c. Cr13Ni17Mo3W3   d. Cr13Ni17Mo3W3

The corrosion resistance of Cr13Ni17Mo3Wx in high temperature sodium under stress is rather better. This may result from lower Cr and higher Ni content in it. Higher Ni and lower Cr is a favourable factor for resistance to cracking of grain boundary and to forming of the grain boundary cavity [8].

When Ni content is high enough, the interlocking potential energy will be increased due to formation of a net structure resulted as a traverse slipping and intercrossing of the dislocation. Therefore, creating and developing of the crack will be inhibited, and the corrosion resistance will be improved. The effect of excessive Cr content is just opposite [8].

The effect of tensile stress on corrosion resistance of materials can be also observed from the surface film of the materials. Figs. 5 a. and b. show the surface film of the SUS 316L SS. specimens without stress and under stress, respectively. The film of the specimen without stress is thinner, and appears in an irregular form. In general, the uniform corrosion is the main corrosion type due to the uniform selection solution of the composition elements of material. The film of the specimen under stress is always thicker and striped at triangle grain boundary zone. EDS demonstrates that the Cr content in this film is rather higher, and the Ni lower. It can be inferred that the grain boundary attack by sodium and impurities in sodium dominates the process, the alloying elements in matrix will transfer and solute quickly along the attacked grain boundary.

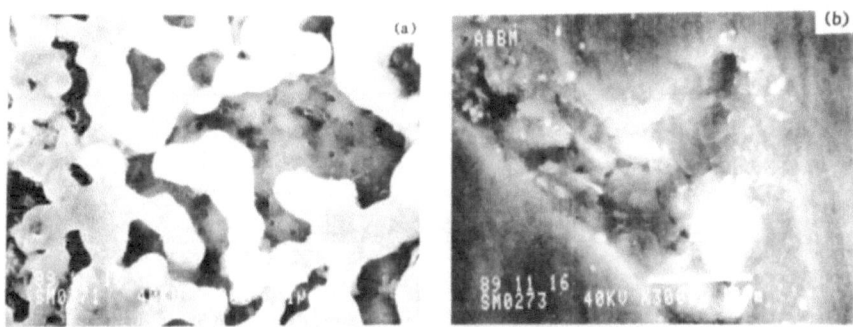

Fig.5. Surface film of the SUS 316L SS. specimens after 550 °C sodium exposure
a. without stress b. under 176 MPa stress

3.2 Corrosion Behaviour of SUS 316L Weld in High Temperature Sodium under Tensile Stress

The weight gain appears for the weld specimens after sodium exposure whether they suffer stress or not. The weight gain is 70.5 $mg \cdot dm^{-2} \cdot a^{-1}$ and 24.0 $mg \cdot dm^{-2} \cdot a^{-1}$, respectively, for the weld under stress and without stress. It is clear that the film on the weld under stress should be thicker than that without stress. This effect of stress on the weld is the same as that on the material itself.

AES analysis for the film shows that it may consists of Cr_2O_3 for the weld under stress, oxide of sodium and carbide of calcium for that without stress (Fig.6).

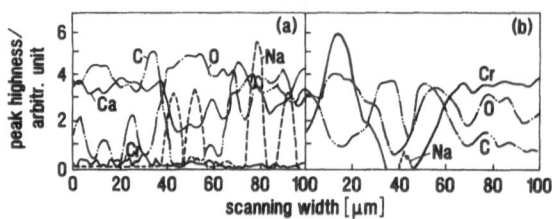

Fig.6. Auger scanning of the weld surface exposed in 550°C sodium
a. without stress   b. under 176 MPa stress

This result indicates that the transfer of Cr from the matrix to the surface of the weld is enhanced by tensile stress. Auger profile analysis of the film on the weld confirms above facts again (Fig.7).

On the other hand, Fig.7 shows, that the contents of alloying elements at the surface film reach the level of those in the matrix for original weld specimen (without stress and no sodium exposure) after Ar ion sparking for three minutes, however, for the weld exposed in sodium without stress seven minutes are needed, while 16 minutes are necessary for the weld which was exposed to sodium under stress. An order about thickness of the film on the weld exposed in sodium can be given as following:

under stress > without stress > non exposed

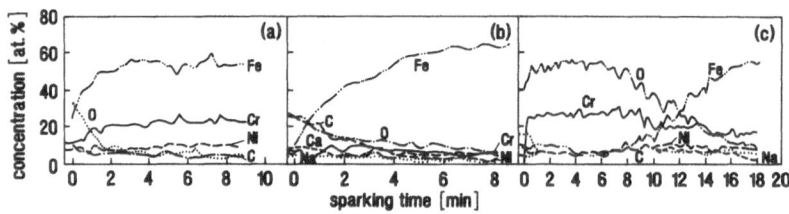

Fig.7. Auger depth profile of the surface film on the weld exposed in 550°C sodium
a. original weld   b. weld without stress   c. weld under 176 MPa stress

High times image of the weld under stress and sodium exposure demonstrates that δ/γ grain boundary and partial δ solute were predominantly attacked, while the γ zone was not affected by sodium (Fig.8). EDS of the γ-δ-γ area was compared with those of original specimen [9], a reduction of high Cr content at δ/γ grain boundary and δ zone was observed. This fact indicates that Cr in the weld transfers from δ/γ grain boundary and δ zone to interface between specimen and sodium.

3.3 Effects of the W Content on the Corrosion Resistance of Cr13Ni17Mo3Wx SS

The corrosion resistance of Cr13Ni17Mo3Wx SS in high temperature sodium under tensile stress seems to be related to the tungsten content in it. The behaviour of Cr13Ni17Mo3W3 was described as above. For Cr13Ni17Mo3W1.5, the changes at the grain boundary of surface were not observed. The grain boundary holes in the matrix were not detected either, and only grains become longer along the tensile direction. For Cr13Ni17Mo3 without tungsten, there are many slip-lines at grains of the matrix, and a ferritic layer appears at its surface. The measurement of microhardness for three kinds of Cr13Ni17Mo3Wx SS. shows that the effect of tensile stress on the hardness of Cr13Ni17Mo3W1.5 seems to be negligible, but considerable effect for Cr13Ni17Mo3W3 (Fig.9). This result indicates that the grain may be strengthened, and the attack of sodium to material may be reduced due to the optimum tungsten content in Cr13Ni17Mo3Wx SS.. However, under stress the excessive tungsten content may weaken the grain boundary, increase the inner defects of the matrix and increase seriously the attack of sodium on the materials. The optimum tungsten content of Cr13Ni17Mo3Wx SS may be 1-2wt.%.

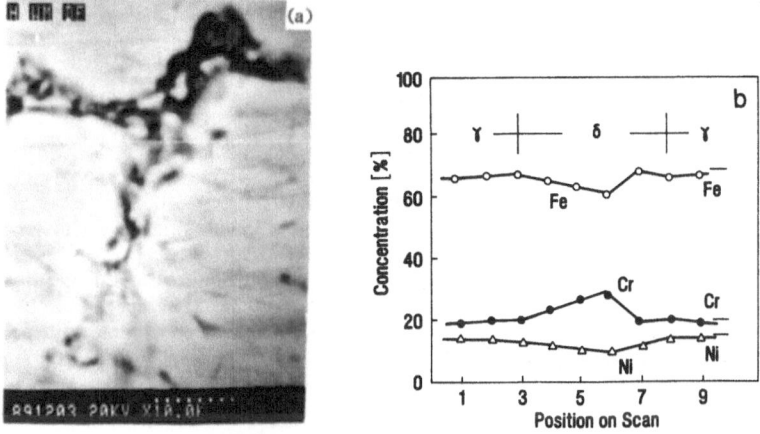

Fig.8. High times image and its EDS of the weld surface under 176 MPa stress and 550 °C sodium exposure
a. high times image of the weld surface   b. EDS of the white line at a.

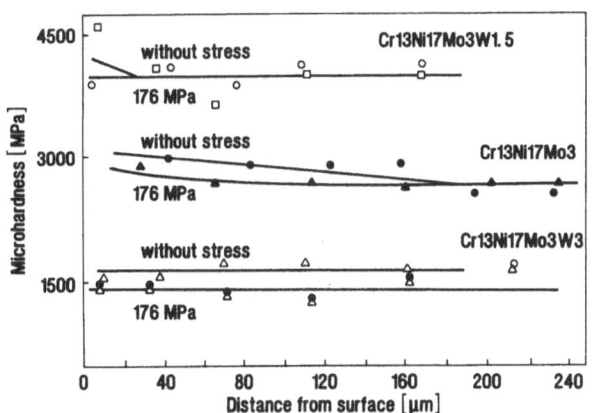

Fig.9. Micro-hardness of Cr13Ni17Mo3Wx SS. exposed in 600 °C sodium for 500 hours

4. CONCLUSIONS

1. The effect of tensile stress on the corrosion behaviour of materials in high temperature sodium is rather considerable, if the stress is high enough. Stresses of less than 117 MPa have no effect on the corrosion behaviour.
2. Higher Ni and lower Cr content in Cr13Ni17Mo3Wx SS. are a favourable factor for resistance to cracking of grain boundary and to forming of the grain boundary cavity. Therefore, its corrosion resistance in high temperature sodium under tensile stress is rather better.
3. The film consists of Cr_2O_3 mainly for the weld in sodium under stress, oxide of sodium and carbide of calcium for that without stress. Tensile stress enhances transfer and selection solution of Cr.
4. The tungsten in Cr13Ni17Mo3Wx SS will strengthen its grain and reduce the attack effect of sodium. The optimum tungsten content is 1-2 wt.%.

References

[1] W. Bennett, K.E. Horton, Metall.Trans. 9A (1978), 143.
[2] K. Natesan, O.K.Chopra, T.F. Kassner, J. Nucl. Mater. 73 (1978) 137.
[3] M.G. Fontana, "Corrosion Engineering", McGraw-Hill, NewYork, (1986).
[4] W.E. Berry,"Corrosion In Nuclear Applications",Wiley & Sons, New York 1997 250-253.
[5] Seiichi Kaga, Katsuhiro Fujii,"Symposium of Welding Society" 7(1), (1989) 33-38.
[6] C.A.P. Horton, B.H. Targett, Proc. 4th Int. Conf. on Liquid Metal Engineering and Technology, Vol.2., Soc.Franc.d'Energie Nucleaire, Paris, France, paper 513 (1988).
[7] H.S. Khatak, H. Shaikh, J.B. Gnanamoorthy, Proc. 4th Int. Conf. on Liquid Metal Engineering and Technology, Vol.2., Soc.Franc.d'Energie Nucleaire, Paris, France, paper 514 (1988).
[8] Hengde Li et al. "Electron Irradiation Test of Cr13Ni17Mo3W3 Stainless Steel", Internal report, (1981).
[9] Haku Tamura, Tadao Shisawa, Akito Takasaki "Symposium of Welding Society", 5(2), (1987) 257-262.

CORROSION BEHAVIOUR OF Nb-BASED AND
Mo-BASED ALLOYS IN LIQUID Na

Satoshi Inoue*, Shigeki Kano**, Jun-ichi Saito*,
Yasushi Isshiki*, Eiichi Yoshida** and Masahiko Morinaga**

* Department of Production Systems Engineering, Toyohashi University of
Technology, Toyohashi, Aichi, 441 Japan
** O-arai Engineering Center, Power Reactor and Nuclear Fuel Development
Corporation, O-arai-cho, Higashi-Ibaraki-gun, Ibaraki, 311-13, Japan

1. INTRODUCTION

The structural materials employed in advanced nuclear power systems (for example, a portable FBR) are exposed under severe environmental conditions [1-3]. Therefore, it is required for the materials to have super-heat resistance at 1473 K, high corrosion resistance to liquid metals and a life time of more than a decade of years. Both Nb-based and Mo-based alloys are ones of the candidate materials for such applications [4]. Since material lives are dominated by corrosion on several occasions, the compatibility of the material with liquid metals is indeed an important factor for the design and development of new materials. The corrosion behavior of stainless steels, which are widely used for the structural applications in nuclear power systems, have been investigated extensively [5-9]. However, the corrosion study is rather limited for Nb-based and Mo-based alloys. The purpose of this study is to obtain basic information on the corrosion in liquid Na for both the Nb-based and the Mo-based alloys. In particular, our attention was directed towards the alloying effects on the corrosion of these alloys.

2. EXPERIMENTAL PROCEDURE

For Nb-based alloys, alloying 3d, 4d and 5d transition elements were selected systematically from the periodic table of elements as shown in Table 1. The concentration of each alloying element was set to be 5 at.%, except for 0.3 at.% of Cr and Cu. The purity of raw materials was better than 99.9 %. For Mo-based alloys, four ternary Mo-Re-W alloys shown in Table 1 were used in the present experiment. For comparison, two specimens of pure Nb having the purity of 99.9 % and 99 % were also prepared. These metals and alloys were melted in a high purity Ar atmosphere using a tri-arc furnace and then heat treated at

Liquid Metal Systems, Edited by H.U. Borgstedt
and G. Frees, Plenum Press, New York, 1995

1323 K for 24 hours (86.4 ks). The specimens for corrosion tests were cut to be 10 mm square and 2 mm in thickness. The corrosion tests were repeated twice for each alloy. The test conditions were summarized as follows;

Table 1:Alloying elements and chemical composition of Nb-based and Mo-based alloys.

Nb-based alloy (Nb-X)	Mo-based alloy /at %
3d: Ti,V,Cr,Fe,Co,Cu	Mo-15Re-5W
4d: Zr,Mo,Ru,Rh,Pd	Mo-15Re-10W
5d: Hf,Ta,W,Re,Ir,Pt,Au	Mo-25Re-10W
X: 5 at% (Cr,Cu: 0.3 at%)	Mo-25Re-20W

Liquid sodium temperature	923 K
Oxygen concentration in liquid sodium	<1 ppm
Testing time	1000 hours (3.6 Ms)
Circulating velocity of liquid sodium in a loop	0 m/s

The corrosion rate was estimated by measuring weight loss or gain of the specimen after the corrosion test. The surface morphology and the structure of corrosion products were examined by SEM observation, EPMA/EDS analysis and X-ray diffraction. X-rays used were Ni-filtered CuKα radiations.

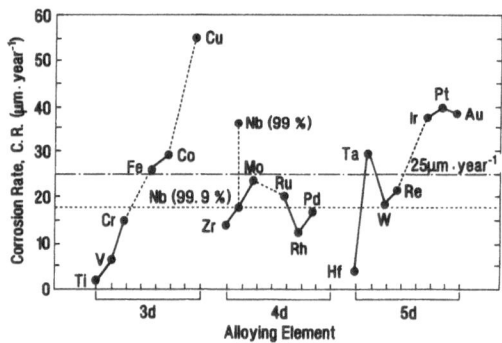

Fig. 1. Corrosion rates of Nb-X binary alloys in liquid Na at 923 K

3. RESULTS

3.1 Corrosion Rate

The resulting corrosion rates are shown in Fig.1 for the Nb-based alloys. All the alloys got a weight loss by corrosion. The corrosion rate of pure Nb with a purity of 99.9 % was about 18 μm/year, which was less than 25 μm/year, an indication value for the corrosion resistance of the alloys for nuclear applications. Compared this result with that of another pure Nb with a lower purity of 99 %, it was found that the lowering of the material purity enhanced the corrosion. The oxygen concentrations were 582 ppm for 99.9 % pure Nb and 1130 ppm for 99 % pure Nb. Thus, impurities including oxygen had an influence on the corrosion behavior of pure Nb. Similarly, corrosion rates changed remarkably with alloying elements. For example, the 4A group elements such as Ti, Zr and Hf decreased the corrosion rate, whereas the 1B group elements such as Cu and Au increased it. These results may be related to the values of heat of formation of oxides for alloying elements, since the values are high for Ti, Zr and Hf but low for Cu and Au [10]. Furthermore, it is interesting to note that the corrosion rate of the alloys containing 3d and 5d elements (except for Ta) changed monotonously with the position of the elements in the periodic table.

In contrast to these Nb-based alloys, corrosion rates of the Mo-based alloys were significantly low, and less dependent on the alloy compositions as shown in Fig.2.

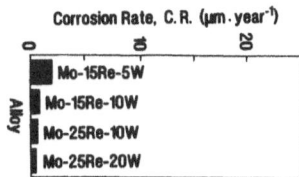

Fig. 2. Corrosion rates of Mo-based alloys in liquid Na at 923 K

3.2 Surface Morphology

The surface morphology of pure Nb and the alloys after corrosion tests was not homogeneous. As shown in Fig.3(a) for 99.9 % pure Nb, the surface region was divided into two parts. The A part is a protuberant part and the B part is a hollow part on the surface. It was supposed that the A part is the residual part where alloying elements may react in some way with Na atoms or O atoms or other impurities in liquid Na. The B part is the dissolved part where solute and/or solvent atoms are dissolved preferably into liquid Na. For alloys, the area fraction of two parts varied with the alloying elements as shown in Fig.3(b)-(g). This change with alloying elements was consistent with the variation in the observed corrosion rates. In other words, the corrosion rates increased with increasing area fraction of the dissolved part. The cross sections of the Ti- and the Cu-containing alloys are shown in Fig.4. Some compounds or adsorbed atom layers seemed to stack on the surface of the A part (the residual part) in the Ti containing alloy.

On the other hand, there was nothing on the surface of the Cu containing alloy. But, the dissolved part was observed only, showing the most active corrosion as shown in Fig.1. The penetration of Na into Nb alloy was not observed in the present experiment. This is partially due to the lower oxygen concentration in the alloy than the threshold concentration (~800 ppm).

As might be expected, the surface state of the Mo-based alloys scarcely changed even after corrosion tests. It kept metallic brilliance and nothing was seen on the surface according to the SEM observation.

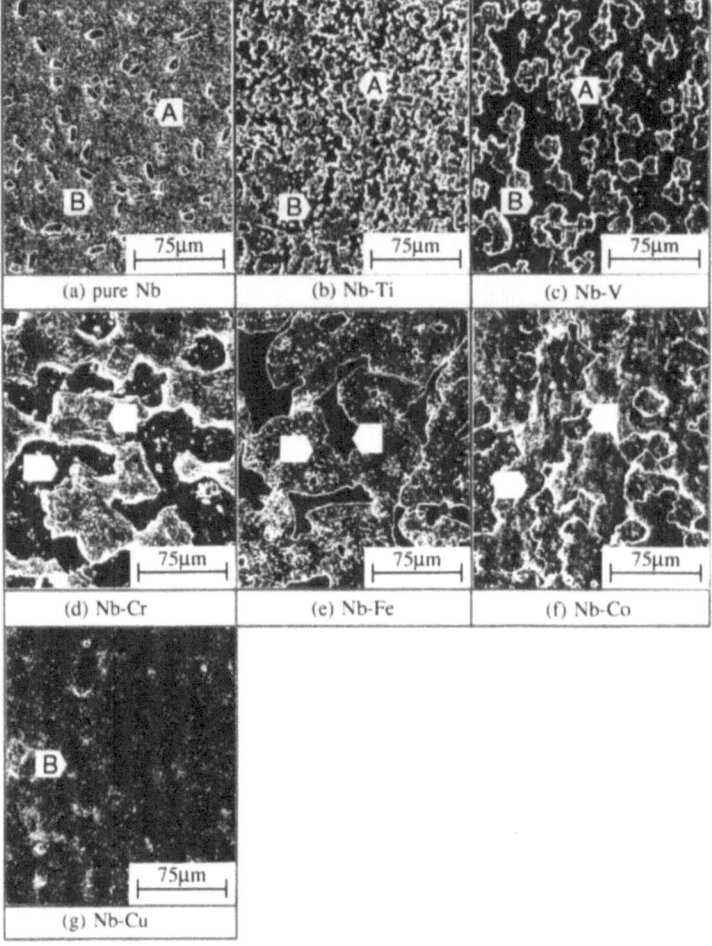

Fig. 3. Surface morphology of pure Nb and Nb-X binary alloys after corrosion tests

3.3 EPMA/EDS Analyses and X-Ray Diffraction

The EPMA/EDS surface analyses were carried out on the both A and B parts in every alloy. Some results were shown in Fig.5 for 99.9 % pure Nb, and the Ti- and the Cu-containing alloys. There were impurity elements such as O, Si and Y even on the surface of 99.9 % pure Nb. A large amount of Si was detected mainly on the A part. Y and O were detected on both parts, but Na was never seen in either of the parts. Most of these impurities might be introduced to the specimen from liquid Na, since there were not appreciable amounts of them in 99.9 % pure Nb. As shown in Fig.5(c) and (d), the Ti containing alloy showed the same trend as pure Nb. Also, the concentrations of Ti and O were significantly higher on the B part than the A part. This was confirmed further from each X-ray image as shown in Fig.6. In the Cu containing alloy, as explained before, only the B part existed and it showed the same feature as the B part of pure Nb.

A similar trend of the O, Si and Y distributions between the A and B parts was observed in the other alloys. Here, it is stressed again that Si was mainly present on the A part, but not on the B part, irrespective of the alloy systems.

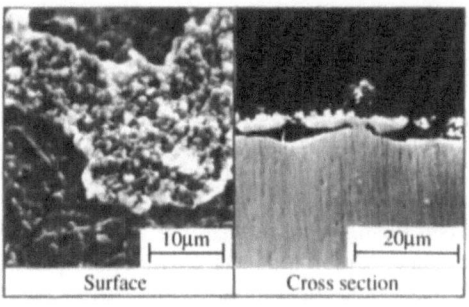

(a) Nb-Ti

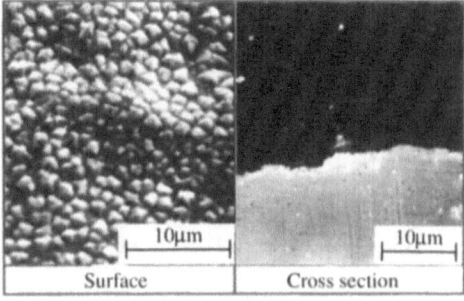

(b) Nb-Cu

Fig. 4. Surface morphology and cross section of
(a) Nb-Ti and (b) Nb-Cu alloys

The X-ray diffraction patterns are shown in Fig.7. These patterns were similar for pure Nb and the Ti containing alloy. This may be related to the coexistence of A and B parts on their surface. On the other hand, the Cu containing alloy showed a different diffraction pattern probably due to the non-existence of the A part on the surface. However Nb oxides such as Nb_2O_5, Nb_2O and NbO, were not observed in any specimens. Also, any simple oxides such as SiO_2 and Y_2O_3 were not observed. Thus, X-ray peaks in the patterns could not be identified despite the extensive search of JCPDS diffraction data.

However, there will be a certain niobium silicate on the A part (the residual part), since the Si, O and Nb contents are high as shown in Fig.5. Also, there is a possibility of the existence of some compound (probably oxide) even in the B part (the dissolved part), since there are some extra peaks observed in the Cu containing alloy. Thus, the detailed structure still remains unknown, but the resemblance in the diffraction pattern indicates that the corrosion mechanism is similar between pure Nb and the Ti containing alloy.

For the Mo-based alloys, there were not any extra peaks in the X-ray diffraction pattern, indicating that nothing was formed on their surfaces.

4. DISCUSSION

4.1 Difference in the Corrosion Rate between Nb-based and Mo-based Alloys

There was an unknown phase (probably oxide) in the highly corrosive Nb-based alloys, but any phases were not observed in the highly corrosion-resistant Mo-based alloys.

Therefore, a main difference between these two alloy systems will be attributable to the formation of oxide phase. According to the phase constitution, oxygen atoms are soluble in Nb, but insoluble in Mo. Also, the heat of formation of oxides at 923 K changes in the order, $MoO_2 < Na_2O < Nb_2O_5$. Thus, the affinity is much weaker between Mo and O atoms than between Nb and O atoms. Needless to say, Mo never reduces Na_2O in liquid Na as MoO_2 is less stable than Na_2O. Therefore, it is supposed that there is only a poor interaction operating between Mo metal and liquid Na, since the catalytic agent of oxygen atoms does not work efficiently between them. In this sense, the Mo-based alloys is considered to be more highly corrosion-resistant to the liquid Na environments, compared to the Nb-based alloys.

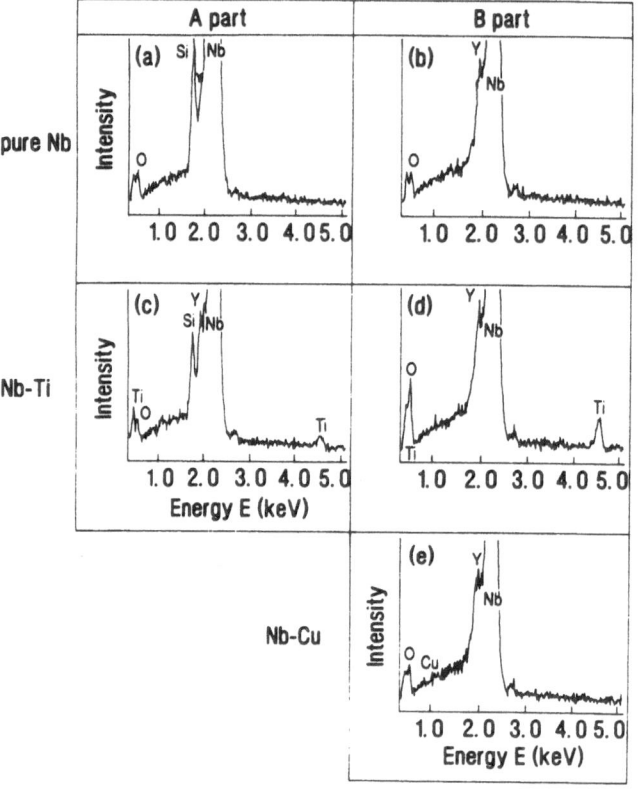

Fig. 5. The results of surface analyses for :
(a,b) pure Nb, (c,d) Nb-Ti and (e) Nb-Cu alloys

4.2 Alloying Effects on the Corrosion of Nb-based Alloys

a) Nb-Ti alloy

Generally, it is known that the oxygen concentration in most alloys increases after corrosion test in liquid Na. It is also reported that the corrosion of pure metals, M, such as Ta, Ti and Zr take place by the formation of a $NaMO_4$ type oxide on the metal surface. These results suggest that oxygen atoms in liquid Na play an important role in the reaction between Nb and Na atoms. In other words, unless oxygen atoms are present, there is little interaction between Nb and Na atoms, because of the non-existence of any compounds and also of the very limited solid solubility in the Nb-Na system. However, once oxygen atoms are introduced into this system, the corrosion proceeds actively through the formation of a stable Na-Nb-O ternary oxide. As the result of this catalytic effect of oxygen atoms, the corrosion will take place preferably in the oxygen-rich region on the alloy surface. In fact, as shown in Fig.6, the oxygen rich region coincided with the dissolved B part, even though as yet the corrosion product was not identified.

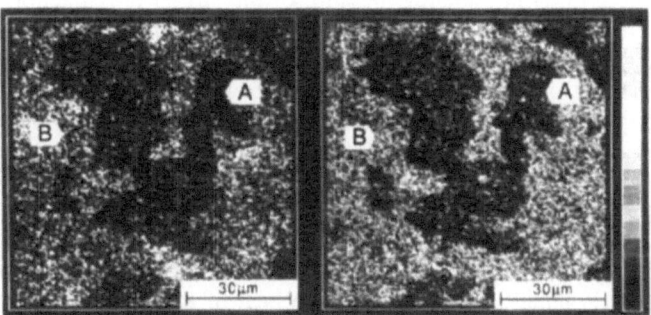

Fig. 6: X-ray images showing the distribution of (a) Ti and (b) O in a Nb-Ti-alloy

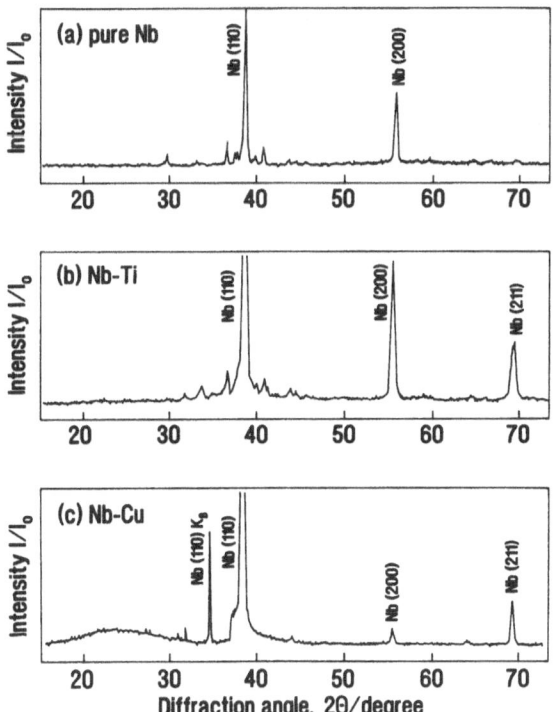

Fig. 7: X-ray diffraction patterns of (a) pure Nb, (b) Nb-Ti, (c) Nb-Cu alloys after corrosion tests

The alloying element which has a high heat of formation of oxides may introduce a large amount of the oxygen atoms into the alloy. This will be understood from the similarity in the distribution of the both Ti and O atoms on the corroded surface as shown in Fig.6. Therefore, active corrosion may be expected to occur in this Nb-Ti alloy. However, this is not the case in the present experiment. The Ti addition into Nb could improve the corrosion significantly. This result was also in contrast to a previous result that the Ti addition into V accelerated the corrosion in liquid Na [11]. However, it is apparent that the presence of considerable amounts of a niobium silicate influenced the corrosion of this alloy. Namely, the niobium silicate existing on the A part seemed to act as a protective film against corrosion, even though its mechanism is unknown. The niobium silicate is probably stabilized in some way by the existence of Ti in the alloy, because of the largest area fraction of the silicate among the 3d series elements. It is interesting that the alloying element which prefers to introduce the oxygen atoms into the alloy tends to promote the silicate formation, too. Thus, the corrosion behavior of this alloy is probably explained by the balancing between the oxygen effect on the dissolution of the B part and the protective effect of the silicate film on the A part.

(b) Nb-Cu Alloy

The affinity between Cu and O atoms is small, because Cu hardly forms binary oxides at oxygen activities as in liquid sodium. In fact, the heat of formation of Cu oxides is much lower than that of Ti oxides. Thus, the amount of oxygen atoms introduced into the alloy is probably very small. However, despite of this weak oxygen effect, the corrosion rate of the Cu-containing alloy was significantly higher than that of the Ti-containing alloy. The difference may be attributable to the non-existence of the niobium silicate on the Cu-containing alloy, since there is a large difference in the surface morphology between these two alloys. In addition, the dissolution of this alloy may be caused by the large solubility of metal Cu into liquid Na. However, it is still unknown why a small amount of Cu (0.3 %) had such a considerable influence on the corrosion of the alloy.

The corrosion rates of Nb alloyed with the other 3d elements were intermediate between these extreme cases of the Ti- and the Cu-containing alloys. Finally, it is interesting to note that the corrosion rate was higher for the Nb-based alloy containing Mo than pure Nb despite the high corrosion resistance of the metal molybdenum.

5. CONCLUSION

A series of corrosion experiments of both the Nb-based and the Mo-based alloys was performed in liquid Na at 923 K. There was a clear contrast in the corrosion behavior between these two alloys. The Mo-based alloys exhibited much higher corrosion resistance than the Nb-based alloys. This result was understood qualitatively by taking into account the difference in the affinity between metal and oxygen atoms. For the Nb-based alloys, it was shown that corrosion rates depended largely on the alloying elements. Furthermore, it was confirmed that impurities such as Si and O in liquid Na affected strongly the corrosion of the alloys. The existence of a niobium silicate on the surface could suppress the corrosion of the Nb-based alloys in liquid Na.

Acknowledgments

We would like to express our sincere thanks to Mr. Y. Hirakawa and Mr. Y. Tachi of the Power Reactor and Nuclear Fuel Development Corporation for their helpful assistance in corrosion experiments.

References

1. L. B. Lundberg, J.of Metals, August (1985),44.
2. R. H. Titran, and T. L. Grogstein, J.of Metals, August (1990), 8.
3. R. Stephens, D. W. Petrasec and R. H. Titran, International Journal of Refractory Metals and Hard Materials, June (1990), 96.
4. Metals Handbook 10th Edition, Materials Park, Ohio, 2 (1990), 557.
5. P. F. Tortorelli and J. H. DeVan, "Liquid Metal Corrosion Considerations in Alloy Development", CONF-840218-6, DE84 008512.
6. Metals Handbook 9th Edition, Metals Park, Ohio, 13 (1987), 56.
7. R. J. Pulham and P. Hubberstey, J. Nucl. Mater., 115 (1983), 239.
8. R. J. Pulham and P. Hubberstey, "Comparison of Chemical Reactions in Liquid Lithium with those in Liquid Sodium", Corrosion, Paper No.13, (1982).
9. B. H. Kolster, J. Nucl. Mater., 55 (1975), 155.
10. Handbook of Chemistry and Physics, 71st Edition, CRC Press, Boca Raton, (1990), 60.
11. R. L. Klueh, Oak Ridge National Laboratory, (1970), Contract No. W-7405-eng-26.

SODIUM COMPATIBILITY OF CERAMICS

S. Kano*, E. Yoshida*, Y. Hirakawa*,
Y. Tachi*, H. Haneda**, T. Mitsuhashi**

* Power Reactor and Nuclear Fuel Development Corporation
4002, Narita, Oarai-machi, Higashi-Ibaraki-gun, Ibaraki, 311- 13, Japan
** National Institute for Research in Inorganic Materials
Namiki, Tsukuba, Ibaraki, 305, Japan

1. INTRODUCTION

As ceramics recently developed possess an excellent heat-resisting property, they are expected to be used as tribological parts and thermal liner to improve the reliability and thermal efficiency of FBR by elevating its coolant temperature. The authors' objective of the research on ceramics is to make clear the points to be improved on chemical compositions, micro-structure and manufacturing process through the investigation of above existing ceramics on their sodium corrosion mechanisms and to create the advanced ceramics which meet sodium environment.

2. TEST METHOD

As the first step of research, sintered ceramics recently catalogued by several companies including C, N, S and K in Japan were tested in sodium. Single crystals and chemical vapour deposition (CVD) ceramics were also tested as reference materials. Generally speaking, sintered ceramics possess grain boundaries and impurities or additives, on the contrary, CVD ceramics possess grain boundaries without impurities and additives, and single crystals do not have any grain boundaries and impurities or additives. As a foundamental approach to clarify sodium corrosion mechanisms, these three kinds of ceramics were mutually compared in the view point of existence of grain boundaries, impurities and additives.

Figure 1 shows the overall research process to create advanced ceramics for usage in fast reactors. Tested ceramics are shown in Table 1.

These ceramics were exposed to 550 and 650°C sodium for 1000 and 4000 hours at an oxygen level of 1 ppm.

As the second step of research, the advanced ceramics were trially manufactured by C based on the countermeasures to improve the resistance to sodium corrosion of the existing ceramics and exposed to 550 and 650°C sodium for 1000 hours to clarify the effect of improvement on corrosion resistance.

Liquid Metal Systems, Edited by H.U. Borgstedt
and G. Frees, Plenum Press, New York, 1995

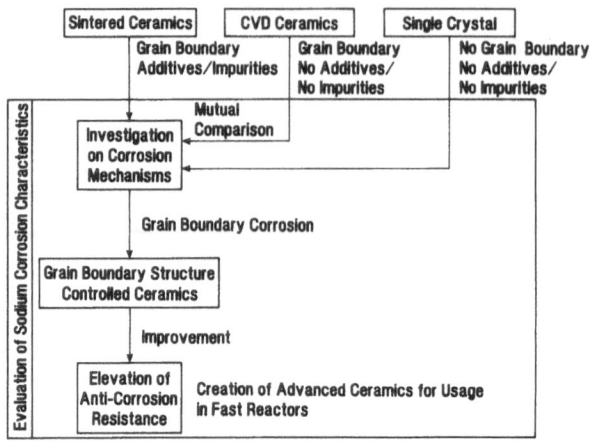

Fig. 1. Research process to create advanced ceramics for the use in fast reactors

Table 1. Tested ceramics

	Sintered		CVD	Single Crystal
	Existing	Advanced		
	Al_2O_3	Al_2O_3		Al_2O_3
	ZrO_2			ZrO_2
	MgO			MgO
	Y_2O_3			
	CaO			
	$MgAl_2O_4$			$MgAl_2O_4$
	$Y_3Al_5O_{12}$			$Y_3Al_5O_{12}$
	SiC		SiC	
	TiC		TiC	TiC
	Si_3N_4	Si_3N_4		
	AlN			
	BN		BN	
	SiAlON	SiAlON		
	AlON			

3. TEST RESULTS OF EXISTING CERAMICS

The weight changes due to sodium exposure are shown in Fig.2 and Fig.3.

Data for Al_2O_3 of 99 - 99.9% purity are plotted in Fig.3. Alumina of lower than 99% purity cracked and fell its grain by grain boundary attack. All ZrO_2 samples increased their weight. In contrast, almost all the other ceramics lost their weight and indicated higher corrosion rate with longer exposure time and higher sodium temperature. Especially, the weight losses of sintered SiC, Si_3N_4, SiAlON and TiC showed the significant time and temperature dependency. Spinels ($MgAl_2O_4$ and $Y_3Al_5O_{12}$), Y_2O_3, AlN and AlON showed the most excellent corrosion resistance among sintered ceramics and no evidence of grain boundary attack by SEM observation. Single crystals and CVD ceramics showed superior corrosion resistance to sintered ceramics. However, CVD BN as well as sintered BN showed the most significant corrosion among all tested ceramics. Single crystal of TiC showed the highest weight loss among single crystals.

In sintered ceramics, a colour changed layer was formed near sodium exposed surface, whose depth depended on exposure time, as shown in Table 2.

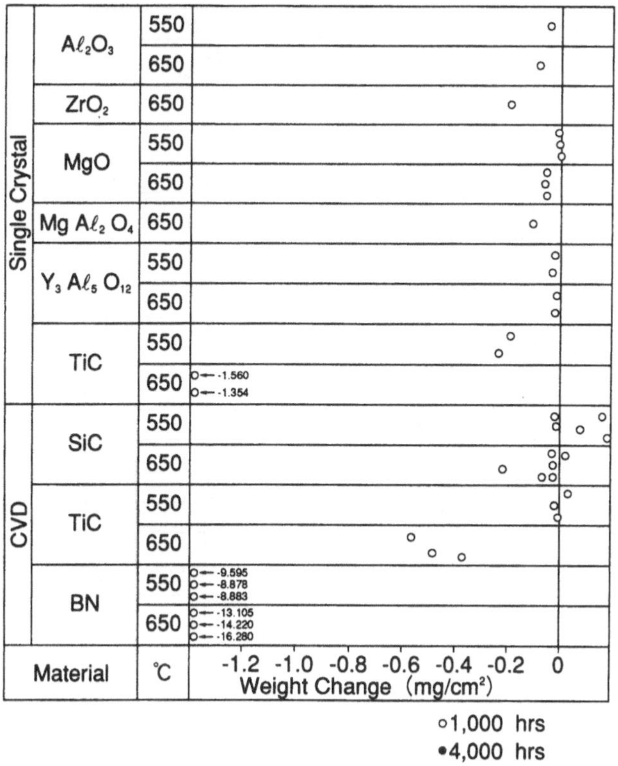

Fig.2 chart:

Material	°C	Weight Change (mg/cm²) notes
Single Crystal		
Aℓ₂O₃	550	
	650	
ZrO₂	650	
MgO	550	
	650	
Mg Aℓ₂ O₄	650	
Y₃ Aℓ₅ O₁₂	550	
	650	
TiC	550	
	650	o← -1.560; o← -1.354
CVD		
SiC	550	
	650	
TiC	550	
	650	
BN	550	o← -9.595; o← -8.878; o← -8.883
	650	o← -13.105; o← -14.220; o← -16.280

Weight Change (mg/cm²) axis: -1.2 -1.0 -0.8 -0.6 -0.4 -0.2 0

○1,000 hrs
●4,000 hrs

Fig.2. Weight change of single crystal and CVD ceramics after sodium corrosion test.

Table 2. Depth of layer with changed colour in sintered ceramics after Na exposure.

	1000 h sodium exposed	4000 h sodium exposed at 650°C / µm
Al₂O₃	Trace	200
ZrO₂	40	400
SiC	Trace	50
Si₃N₄	80	300
SiAlON	10	50

In ZrO₂ and Si₃N₄, the depth of colour changed layer reached 400 µm and 300 µm, respectively, after sodium exposure at 650°C for 4000 h. Sodium was detected in this layer by TEM-EDS analysis and EPMA line analysis. The sodium detected depth corresponded to the colour changed width.

The observation of cross-sectional micro-structure and sodium surface analysis by EPMA for ZrO₂ (stabiliser : MgO), which was exposed to 650°C sodium for 1000 h. showed that much sodium was detected at pores in ceramics. This suggests that sodium penetrated into pores through grain boundaries.

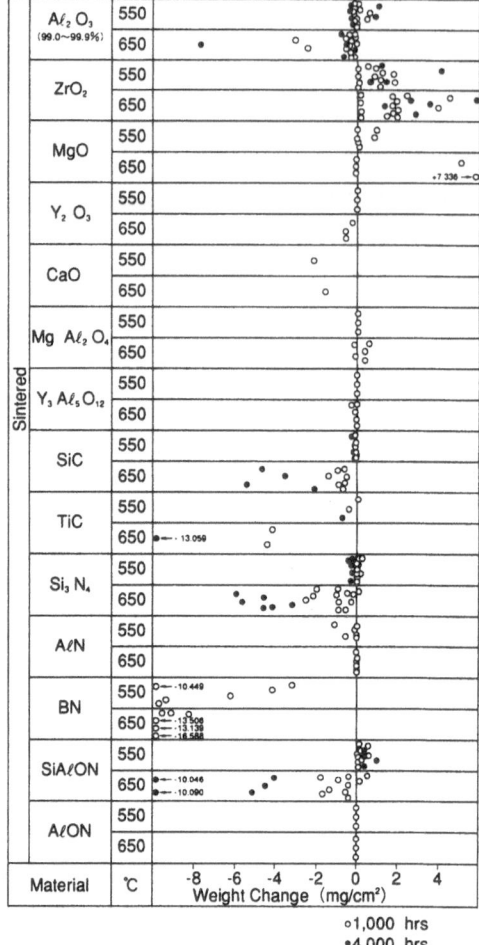

Fig.3. Weight change of sintered polycrystal ceramics after sodium corrosion test.

The observation of sodium exposed ZrO_2 surface by electron diffraction showed that Na_2ZrO_3 generated on the surface. It is considered that sodium penetration and Na_2ZrO_3 generation caused weight gain shown in Fig.3. The effect of stabilisers in ZrO_2 on sodium corrosion was not observed between MgO and Y_2O_3.

Figure 4 shows cross-sectional micro-structures observed by SEM for Al_2O_3, which was exposed to 650°C sodium for 1000 h. Low purity Al_2O_3 shows the significant grain boundary attack and its grains fell off partially. Figure 5 shows the relation between purity and weight loss for Al_2O_3. Relatively speaking, the higher the Al_2O_3 purity is, the lower the weight loss is. Therefore, it is understood that the corrosion rate depends on the impurity level in Al_2O_3 raw material.

Silicon carbide showed the grain boundary attack near sodium exposed surface. Silicon nitride showed the similar grain boundary attack. As for SiAlON and TiC, an aspect of grain boundary attack was not apparent by SEM observation

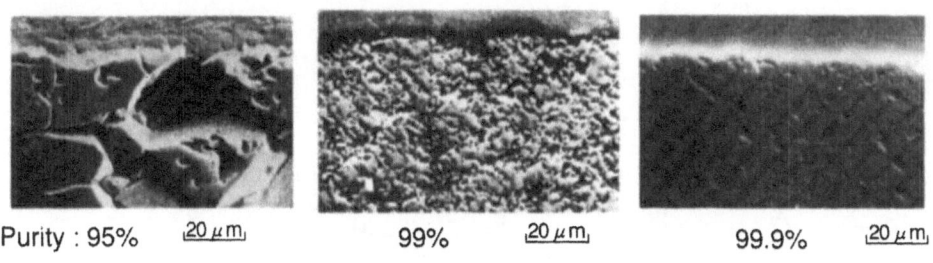

Purity : 95% ⎣20 μm⎦   99% ⎣20 μm⎦   99.9% ⎣20 μm⎦

Fig. 4. Scanning electron micrographs of cross section of Al_2O_3
after sodium exposure at 650° for 1000 hours.

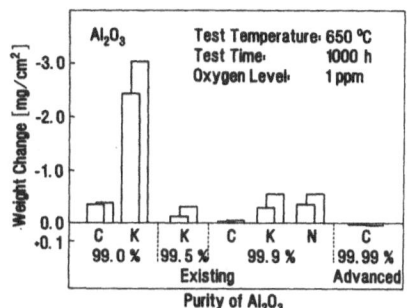

Fig. 5. Effect of Al_2O_3 purity on weight change.

Figure 6 shows the compositional analysis by TEM-EDS for Si_3N_4, which was exposed to 650°C sodium for 1000 h. At grain boundary which was glassy structure, not only Al_2O_3 and Y_2O_3 as additives but also sodium were detected. Sodium attack of Si_3N_4 grain was not observed by high resolution TEM.

Two silicon nitrides having the same compositions, one was low pressure sintered and the other was thereafter treated by hot isostatic pressing (HIP), were exposed to 550°C and 650°C sodium for 1000 h. However, the difference in manufacturing process is not significant on sodium corrosion

4. DISCUSSION ON CORROSION CHARACTERISTICS OF EXISTING CERAMICS

It is estimated that sodium in colour changed layers observed in sintered ceramics penetrated into the inside mainly by grain boundary diffusion. Though these colours changed layers show minute micro-structures, their effect on the integrity of sintered ceramics is not yet examined. It is assumed that the higher corrosion rates of sintered ceramics than those of single crystals and CVD ceramics were presumably attributed to the existence of additives and impurities in grain boundaries, resulting from the selective sodium penetration through them along grain boundaries by diffusion and grain boundaries attack based on the reaction with them.

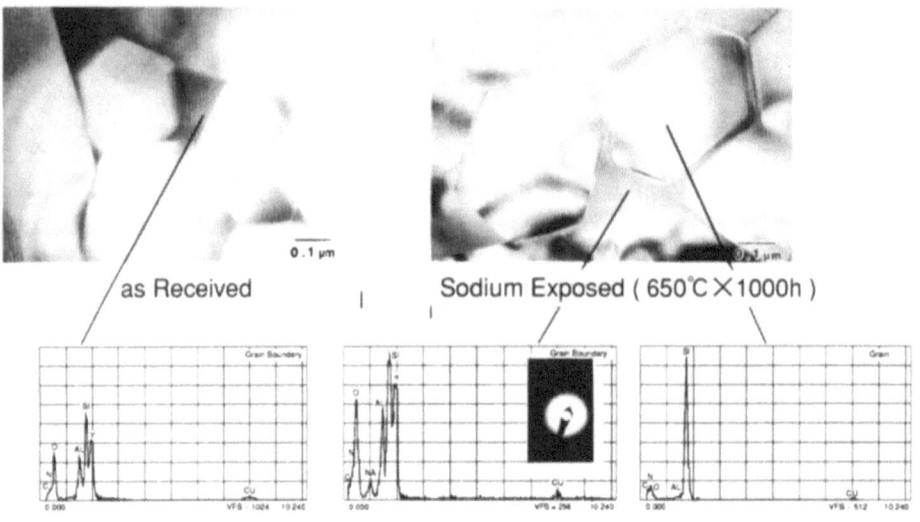

Fig. 6. Results of TEM and EDS analysis on Si_3N_4.

Particularly, it was detected by electron diffraction analysis for SiC, Si_3N_4 and SiAlON that their grain boundaries were glassy. In general, silicon ceramics produce SiO_2 partially at their grain boundaries during sintering process. In accordance with the phase diagram of SiO_2 and Na_2O, these two chemical compounds produce composite oxides ($Na_xSi_yO_z$) [3] Further, Na_2SiO_3 is thermodynamically more stable that SiO_2 and Na_2O. Therefore, it is considered that one of main causes of grain boundary attack is what the glassy SiO_2 existing in the grain boundaries produces soda glass by the reaction with Na_2O in sodium.

Figure 7 shows that the grain boundary of 95% Al_2O_3 is glassy structure. In high purity Al_2O_3, glassy structure was not detected at the grain boundary by electron diffraction analysis.

Figure 8 shows depth profiles of sodium by SIMS analysis in single and polycrystal Al_2O_3. [4,5] The depth profile of polycrystal has a long tail at deeper part, which indicates the grain boundary diffusion of sodium ions. The profile did not depend on the depth at deeper position in the sample treated at higher temperature.

On the other hand, although sodium ions penetrated along grain boundary in the transparent YAG, shown in Figure 9 [4], the depth profile is little affected by sodium grain boundary diffusion, because of the grain boundary diffusivity in polycrystal YAG much lower than polycrystal Al_2O_3 shown in Fig.8. The difference in grain boundary diffusion characteristics among similarly transparent ceramics is believed to reflect grain boundary characteristics, such as structures and compositions.

As shown in Figure 10 [4], Mg ions as sintering aid additive segregated near grain boundary in the transparent Al_2O_3. On the other hand, the segregation of additive was not observed in the transparent YAG. Furthermore, the observation by high resolution TEM showed that the range of lattice irregularity at grain boundary was under 1 nm in YAG ceramics.

It is suggested that the additive and impurity in Al_2O_3 polycrystal affected the grain boundary diffusivity and they formed a liquid phase at grain boundaries with sodium ions during sodium exposure at higher temperature (650°C), and then the sodium ions rapidly diffuse in this liquid phase at grain boundaries. The mechanism of high diffusivity along grain boundary was observed in Bi or Na ion diffusion at grain boundary of a strontium titanate ceramics, where a liquid phase was believed to exist [6].

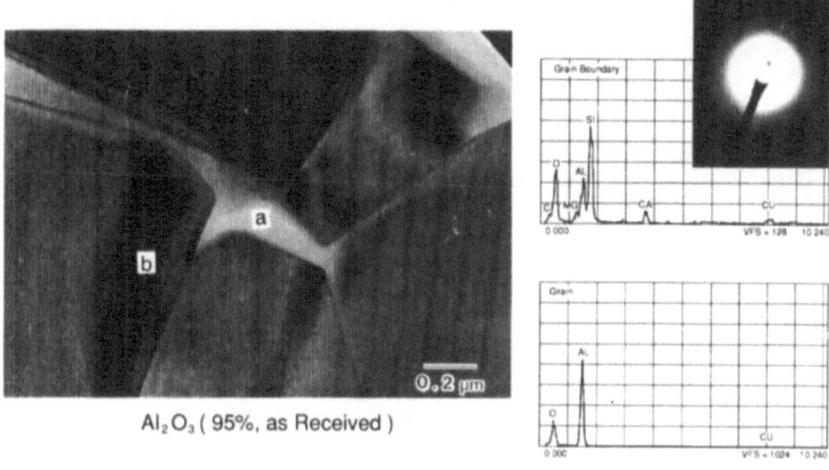

Al₂O₃ (95%, as Received)

Fig. 7. Result of TEM and EDS analysis on 95 % Al$_2$O$_3$.

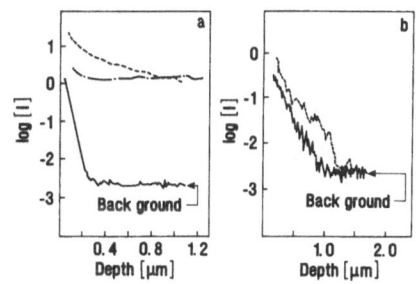

Fig. 8. Depth profile of Na in single and polycrystals.

a) for Al$_2$O$_3$ (solid line: single; broken line: polycryst. at the low temp.; dashed and dotted line at the high temperature, respectively)

b) for YAK (solid line: single; broken line: polycryst. at the low temperature)

Further, Al$_2$O$_3$ is thermodynamically more stable than Na$_2$O and SiO$_2$. Therefore, it is considered that Al$_2$O$_3$ possesses more excellent corrosion resistance than the other sintered ceramics. It is assumed that the grain boundary attack of low purity Al$_2$O$_3$ is attributed to the glassy structure and the corrosion of SiO$_2$ existing much as major impurity in Al$_2$O$_3$ raw material.

Figure 11 shows the schematic drawing of sodium corrosion mode on sintered ceramics.

Although TiC was in a form of single crystal, phenomena like grain boundary diffusion were observed [7]. It was reported that the sub-grain boundary existed in this single crystal and the misfit degree of sub-grain was stayed in the range of some seconds to some minutes [8]. The sodium image of TiC by SIMS [7], therefore, reflected the diffusion of sodium ions along these sub-grain boundaries. The oxygen ions were also observed at the same position as sodium ions by SIMS [7]. It is believed that the oxygen impurity in sodium firstly attacks at sub-grain boundary and then forms a liquid phase with sodium ions, so the high diffusivity of sodium ions at sub-grain boundary is observed.

The improvement of corrosion resistance to sodium is inevitable for the application of ceramics to sodium environment in the fast reactors. Based on the above discussion, it is necessary for the improvement that the micro-structures of grain boundaries must be controlled. To satisfy this, the following considerations can be proposed:

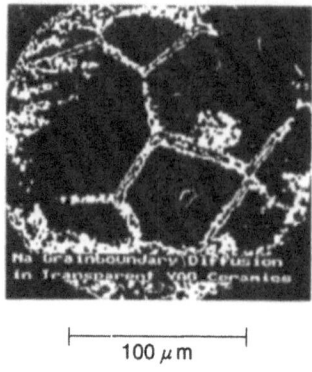

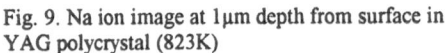

100 μm

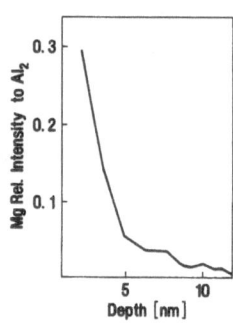

Fig. 9. Na ion image at 1μm depth from surface in YAG polycrystal (823K)

Fig. 10. Mg segregation profile near grain boundary in alumina.

the change of grain boundary from glassy structure to crystalline structure,
the selection of additives chemically more stable in sodium,
the reduction of grain boundary area,
the decrease of SiO_2 production at grain boundary,
the improvement of raw material purity,
the related improvement of manufacturing processes, etc..

5. CORROSION CHARACTERISTICS OF ADVANCED CERAMICS

Advanced ceramics were trially manufactured based on the countermeasures proposed to improve the resistance to sodium corrosion. The effects of improvements on the corrosion resistance of advanced ceramics have been evaluated by sodium exposure test.

Figure 12 shows the weight change of advanced Al_2O_3 with higher purity of 99.99% comparing with those of existing Al_2O_3 manufactured by the same company, which all were exposed to 650°C sodium for 1000 h. The corrosion of advanced Al_2O_3 by sodium was negligible and SEM observation showed no evidence of grain boundary attack. As the above discussion, Al_2O_3 possesses more excellent corrosion resistance with higher purity.

Figure 13 shows the weight change of advanced Si_3N_4 with crystallised grain boundary comparing with those of existing Si_3N_4 with glassy grain boundary, which all were exposed to 650°C sodium for 1000 h. Advanced Si_3N_4 relatively shows the tendency to have higher resistance to sodium corrosion than existing ones. This will be attributed to the crystallisation of grain boundary.

Figure 14 shows the weight change of advanced SiAlON with narrower grain boundary comparing with those of existing SiAlON, which all were exposed to 650°C sodium for 1000 h. Advanced SiAlON relatively shows the tendency to have higher resistance to sodium corrosion than existing ones.

6. CONCLUDING REMARKS

It was clarified through the detailed evaluation of sodium corrosion characteristics on three kinds of existing ceramics, those are single crystals, CVD and sintered ceramics that the

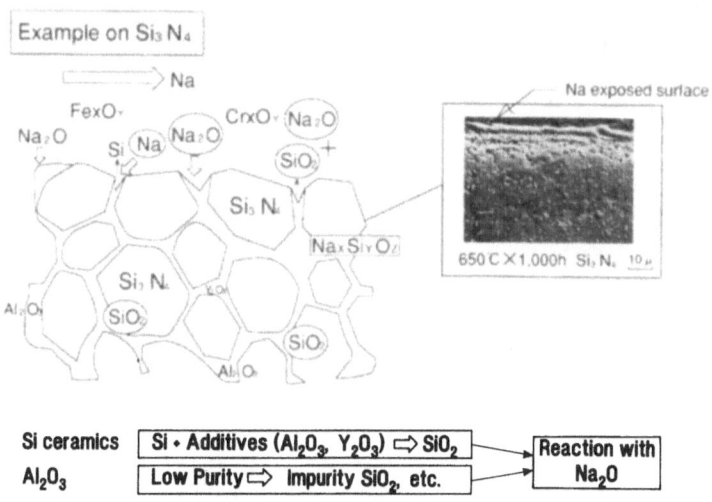

Fig. 11. Schematic drawing of sodium corrosion mode on sintered ceramics.

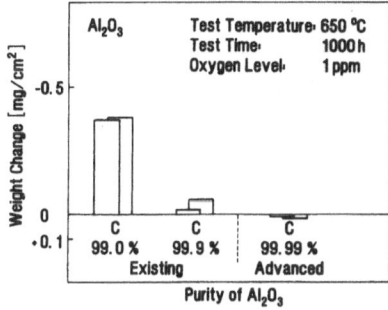

Fig. 12. Comparison of weight change between existing and advanced Al$_2$O$_3$.

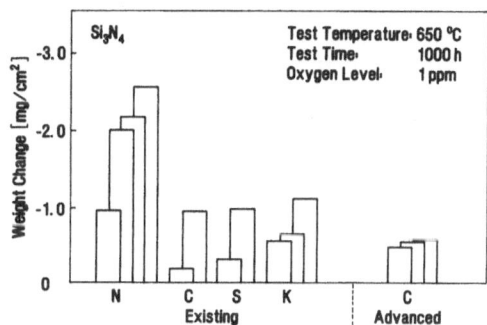

Fig. 13. Comparison of weight change between existing and advanced Si$_3$N$_4$.

grain boundary attack was a dominant corrosion process mainly on sintered ceramics, thus the micro-structures of grain boundaries should be controlled to improve corrosion resistance to sodium. The countermeasures were proposed to create the advanced ceramics having excellent sodium corrosion resistance.

The advanced ceramics were trially manufactured based on these countermeasures and tested in sodium. The effects of improvements on the corrosion resistance of advanced ceramics were evaluated and clarified.

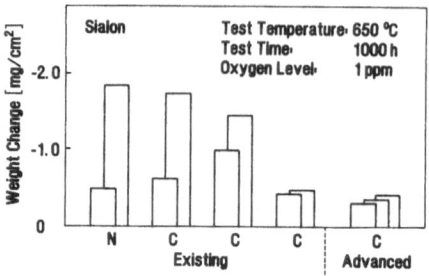

Fig. 14. Comparison of weight change between existing and advanced SiAlON.

REFERENCES

1. S.Kano, E.Yoshida, M. Inoue, S. Nomura, Y. Hirakawa, K. Saka, H. Haneda, T. Mitsuhashi, and Y. Kimura, Proceedings of the Intl. Conf. on Fast Reactors and Related Fuel Cycles, Vol. 4, P4-11, Tokyo, 1991.
2. E.Yoshida, S. Kano, Y. Hirakawa, Y. Wada, Y. Himeno, S. Saka, H. Haneda, and T. Mitsuhashi, Proceedings of the Intl. Symp. on Material Chemistry in Nuclear Environment, p. 305, Tokyo 1992.
3. J.Loeffler, Glastech. Ber., 42, No.3(1969) 92.
4. H.Haneda, S. Otani, T. Mitsuhashi, S. Shirasaki, E. Yoshida, and S. Kano, Proceedings of the Intl. Symp. on Material Chemistry in Nuclear Environment, p. 295, Tokyo 1992
5. H.Haneda, H. Toshima, Y. Miyazawa, T. Mitsuhashi, E. Yoshida and S. Kano, Proceedings of the 6th Intl. Conf. on Intergranular and Interphase Boundaries in Materials ,Held on June 22-26, 1992 at Thessaloniki, Greece
6. H.Haneda, J. Tanaka, T. Mitsuhashi, K. Ito, and S. Shirasaki, Proceedings of the Intl. Symp. on Material Chemistry in Nuclear Environment, p. 287, Tokyo 1992
7. T.Mitsuhashi, H. Haneda, S. Otani, S. Kano, and E. Yoshida, Proceedings of the Intl. Symp. on Material Chemistry in Nuclear Environment, p. 109, Tokyo 1992
8. S.Otani, S. Homa, T. Tanaka and Y. Ishizawa, J. Crys. Growth, 61, 1 (1983)

COMPATIBILITY TESTS OF INSULATING
MATERIALS IN Pb-17Li ALLOY

H. Glasbrenner, H.U. Borgstedt, Z. Peric

Kernforschungszentrum Karlsruhe GmbH.
Institute of Materials Research III
D-76021 Karlsruhe, Germany

1. INTRODUCTION

The liquid alloy Pb-17Li is used as breeder material and as coolant in the concept of a self-cooled liquid metal blanket of a fusion reactor [1]. The circulation of the eutectic has to be performed at relative high flow velocity in order to get a good heat-exchange. One serious problem of this concept is the pressure drop due to the magneto-hydrodynamic (MHD) forces in the flowing liquid metal under the influence of the magnetic field [2]. This pressure drop has to be drastically reduced. An electrical insulation of the flowing liquid alloy against the metallic tube walls may suppress the effects of MHD forces. There exist three different methods to insulate the tube walls. The so called "laminated walls" and the "flow channel inserts" are described elsewhere [3]. The direct insulating of the tube walls by means of a coating is a third method. The insulating coating on the wall surface performs the easiest way to reduce the MHD pressure drop. In this case electrical currents could only flow inside the liquid metal itself, and the result is a reduction of the MHD pressure drop by some orders of magnitude. However, the insulating coating has to be compatible with Pb-17Li over the whole lifetime of the blanket and has to withstand wear, crack formation, or peeling off. The electrical resistance should remain above a certain limit even under irradiation.

Pilot tests indicate the excellent compatibility of Pb-17Li with sintered corundum, Al_2O_3 up to 650 °C. If the insulating layers would have the ability of self-healing after a damage, their insulating character would be ensured all the time. For that purpose a layer with the potential to supply the insulating layer with atoms of the metallic compound has to be placed between the substrate and the insulating coating. A realistic construction of a multilayer system seems to be a coating of the substrate with an intermediate existing of iron-aluminium intermetallics, which might be covered with an insulating overlayer. This structure should fulfil the desired demands.

The preparation of such multilayer systems can be realised by means of the aluminising process. Aluminising is a well established and widely applied method which is used to

Liquid Metal Systems, Edited by H.U. Borgstedt
and G. Frees, Plenum Press, New York, 1995

produce a surface layer on steel, consisting of various iron aluminides. Subsequent exposure to oxygen at high temperature produces an adherent layer of Al_2O_3 on the surface of the intermetallics. Additional to the well established pack-aluminising [4] the so-called hot-dipping aluminising was recently developed. Examinations on such layers prepared by hot-dipping and a following heat-treatment and high temperature oxidation will be presented in this paper. The influence of flowing Pb-17Li on these covered steel specimens after 4000 h at 450 °C was studied and will be also discussed.

2. EXPERIMENTS

2.1 Preparation of the Samples

The initial surface scale on the plate of the martensitic steel X18CrMoVNb 12 1 with a composition given in [5] was removed by sand blasting. After this process the cleaned samples were heat-treated for 0.5 h at 1075 °C in vacuum and cooled within 2 h to 25 °C by blowing with compressed air in order to produce the martensitic structure. An Al melt was heated up to 800 °C in a crucible which was placed in a muffle furnace. The samples were dipped a few times into the molten Al for 2 min to get a uniform coating on the steel. The whole procedure was performed in a glove-box under purified Ar atmosphere ($p_{O_2} \leq 10^{-6}$ bar, $p_{H_2O} \leq 10^{-6}$ bar). Afterwards the coated samples were heat-treated at 610 °C for 2 h under vacuum. This heat-treatment improved on one hand the adhesion of the Al layer on the steel and secondly it promoted the formation of the intermetallic compounds by diffusion of Fe out of the steel and Al into the steel. Such layers are necessary for the required self-healing mechanism.

The oxidation of these samples was carried out at 950 °C for 30 h in air. The prepared samples were then exposed to flowing Pb-17Li for 1000, 2000, 3000, and 4000 hours at 450 °C in the PICOLO loop. After the exposure to Pb-17Li the surface of the specimens was covered with adsorbed alloy. Therefore the samples were washed in a solution of acetic acid / hydrogen peroxide, neutralised, washed, and dried to remove the remaining traces of the alloy.

2.2 PICOLO Loop

The compatibility of insulating layers on MANET is investigated in flowing lead-lithium alloy in the test rig PICOLO. It consists of a figure-of-eight type loop with a central recuperative heat-exchanger, a hot leg for specimen exposure and a cold leg for pumping, corrosion product deposition and impurity removal. The tubing in the hot-section is made of ferritic steel X10CrAl 17, whereas in the cold section austenitic material (AISI 316Ti) is used. The volume of the loop is in the order of 10 l. The cylindrical samples have a diameter of 8 mm and a length of 50 mm. Six of these samples were stacked together using an adapter and inserted into the test section, a tube of 440 mm length and 15 mm inner diameter. The pump attains a flow of 120 l/h, which corresponds to a flow velocity of 0.30 m/s in the assembled test section, and a Reynolds number of 2.1×10^4. More details of the loop have been given in [6].

The eutectic alloy was delivered by Métaux Spéciaux S.A., Paris, France. Its composition was 0.68 at % Li, which corresponds well to the eutectic point.

2.3 Test Procedures

The Li content in Pb-17Li was analysed by means of atomic absorption spectroscopy (AAS) or optical emission spectroscopy (OES) with inductive plasma excitation [7].

The samples were examined on the surface and the polished cut by combination of scanning electron microscope with energy dispersive analysis by X-ray (EDAX) capabilities. Every sample was metallographical examined. The original sample was investigated with Auger electron spectroscopy (AES).

3. RESULTS

3.1 Characterisation of the Original Material

Two layers could be found on the surface of the steel (Fig. 1) after the high temperature oxidation of the HDA MANET steel for 30 h at 950 °C. The inner diffusion zone was characterised by a thickness of about 220 µm, and the outer diffusion zone, called overlayer, was about 80 µm thick. The adhesion of the inner diffusion layer with the steel is excellent, no cavities or pores can be observed. The intermediate and the overlayer are separated by a band of pores.

Hot-dip aluminised MANET steel

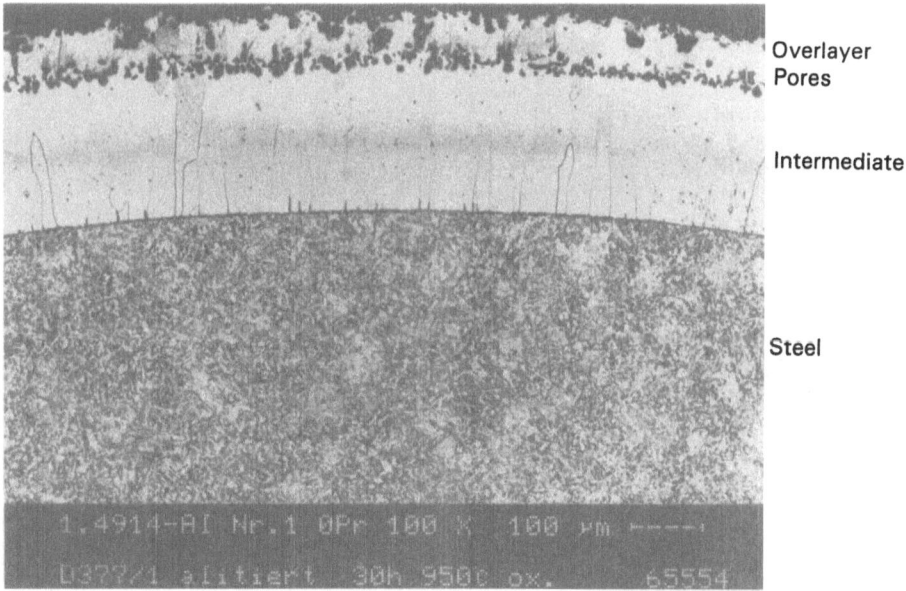

Fig. 1. The micrograph of the cross-section shows the HDA MANET steel oxidised for 30 h at 950 °C. The two diffusion zones and the substrate are well determined.

Fig. 2 a and b show SEM micrographs of the same sample with different magnification. The substrate, the two layers, the band of pores between them and the pores on the overlayer can be well distinguished. In Fig. 3, characteristic K_α X-ray maps of oxygen (a), aluminium (b), and iron (c) were taken of the area of Fig. 2 b. The brighter region corresponds to higher concentration of the element in question. The X-ray image (Fig. 3 a) shows the element distribution of oxygen. Oxygen is enriched in the pores and on the surface. Oxygen perhaps diffused along grain boundaries into the material. No oxygen could be found in the two diffusion zones.

The element concentration pattern of Al (Fig. 3 b) shows that Al is distributed continuously in both diffusion zones. The concentration of the element is higher in the over-layer than in the intermediate. The content of aluminium is quite low in the pores and on the surface, but does not reach zero.

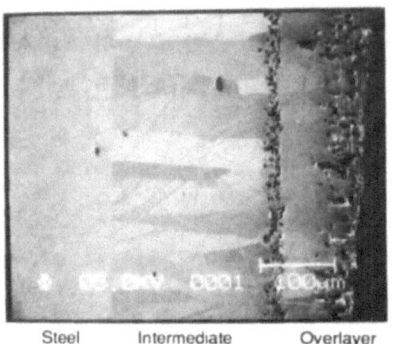

| Steel | Intermediate | Overlayer | | Intermediate | Pores | Overlayer |

Fig. 2 a. The SEM microgaphs show the matrix steel, the two diffusion zones, the band of pores between them and the pores in the overlayer.

Fig. 2 b. Enlarged upper region of Fig. 2 a.

Characteristic K_a X-ray maps

a) K_α O

b) K_α Al

c) K_α Fe

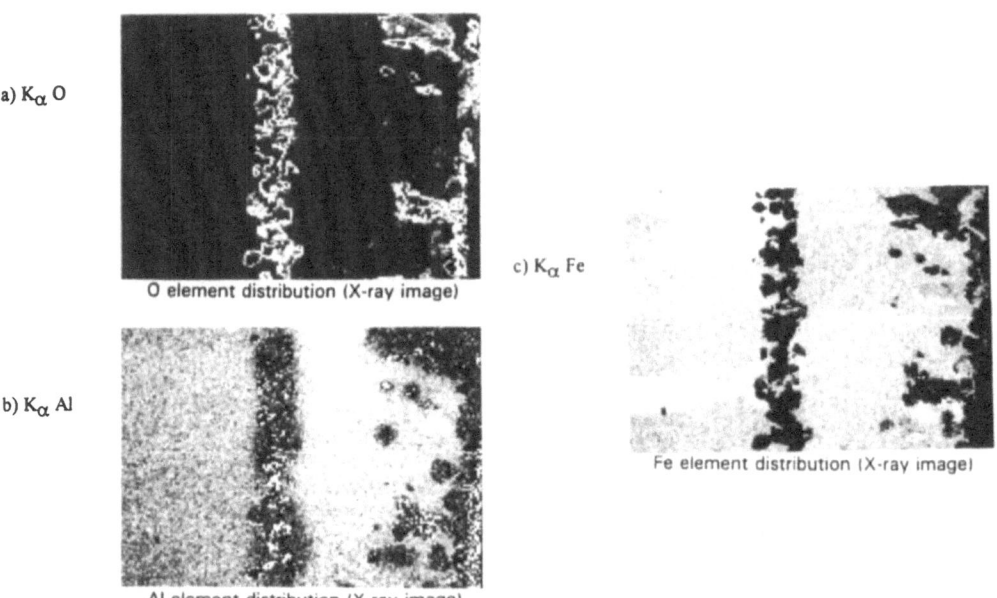

O element distribution (X-ray image)

Fe element distribution (X-ray image)

Al element distribution (X-ray image)

Fig. 3. The X-ray images of HDA MANET steel after high temperature oxidation for 30 h at 950 °C.
a) K_α O   b) K_α Al   c) K_α Fe

Fig. 3 c shows the X-ray map of iron. No Fe can be observed on the surface and in the pores. The intensity in the overlayer is lower than in the intermediate, that means the concentration is lower in the overlayer than in the intermediate. The element distribution in both diffusion zones is continuous, and no enrichment in any region can be determined.

EDAX measurement were carried out on the surface, in the two diffusion zones and in the steel matrix. The contents of aluminium, iron, chromium, and manganese were measured and the values were calculated half-quantitatively with a standard program which is integrated in this system. This method is only usable for elements with an atomic number higher than 12, therefore, the content of oxygen could not be measured. In Fig. 4, the contents of the elements are plotted versus the depth of the sample.

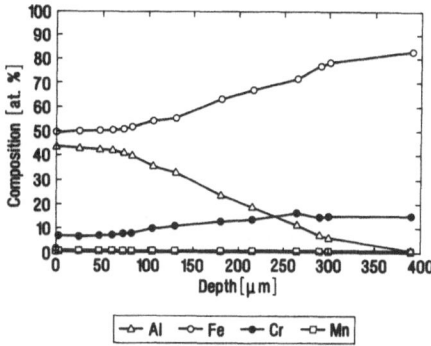

Fig. 4. The depth profile of HDA MANET steel after high temperature oxidation for 30 h at 950 °C, which was measured by EDAX on the polished cut of the specimens.

The content of aluminium on the surface is about 95 at %. Iron, manganese, and chromium can only be detected in small concentrations on the surface. The depth profile shows, that the Al content decreases with increasing depth. The iron and the chromium content show the opposite trend, they increase with increasing depth. The concentration of manganese is constant on a low level over the whole measured area. In the overlayer with a thickness of about 80 μm the concentration of the elements is nearly constant. The value for aluminium decreases from 40 to 35 at %, the Fe increases from 50 to 56 at %, and the Cr from 9 to 12 at %. The change of the concentrations of each of the elements is more significant in the diffusion layer. The aluminium content decreases from 30 to 5 at % in this 220 μm thick layer, and the Fe and Cr content have nearly reached the steel composition.

3.2 Characterisation of the Material after Exposure to Pb-17Li

Metallographic examination (Fig. 5 a and b) shows the behaviour of the aluminised steel after 1000, and 2000 h in flowing Pb-17Li. The thickness of the overlayer and the intermediate is still of the same order as in the original sample. Small differences in the thickness might be due to the not yet optimised preparation process. No visible damage can be observed after the interaction of the flowing eutectic with the overlayer. The adhesion between the intermediate and the steel remains well. The band of pores, which separates the two diffusion zones does not show any change, neither in the size nor in the number of pores. Hence no attack has been evidenced on the specimens for the increasing exposure time by the means of this method.

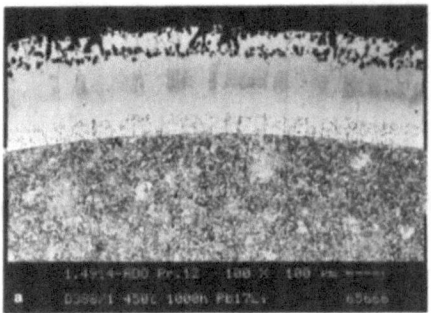 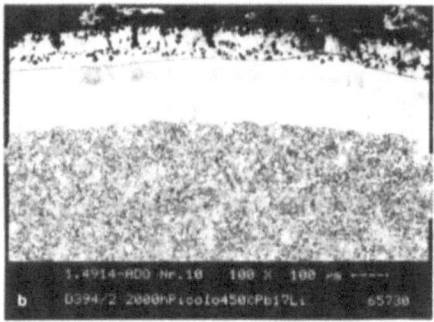

Fig. 5. Metallographic examination of HDA MANET steel after the exposure to Pb-17Li for
a) 1000 h and for   b) 2000 h.

The surface composition is plotted in at % versus the exposure time in Fig. 6. The values are evaluated by EDAX measurements of the surface of the specimens after the exposure to Pb-17Li for 1000, 2000, and 3000 h. For a better comparison, the values of the original samples were also added in the plot. Before the measurements, the samples were cleaned of the traces of the Pb-17Li alloy.

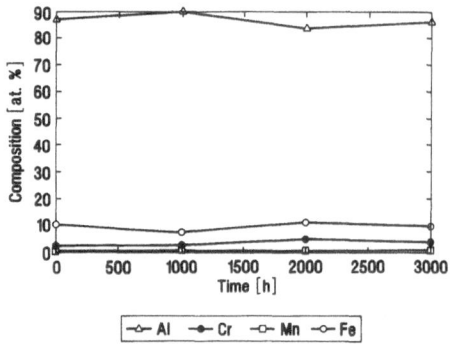

Fig. 6. The surface composition of the specimens is plotted after the exposition to the eutectic alloy for different periods of time.

The value of manganese is constantly low over the whole time. The aluminium content decreases slightly with raising exposure time. The elements Fe and Cr show the opposite run, their contents increase. The interpretation of these small changes is very difficult. The reasons for these tendencies could be attributed to corrosion by the Pb-17Li alloy, however, they might also be due to a reaction with the purification liquid (acetic acid / hydrogen peroxide). Other possible influencing parameters could be the tolerance by the measurement, or the calculation or the not yet optimised preparation process of the specimens.

To get more information about the discussed process, depth profiles were measured with EDAX on each of the exposed samples. The depth profile is plotted for the sample exposed for 3000 h in Fig. 7. The Al content decreases, the Fe and Cr content increase with increasing depth. Mn is constant in agreement with the results of the standard sample. In the overlayer Al, Fe, and Cr are nearly constant over this span of 80 µm. The concentrations of the elements show more significant changes in the intermediate. Aluminium decreases to 5 at

%, and iron and chromium nearly raise up to their concentration of the steel. Comparing these results with the original sample, no differences can be observed. The concentration profiles for the elements show always the same trend. Fig. 8 shows the concentration profiles of Al for each of the samples across the two diffusion zones and the steel. Generally, the same trend has been found in all samples. High concentration of aluminium on the surface, nearly constant concentration in the overlayer and a strong decrease in the intermediate. The results of the intermediate are comparable. The highest deviation in the overlayer can be observed on the 1000 h sample. In this case the concentration is lower than in the other three samples. The original sample and the 3000 h sample show the same results. Hence, it does not seem to be a corrosion problem, but to be related to production process.

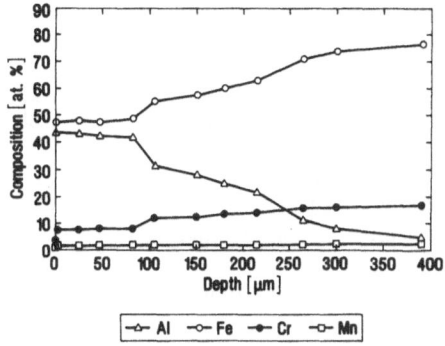

Fig. 7.   The depth profile shows the trends of Fe, Al, Cr, and Mn after 3000 h in Pb-17Li.

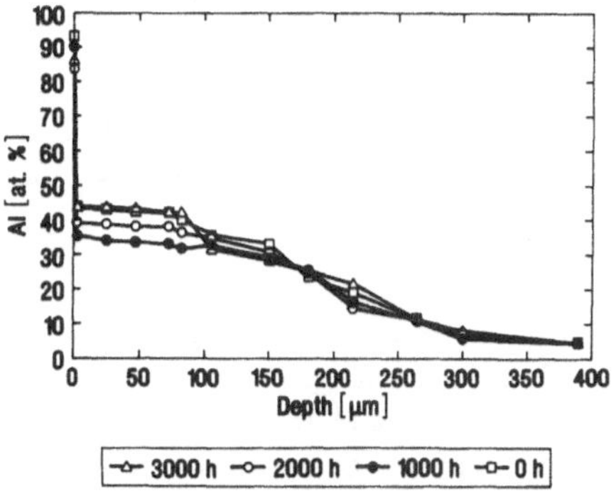

Fig. 8.   Concentration profiles of Al, plotted versus the depth across the two diffusion zones and the substrate.

The Pb-17Li in the PICOLO loop was periodically controlled in respect to impurities and the lithium content. No increase of impurities as iron, aluminium, or other metals could be measured in the Pb-17Li over the whole time of operating. The lithium content decreased with increasing time of operating, the values are summarised in Tab. 1.

Table 1. The lithium content decreases with increasing time of operation.

Time of operation / h	Lithium content / at %
0	0.681
500	0.638
1500	0.623
2500	0.493
2644	0.614
3120	0.584
3500	0.611

The reason of the decrease of Li in the alloy may be the reaction with traces of oxygen and nitrogen in the Ar gas. The extremely small value for lithium at 2500 h can be explained with a leakage in the box. The oxygen level raised up in the glove box, and lithium of the alloy could react to Li_2O, which was deposited on the surface of the melt. Hence the Li content in the alloy decreased. Fresh Pb-17Li was added to the PICOLO loop after the leakage was fixed to compensate the loss of lithium. Afterwards the lithium contents reached again their expected values for the corresponding time of operating. There might be an influence of the segregation of the Pb-17Li alloy in a semi-stagnant region of the loop [8].

4. DISCUSSION AND CONCLUSIONS

A multilayer system existing of MANET steel, intermetallics, and an oxide layer can be prepared by means of a hot-dipping aluminising process. The results are summarised in Table 2.

Table 2. The summarised results for the HDA MANET steel after the diffusion and oxidation processes.

Region	Thickness / μm	Al / at %	Fe / at %	Cr / at %
Surface	~ 5	93 - 85	5 - 10	2 - 5
Overlayer	~ 80	35 - 40	50 - 56	9 - 12
Intermediate	~ 220	5 - 30	55 - 78	12 - 17
Steel	Bulk	--	85	15

The adhesion between the intermetallic layer and the steel is good. The thickness of the two diffusion zones is sufficient and the Al content should be high enough for the desired demands of self-healing.

The thickness and the structure of the oxide layer is not yet optimised. Alternative procedures have to be tested for the oxidation of the intermediate to get uniform α-Al_2O_3 layers without any impurities and pores.

No corrosion attack could be observed on the aluminised samples at least after 4000 h in flowing Pb-17Li. The element concentrations in the different layers and the thickness did not change with increasing exposure time in the PICOLO. The composition on the cleaned surfaces of the samples shows a small increase of the steel components Fe, Cr, and Mn with increasing exposure time. This might be an effect of the cleaning procedure or of the corrosion attack.

It could be shown that the hot-dipping aluminising is an excellent method to form intermetallic layers on steel. The interconnection between these layers and the steel is of high quality. The intermetallic layers are compatible with the eutectic alloy under the conditions of the self-cooled liquid metal blanket.

The lithium content in the Pb-17Li alloy decreases with increasing time of operating. Therefore Li should be added to the alloy to reduce this decrease, and to be sure to work with the composition of the eutectic.

Acknowledgements

We gratefully acknowledge the metallographic work of Mr. H. Zimmermann and Mr. P. Graf, the AES measurements of Mr. E. Nold, and the analytical contribution of Pb-17Li of Dr. Ch. Adelhelm.
This work has been performed in the framework of the Nuclear Fusion Project of the KfK and is supported by the European Communities within the European Fusion Technology Program.

REFERENCES

1. S. Malang, H. Deckers, U. Fischer, H. John, R. Meyder, P. Norajitra, H. Reiser, K. Rust, Fusion Technol. Des. 14 (1991) 373-399.

2. L. Barleon, V. Casal, L. Lenhart, A. Sterl, KfK-Nachrichten <u>21</u> (1989) 224-231.

3. S. Malang, K. Arheidt, L. Barleon, H.U. Borgstedt, V. Casal, U. Fischer, W. Link, J. Reimann, K. Rust, G. Schmidt, Fusion Technol. <u>14</u> (1988) 1343-1356.

4. K.S. Forcey, D.K. Ross, C.H. Wu, J. Nucl. Mater. <u>182</u> (1991) 36-51

5. H.U. Borgstedt, G. Frees, M. Grundmann, Z. Peric, Fusion Engng. Des. <u>14</u> (1991) 329-334.

6. H.U. Borgstedt, G. Drechsler, G. Frees, Z. Peric, J. Nucl. Mater. <u>155-157</u> (1988) 728-731.

7. Ch. Adelhelm, H.U. Borgstedt, D. Linder, G. Streib, Fusion. Engng. Des. <u>14</u> (1991) 235-239.

8. H.Feuerstein, L. Hörner, J. Oschinski, S. Horn, to be published.

DEPOSITION OF THE CORROSION PRODUCTS IN Pb-17Li

M. G. Barker and M. J. Capaldi

Department of Chemistry
The University of Nottingham
Nottingham NG7 2RD, United Kingdom.

1. INTRODUCTION

Both liquid lithium and the eutectic lithium-lead alloy, Pb-17 at%Li have been proposed as the tritium breeders as well as the coolant for fusion reactors operating on the D-T fuel cycle. The choice of Pb-17Li rather than pure liquid lithium was based on the enhancement of tritium breeding due to the neutron multiplication properties of lead and the extremely low reactivity of the alloy towards air and water relative to liquid lithium [1,2,3]. Successful nuclear fusion has been demonstrated in the Joint European Torus (JET) and the next generation of fusion reactors are envisaged as having a Pb-17Li blanket. The molten alloy is likely to be contained within either 316L stainless steel or the newer German type 1.4914 martensitic steel. The results of preliminary studies have shown that corrosion product mass transfer and deposition may affect the maximum operating temperature in some liquid metal blanket systems.

A knowledge of the chemical composition and deposition behaviour of dissolved corrosion products is essential for the safe operation of fusion reactors. The deposition of activated elements has been the subject of much conjecture and controversy over the past few years yet no comprehensive data exist which will enable accurate predictions of the deposition behaviour of such material to be made. We have addressed the problem firstly by the determination of the solubilities in Pb-17Li of the elements likely to be used in a fusion reactor namely Fe, Cr, Ni and Mn. This has been followed by a simple study of the deposition of corrosion products from Type 316 and 1.4914 steels in Pb-17Li under a temperature gradient. To date some six tests have been carried out and the results indicate a consistency in the behaviour of the deposition process.

2. EXPERIMENTAL

All tests were carried out in similar apparatus comprising of a single 25cm length of steel tube, either Type 316L stainless steel which had an internal bore of 10mm or 1.4914 Martensitic steel with a bore of 25mm. The tubes were welded closed at one end and filled

Liquid Metal Systems, Edited by H.U. Borgstedt
and G. Frees, Plenum Press, New York, 1995

with liquid Pb-17Li alloy in an argon filled evacuable glove-box. The open ends of the tubes were then sealed either with a Dralim coupling or by welding, also in an inert atmosphere to prevent conta ation of the alloy. Chromel-Alumel thermocouples were silver soldered at centimetre i .ervals along the tubes and the entire length of the tubes heated to 600°C for 1 hour to ensure efficient wetting of the steel surface by the lithium-lead. The wetted tubes were then placed in a furnace where a temperature gradient, approximately linear from 650°C to 250°C was set up. The gradient was arranged so that the hot end of the tube was at the top in order to minimise the effect of convection. The tubes were left in this state for the duration of the tests which ran for between 672 and 1600 hours. The temperatures were noted at regular intervals.

On completion of the tests the tubes were cooled, sectioned longitudinally and cut into two centimetre lengths. These samples were mounted in conducting bakelite and polished through successively finer carborundum paper and finally 0.05 micron gamma alumina. The prepared samples were then examined by both optical and electron microscopy using a Jeol 35C electron microscope and compositional information was obtained from EDX studies.

In addition to the basic configuration as set out above, each test had additional metal in the hot region to increase the surface area available for corrosion and hence increase the amount of material entering solution. This metal was the same as the tube material and took the form of bundles of wire for 316L and small plates for 1.4914. The exception to this was a test designed to study the mass transfer of nickel where the added metal to a 316L system took the form of a nickel foil liner in the hot zone. Also in two of the tests a ceramic rod was inserted into the centre of the tube on filling in an attempt to preferentially deposit corroded material onto an inert surface (Fig. 1). In one test a magnetic field was placed around a region of the tube where deposition was thought likely to occur in the hope of trapping out any ferromagnetic material. The details of all tests carried out are shown in Table 1 below.

Table 1. Details of tests

Test	Tube metal	Added metal	Temperature gradient	Duration /hours	Ceramic /magnet
1	316	316	600-300°C	672	-
2	316	Ni	650-250°C	1600	Al_2O_3
3	316	316	650-250°C	1600	-
4	1.4914	1.4914	650-250°C	1500	-
5	316	316	650-250°C	1600	ZrO_2/CaO
6	316	316	650-250°C	1600	Magnet

3.RESULTS AND DISCUSSION

3.1. Corrosion

3.1.1 Type 316 systems

In tests 1,3,5 and 6 above, the corrosion behaviour followed the standard pattern for type 316 stainless steel at high temperatures. There was formation of a porous layer through the complete depletion of nickel and considerable chromium depletion. This causes a destablisation of the austenite structure and this layer is ferritic in nature. EDX analysis was used to determine the composition of this layer which is marked B in figure 2, this shows the depletion of nickel and chromium from bulk steel which is marked C in Fig. 2. The depth and appearance of this layer were concurrent with published results [4,5,6,7].

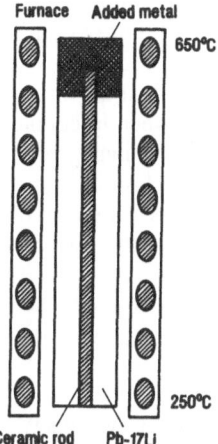

Fig. 1. Schematic of tube. Ceramic: Al_2O_3, CaO stabilised ZrO_2. Added metal: 316L wire, nickel foil, 1.4914 plates.

In test 2 above, there was a difference in the corrosion behaviour from the standard type 316. This was apparent by the absence of a porous ferritic zone even at very high temperatures. This was due to the presence of the nickel foil in the hottest region of the test. As the solubility of nickel in Pb-17Li is very high [8] the solution in the high temperature zone was saturated with respect to nickel and thus prevented the dissolution of nickel from the 316 matrix. This is a step which is responsible for formation of a porous layer.

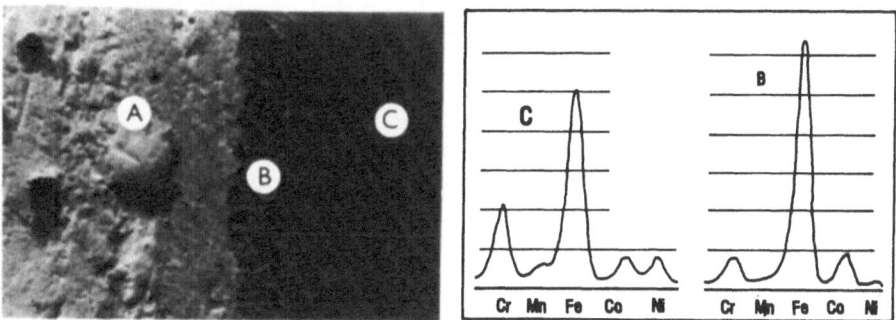

Fig. 2. Porous corrosion layer formed on 316L stainless steel at high temperature. EDX analysis shows depletion of nickel.

Neither the alumina ceramic in test 2 or the calcia stabilised zirconia ceramic in test 5 had any noticeable effect on the corrosion of the 316 steel. The presence of a magnetic field in test 6 also did not alter the corrosion behaviour.

3.1.2 1.4914 Martensitic systems

Test 4 using a 1.4914 Martensitic steel tube of larger dimension than in the 316 tests showed no visible signs of corrosion both under optical and electron microscopy. We feel that it is not possible for there to be no corrosion at all of this material at such elevated temperatures and must therefore conclude that the main constituents of this steel, namely iron and chromium, corrode at a similar rate when in contact with molten Pb-17Li and the only evidence for corrosion would be a gradual thinning of the walls of the tube or the formation of material deposited from solution at lower temperatures. Our specimen preparation techniques did not allow for an accurate measurement of wall thicknesses to be made. These observation are in agreement to those reported by M. Broc et al. [9].

3.2. Deposition

3.2.1 Type 316 systems

Tests 1,3 and 5 described in Table 1 were systems where the material to be corroded was solely type 316 stainless steel either in the form of wire or the walls of the tube. There were two distinct types of deposits found in these tests. The first of these was found in regions of the tube which had been at a temperature of 550°C during the test and extended to regions which had been at 450°C. These were large and highly crystalline (Fig.3) and had a dendritic appearance. Using EDX the composition of this material could be determined. It was found to be a mixture of iron and chromium. Crude quantitative data were obtained by calculating the ratio of peak areas from the EDX traces. This gives an Fe/Cr ratio of approximately 3 for bulk 316 steel. The dendritic material observed at 550°C produced an Fe/Cr ratio of between 7 and 9 which implies an iron rich deposit whereas at 450°C this ratio falls to below 1 indicating a chromium rich phase. There was no nickel detected in any of this deposit. Some of this material was found adhered to the walls of the tube and in test 5 adhered to the ceramic surface, but the bulk of the deposit was found suspended in the frozen alloy.

The second type of deposit was found in regions of the tube which had been at temperatures of below 350°C during the test. This material was much less abundant and less crystalline in appearance (fig.4). EDX analysis showed the composition to be nickel rich with traces of iron and chromium at higher temperatures. This composition changed to nickel and manganese in a 1:1 ratio at lower temperature. As for the dendritic iron/chromium deposits this low temperature deposit was found adhered to the walls of the tube and the ceramic surface in test 5 but was mainly found in the bulk of the alloy.

3.2.2 316 Systems with a magnetic trap

It was noticed from the results of the 316 tests described above that the dendritic iron/chromium deposits were ferromagnetic. Test 6 was planned to exploit this by placing a permanent magnet around the region of the tube where deposition was expected to occur on the basis of the temperature gradient. This it was hoped would trap out the magnetic material.

Again as with the other 316 tests two distinct types of deposit were found. The first of these was the dendritic iron/chromium material described above. This however was found at a lower temperature than in previous tests being first evident at 500°C rather than 550°C and did not extend beyond 450°C. This was a much narrower range than in previous tests and corresponded to the region of the tube which was directly in the magnetic field. There

Fig. 3. Dendritic iron/chromium crystals

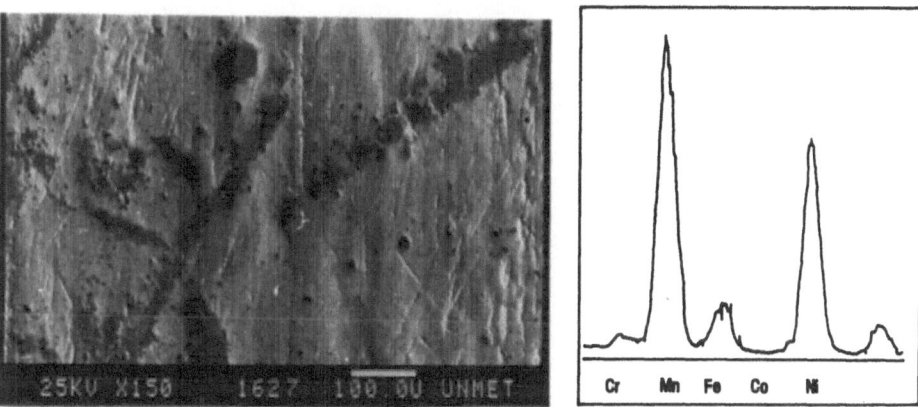

Fig. 4. Non-dendritic manganese/nickel deposits found below 350 °C.

was also no temperature dependence on composition as before with both iron rich and chromium rich deposits found in the same region. EDX analysis also showed traces of nickel associated with this material. Further evidence for the effect of the magnetic field was the fact that the deposition was only found on one side of the tube (fig.5). The region of the tube immediately adjacent to the poles of the magnet showed a large cluster of crystals whereas the opposite side was completely free from deposition. The second type of deposit found was as for previous 316 tests, small, non-crystalline and nickel rich. This was evident below 350°C and was composed of nickel and manganese only at very low temperature. This deposit was unaffected by the presence of the magnetic field.

3.2.3 Nickel mass transfer

Test 2 described above was designed to study the mass transfer of nickel in a lithium-lead environment. As there was no evidence of any corrosion of the 316 steel in the hottest regions of this test the only deposits found were almost entirely composed of nickel. This material was found at low temperatures, below 300°C, and was either in the form of sparsely distributed dendritic deposits which contained traces of iron, showing some corrosion of the 316 matrix but not sufficient to produce a porous zone, or as large regular crystals of pure nickel which were found in large quantities on the walls of the tube, on the surface of the alumina rod and in the bulk of the Pb-17Li alloy (figs.6a,6b).

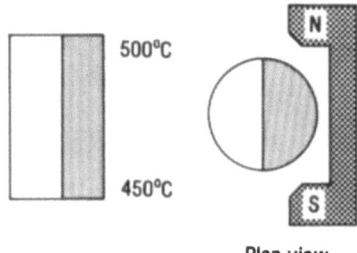

Plan view

Fig. 5. Section of tube from test 6. Shaded area represents crystalline Fe/Cr deposit.

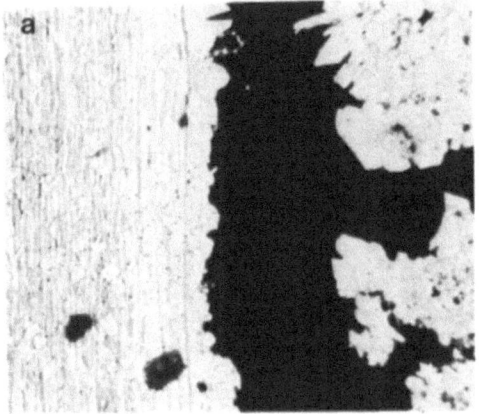

Fig. 6a. Nickel crystals deposited on steel surface and in bulk Pb-17Li.

3.2.4 1.4914 Systems

No deposition was found in test 4. This we suggest may have been due to the large volume of alloy in this test which did not allow the concentration of the corroded elements to reach a value where precipitation from solution was favourable.

4.3. CONCLUSIONS

1. In systems entirely composed of type 316 stainless steel two types of deposition occur.

The first type of deposit, type A. takes the form of large, dendritic crystals which are composed of iron and chromium only. This material deposits in the temperature range between 550 and 400°C. There is a temperature dependence on composition with a change from iron rich to chromium rich with decreasing temperature. Neither iron nor chromium are ever deposited as single elements.

The second type of deposit, type B. is smaller and non-crystalline. This material is more sparsely distributed than the type A deposit and is found at temperatures below 350°C. The composition of this deposit is generally nickel rich with some iron and chromium associated at the higher end of the deposition temperature range. This changes to nickel and manganese only in a 1:1 ratio below 300°C. A large percentage of the small amount of nickel which deposits in 316L stainless steel systems is found associated with manganese, pure nickel is never seen. The remainder of the nickel dissolved from the 316 matrix must remain in solution in the Pb-17Li due to the high solubility. This may constitute a hazard in a working reactor because of the long lived activation products produced from nickel on exposure to a high neutron flux.

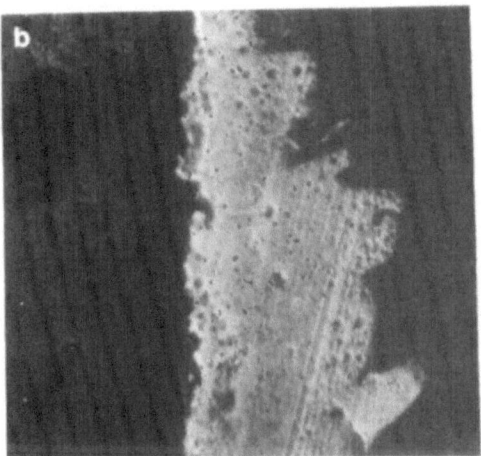

Fig. 6b. Nickel crystals deposited on alumina rod.

2. The presence of a magnetic field destroys the temperature dependence on the composition of the iron/chromium dendrites, type A deposits, and also compresses the deposition range. It has no effect on the nickel/manganese deposits. It has been shown that magnetic trapping would be an effective method of removing this iron/chromium deposit from a conceptual fusion reactor using a lithium-lead blanket.

3. Nickel will readily mass transfer in Pb-17Li under a thermal gradient and will precipitate out of solution at low temperatures, ca.300°C,but only if present in solution at sufficiently high concentrations. The presence of a large amount of nickel in the hottest regions of a Pb-17Li system will inhibit the corrosion of the 316L stainless steel containment.

4. A ceramic surface either alumina or zirconia does allow the deposition of all the types of deposits described above but there seems to be no preference for this surface over the walls of the steel tube or the bulk Pb-17Li.

Acknowledgements

The authors would like to thank the Joint Research Centre, Ispra for financial support. The work was instigated by the late Dr. Vittorio Coen whose inspiration and friendship will be sorely missed by all who knew him.

References

1. G. Kuhlborsch and F. Reiter, Nucl. Enginerring and Design/Fusion 1 (1984) 195.
2. D. W. Jeppson and L. D. Muhlestein, Fusion Technology 8 (1985) 1385.
3. V. Coen, A. T. Dadd, H. Kolbe and L. Orecchia, 3rd Int. Conf. on Liquid Metal Engineering and Technology, BNES, London, (1984) Vol. 1, 347.
4. V. Coen, P. Fenici, H. Kolbe, L. Orecchia and T. Sasaki, J. Nucl. Mater. 110 (1982) 108.
5. M.G. Barker, V. Coen, H. Kolbe, J.A. Lees, L. Orecchia, and T. Sample, J. Nucl. Mater. 155-157 (1988) 732.
6. M. Broc, P. Fauvet, T. Flament and J. Sannier, J. Nucl. Mater. 141-143 (1986) 611.
7. M. G. Barker, J. A. Lees, T. Sample and P. Hubberstey, J. Nucl. Mater. 179-181 (1991) 599.
8. M. G. Barker and T. Sample, Fusion Engineering and Design 14 (1991) 219.
9. M. Broc, T. Flament, P. Fauvet and J. Sannier, J. Nucl. Mater. 155-157 (1988) 710.

AN INVESTIGATION OF CHROMIUM AND NICKEL MASS TRANSFER IN NON ISOTHERMAL SODIUM LOOP

V.V. Alekseev, F.A. Kozlov and Ju.I. Zagorulko

Institute of Physics and Power Engineering,
249020 Obninsk, Russia

1. INTRODUCTION

The formation of depositions of impurities on the surfaces of sodium loops may influence hydrodynamic and heat exchange conditions of the loop units, leads to localised accumulation of radioactive impurities and to occurrence of seizing phenomena in the equipment with movable parts. In this connection the development of the calculation methods for impurities in non isothermal mass transfer is of exceptional importance in estimation of depositions accumulation rates and their distribution in the circuit channels.

2. MODEL OF IMPURITIES MASS TRANSFER IN THE COOLED CHANNEL

Non isothermal mass exchange of impurities is essentially a set of simultaneous hydrodynamic and physico - chemical processes.

The main principles of the impurities mass exchange model, presented here, are as follows:
- the coolant, containing dissolved impurities, forms a supersaturated solution, that results in volume nucleation of spontaneous impurities (origin of the impurity solid phase nuclei);
- the growth of the originated particles proceeds for the sake of crystallisation of dissolved substances on the particles surfaces and due to aggregation of the colliding particles;
- depositions on the channel wall are formed for the sake of dissolved substance crystallisation and particles transfer to the channel wall (at high rate of impurities dissolution the depositions on the channel wall are not formed at all);

Main assumptions:
- the formation of the nuclei takes a negligible small time and is uniform over the channel cross section;
- particles and dissolved substance are uniformly distributed over the channel cross-section, except of sodium boundary layer near the wall surface;

Liquid Metal Systems, Edited by H.U. Borgstedt
and G. Frees, Plenum Press, New York, 1995

- the particles are spherical ones and of the same dimension for some particular cross section. After secondary nucleation occurrence in the flow are present the particles of two different dimensions. Particles dimensions are negligible in comparison with channel diameter;
- the velocity of colliding particles is the sum of that of the components, determined Brownian and turbulent diffusion of the ; at that, all collisions are effective ones, if the particles dimensions are not larger than some limited value. The aggregates with larger dimensions are disintegrated due to the action of sodium velocity pulsations;
- temperature distribution over the channel cross section is uniform, and sodium temperature in the boundary layer is taken to be equal to the temperature of the channel wall;
- sodium flow regime is turbulent.

The description of mass exchange after the formation of particles in the sodium flow is given by a set of differential equations, determining a rate of change with time (τ) of the following parameters:

impurity particle dimension (1):

$$dl/d\tau = 2\,\beta_p(c_d^f - c_s^f)/\rho_p + \xi(6c_p/(\pi\,\rho_p\,n_0))^{1/3}(1+\xi\tau)^{-2/3}/3 \qquad (1)$$

where

c_d^f, c_s^f concentration of impurities in sodium and concentration of saturation, respectively;

c_p = concentration of particles in sodium;

n_0 = volume concentration of solid phase nuclei;

β_p = coefficient of mass exchange between the particles and sodium flow for impurity being in solution;

ρ_p = density of particles substance;

ξ = parameter, characterising the particles coagulation and calculated according to (1);

particles concentration in the sodium flow:

$$dc_p/d\tau = c_p(6\,\beta_p\,(c_d^f - c_s^f)/(\rho_p l) - 4K/D_c \qquad (2)$$

where

D_c = diameter of channel

K = coefficient, accounted the particles deposition on the channel wall;

concentration of impurity, dissolved in sodium:

$$dc_d^f/d\tau = 4\,\beta_t(c_d^w - c_d^f)/D - 6\beta_p c_p(c_d^f - c_s^f)/\rho_p l \qquad (3)$$

where

in the case of the formation of depositions on the channel wall, the boundary layer concentration of impurity, dissolved in sodium, is calculated according to equation:

$$c_d^w = (\beta_t c_d^f + K_r\beta_k c_s^w)/(\beta_t + K_r\beta_k) \qquad (4)$$

where

c_s^w = saturation concentration of impurity in sodium in channel boundary layer;

K_r = coefficient accounted for wall surface roughness;

β_k = crystallisation rate constant;

β_t = coefficient, accounted for dissolved impurity mass transfer from the sodium flow to the channel wall;

In the absence of the formation of depositions, when impurity is intensively dissolved in the wall material, expression (4) is substituted for the following one:

$$c_d^w = (N\,c_d^f + c_w)/(N + 1/c_s^w) \qquad (5)$$

$$N = \beta_t\,(\pi\tau/D_w)^{0.5}/(2\,\rho_w)$$

where: c_w = impurity initial concentration in the wall material;

D_w = impurity diffusion coefficient in the wall material;

ρ_w = density of the wall material.

Coefficient K at $K < 0.2\,U^*$ calculated by the formula:

$$K = U^*(2,08.10^{-4}\,l^2\,U^{*2} + 5,35.10^{-25}T/(l\,\rho))/\upsilon^2 \qquad (6)$$

c_w = impurity initial concentration in the wall material;

in other cases it is assumed, that

$$K = 0.2 \, U^*$$ (7)

where $U^* = U/(5.15 \lg Re - 4.64)$ dynamic velocity ;

Re = $UD_c \, 7 \, \upsilon$ (Reynolds number)

U = sodium average velocity in the channel;

T = absolute temperature

ρ = density of sodium

υ = sodium viscosity coefficient.

The impurity concentration in sodium, at which the solid phase nuclei are formed, is calculated by the Thompson equation:

$$c_d{}^f/c_s{}^f = \exp(4 \, \sigma \, M / \rho_p \, l_0 \, RT)$$ (8)

where

l_0 = solid phase nuclei diameter;

M = impurity molar mass;

σ = specific energy of interface surface between nuclei and sodium

R = gas constant.

Alteration of the determined parameters along the channel inlet part, before the impurity nucleation is occurred, has been calculated using an equation:

$$dc_d{}^f/d\tau = 4 \, \beta_t \, (c_d{}^w - c_d{}^f)/D_c$$ (9)

Solution of equations (8) and (9) gives the value of the parameters, corresponding to the onset of impurity nucleation in the sodium flow.

The impurity flux to the channel wall is calculated by equation:

$$I = K \, c_p + \beta_t \, (c_d{}^f - c_d{}^w)$$ (10)

3. DESCRIPTION OF EXPERIMENTS

The flow chart of experimental unit is given in Fig. 1. It incorporates a heater, a heat exchanger with mass exchange tube 3, a chamber with impurity source 2 and a chamber with a stainless steel sintered filter 4 for particles trapping. The mass exchange tube was made of stainless steel X18H10T with a diameter of 10 mm and a length of 1 m. As impurity sources nickel chips and chromium powder were used, which have been loaded into the chamber 2. Characteristics of the impurities sources and experimental conditions are listed in the Table 1.

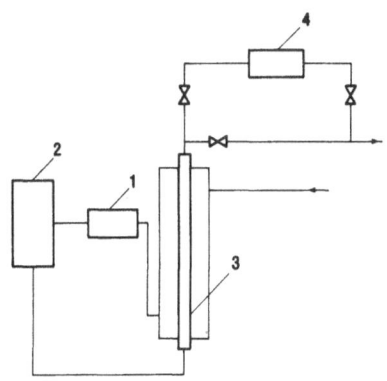

Fig. 1. Flow- chart of experimental part
1-heater, 2-source of impurity,
3-cooled tube,   4-chamber for particles trapping

An experimental procedure was as follows: Sodium, supplied from the sodium loop, entered into interbank region of the heat exchanger with temperature of 350 - 450 °C, where it was heated up to 630 - 660 °C and then supplied into heater 1, where the temperature was raised to 750 - 770 °C. It was directed to the impurity source chamber 2 and after that into the mass exchange tube 3, where the impurity mass exchange between sodium flow and cooled tube wall took place. Temperature conditions on the mass exchange tube and outlet are given in the Table 1.

After the accumulation of impurity has been finished sodium was drained from the experimental unit. After unit has been cooled the heat exchanger was cut off and mass exchange tube removed. The latter was cut for 20 equal parts (with the length of individual part of 50 mm).

Table 1. Parameters of Mass Transfer Tests

Experiment	1	2	3
Impurity source: surface, (m$^{2)}$	Nickel chips	Nickel chips	Chromium powder
weight, (kg)	0.47	0.47	0.3
	0.575	0.575	0.641
Sodium temperature °C			
in the impurity source	750	750	760
on the mass exchange			
tube inlet	730	730	740
on the mass exchange			
tube outlet	520	420	530
Sodium velocity in the channel, u			
(m. s^{-1})	1,0	0,58	1,0
Duration of impurity	304	321	325
accumulation, (h)			

In the studies of nickel mass transfer the inner surfaces of the tube parts have been washed with a solution of nitric acid, which further was gathered and analysed for nickel. By this method the depositions on the tube surface were determined. Nickel dissolved in the tube material was determined by means of X-ray spectrometric analysis.

In the studies of chromium mass transfer the method of X-ray spectrometry was used. In order to estimate the sum quantity of chromium transferred the inner layers of every tube part with a thickness of 0.5 - 1.0 mm were cut off and chips analysed for the metals by means of chemical determinations.

4. EXPERIMENTAL RESULTS; COMPARISON WITH CALCULATION DATA ON NICKEL AND CHROMIUM MASS TRANSFER

Impurity mean flux to the surface of every tube part, calculated for experiment whole duration, was estimated by formula:

$$W = m/(s \tau) \tag{11}$$

where
m = mass of impurity deposited on the surfaces and dissolved in the wall material

s = surface of mass exchange for particular part,

τ = duration of impurity accumulation.

Flux values of impurities deposited on the channel surface in the experiments 1 and 2 and the sum of fluxes (diffusion in the wall material + deposition) in the experiment 3, calculated using the experimental data, are given in the Table 2 (units of 10^{-9} kg/m^2 s).

In order to estimate fluxes of penetration of the impurities into the wall material due to diffusion, their content in the surface layers was determined, using X-ray spectrometry of these experimental data diffusion coefficients impurities in the wall material were estimated by formula:

$$D = \delta^2 / (16\ \tau) \qquad (12)$$

where

δ = depth of impurity diffusion in the wall material.

Values of D obtained in this way were further used for calculations of coefficients K_d and A_d in the equations, applied in estimations impurity mass transfer :

$$D = K_d\ exp(-A_d / T) \qquad (13)$$

Results of X-ray spectrometric analysis for impurities content in the wall are given in the Table 3.

Table 2. Flux values of impurity deposition in Experiments 1 - 3

Distance from the channel inlet (cm)	Samples number	No of experiment		
		1	2	3
95-100	1	1,28	0,94	0,52
90-95	2	-	1,21	11,2
85-90	3	1,81	1,21	18,8
80-85	4	4,09	1,38	12,1
75-80	5	4,44	0,88	0,0
70-75	6	4,38	0,96	14,2
65-70	7	4,61	1,65	10,7
60-65	8	3,56	2,81	4,7
55-60	9	1,81	1,87	12,8
50-55	10	1,34	1,24	11,5
45-50	11	1,31	0,88	13,7
40-45	12	2,39	1,21	0,0
35-40	13	0,82	0,55	5,83
30-35	14	0,47	7,55	1,98
25-30	15	0,29	3,09	2,25
20-25	16	0,18	2,20	17,7
15-20	17	0,06	-	17,7
10-15	18	0,04	0,83	18,
5-10	19	0,0	0,0	15,5
0-5	20	0,0	0,0	19,7

Table 3. Results of analyses by X-ray spectrometry

No. of experiment	1		2			3		
No. of samples	20	11	20	15	10	20	16	11
An temperature (°C)	730	630	730	670	610	740	690	640
Channel wall temperature (°C)	710	610	710	650	590	720	670	620
Maximum concentration in the surface layer (%)	30	-	25,1	30,2	26	58	59	43
Depth of impurity diffusion (µm)	20	5	28,5	22	19,4	7,1	8,6	9,2
Diffusion coefficient (10^{-7} m^2/s)	2,28	0,14	4,39	2,62	2,03	0,27	0,40	0,09

Dependencies, obtained for impurities diffusion coefficients are given in Table 4.

Table 4. Diffusion coefficients in experiments 1 - 3

No. of experiment	Expression for diffusion coefficient, (m^2/s)	
1	$9{,}74 \cdot 10^{-9} \exp(-19475 / T)$	at T > 850 K
	0	at T < 850 K
2	$4 \cdot 10^{-17}$	at T > 850 K
	$10^{14} \exp(-59533 / T)$	at T < 850 K
3	$3{,}5 \cdot 10^{-18}$	at T > 943 K
	$1{,}18 \cdot 10^{-7} \exp(-22860 / T)$	at T < 943 K

Note : T value varies in the range indicated in the Table 1.

Due to the lack of exact data for the physical constants the calculations of nickel and chromium mass transfer were performed by a successive approximation method. At this, parameters β_k, l_o, C_s, K_r and D_{Na} (diffusion coefficient of impurity in sodium) were varied.

Distributions of the impurities fluxes densities along the channel, obtained in calculations, were compared with experimental data. Thereby it was assumed that constants values were correct in the case of coincidence of experimental and calculated data. Constants values, obtained in this manner, are given in the Table 5.

The value of the crystallisation rate of nickel and chromium for all three experiments was taken to be $3 \cdot 10^{-4}$ m/s.

The concentration of impurities in sodium on the channel inlet was taken to be equal to saturation concentration at temperature of 730 °C.

In the Fig. 2 are given histograms of nickel fluxes to the channel wall, obtained on the basis of mechanical analysis (curve 1) and X-ray spectrometric analysis (curve 2), as well as integral (curve 3) for experiment No. 1.

The solubility expression for nickel in sodium, represented in the Table 5, gives its saturation concentration 40 times lower in comparison with literature data [3].

Table 5. Results of the calculations

No. of experiment		1,2	3
Impurity concentration on the channel inlet,	10^{-5} kg/m^3	6,2	17
Solid phase nucleus - dimension	m	$4 \cdot 10^{-9}$	10^{-9}
Specific energy of nucleus - sodium interface surface formation	J/m^2	0,255	0,274
Coefficient accounted for surface increase due to roughness	m^2/s	1,5	2
Diffusion coefficient of impurity in sodium		$2 \cdot 10^{-9}$ at T <800K $9 \cdot 10^{-11} \cdot T - 7 \cdot 10^{-8}$ at T>800 K	$3 \cdot 10^{-9}$
Impurity solubility in sodium	ppm	exp.(1.077-3615/T)	exp.(19.2-20746/T)

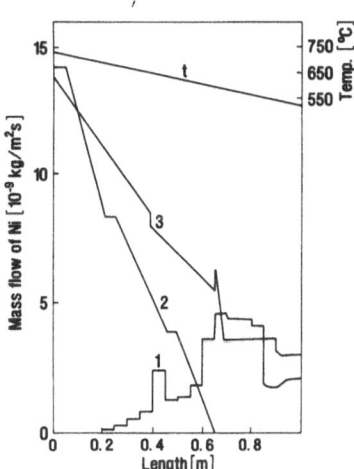

Fig. 2. Distribution of the mass flows of nickel, deposited on the channel wall at U = 1 m/s^{-1} (experiment 1):
1 according to chemical analysis of deposition substrate on the channel wall;
2 according to X-ray-spectrometric analysis of the nickel content in the surface layer of the channel wall material;
3 calculated curve

The distribution of chromium fluxes over the channel length in the experiment N 3 is characterised by the presence of four clearly distinguishable maxima, see Fig. 4. The peaks on the calculated curve 2 are caused by chromium flux sharp increase in the form of particles, originated in sodium, supersaturated by chromium. Sharp decrease of chromium fluxes in the pikes regions is related to the rapid enlargement on the particles (thereby the coefficient of their deposition on the channel wall is decreased).

On the channel part, approximately up to 0.6 m from inlet the main component of nickel flux penetrates the wall material due to diffusion.

The first maximum on the curve 1 is caused by critical supersaturation of sodium by nickel, when in the sodium flow the solid phase nuclei are originated and the particles begin to deposit on the channel wall (at the distance of 0.4 mm from the channel inlet). On the curve 3 at the same distance jump-like drop is noted. Presumably this behaviour is related to the fact that concentration of impurity, being in sodium solution, sharply dropped due to formation of particles.

At the distance of approximately 0.65 m from the channel inlet is seen the second drop of concentration of nickel, dissolved in sodium, that is related to a secondary supersaturation of sodium by nickel with following nucleation.

The third drop on the curve is noted at a distance of 0.85-0.98 m from the channel inlet.

The sodium velocity being diminished and unaltered conditions at the channel inlet, the fluxes drops are shifted towards the channel inlet and the distance between the drops are decreased (see experiments No. 2, Fig. 3).

Solubility expression for chromium in sodium, represented in the Table 5, gives the saturation concentration 10 times lower as compared with literature [3].

Obtained values of the coefficients, characterising mass transfer of nickel and chromium in non isothermal sodium loop correspond to the case when one impurity is present in sodium flow at larger concentration.

In the loop, made of chromium-nickel stainless steel, both components may be present in sodium flow as impurities. Their mutual influence can cause the change of corresponding constants values. The data, which may permit account for impurities interrelations in their mass transfer phenomena should be obtained in further experiments.

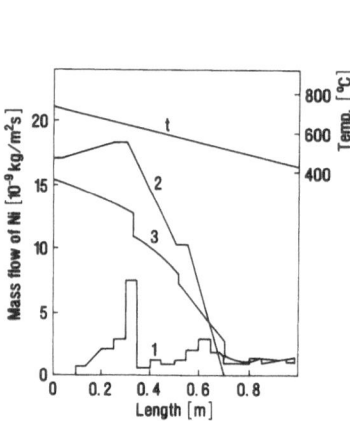

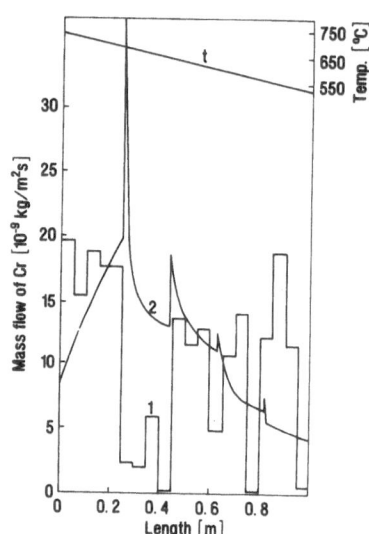

Fig. 3. Distribution of the mass flows of nickel, deposited on the channel wall at U = 0.58 ms^{-1} (experiment 2):
    1 according to chemical analysis of deposition substrate on the channel wall;:
    2 according to X-ray-spectrometric analysis of the nickel content in the surface layer of the channel wall;
    3 calculated curve

Fig. 4. Distribution of the mass flows of chromium along the channel length (experiment 3):
    1-experimental data;
    2-calculated curve

References

[1] S.K. Beal, "Agglomeration of particles in turbulent flow", WARD-TM-904, 1969
[?] A. Ubellode, "Melting and crystalline structure", Moscow, Mir ed., 1969
[3] T.D. Claar, "Solubilities of metallic elements in liquid sodium", Reactor Technology, 13, (1970), 124

VARIATION IN THE TENSILE PROPERTIES OF AISI 316 STAINLESS STEEL ON EXPOSURE TO HIGH CARBON DYNAMIC SODIUM AT 723 K.

H.S. Khatak*, J.B. Gnanamoorthy* and P. Rodriguez*
R.D. Kale[#], K. Swaminathan[#], M. Rajan[#] and K.K. Rajan[#]

* Metallurgy and Materials Group,
[#] Engineering Development Division,
Indira Gandhi Centre for Atomic Research,
Kalpakkam-603102, India

1.INTRODUCTION

The liquid sodium coolant of fast reactors is an effective mass transfer medium for both metals and non metals, and the resulting compositional and microstructural changes adversely affect the mechanical properties of structural and cladding alloys. The purity of sodium is an important parameter which determines the extent of damage. Tensile, creep and fatigue properties have been found to be affected by sodium [1-7]. Carbon transport is of particular interest in this respect. [1,3] and experience has shown that the carbon movement across the sodium/steel interfaces occurs in the direction of decreasing carbon activity. The sodium used in our experiments contains about 25 ppm of carbon which is much higher than the carbon content of sodium used in other countries. Therefore, an in-sodium material testing programme has been initiated. The first phase of the programme was studies in thermal convection loops [8,9] (which are inexpensive and easy to fabricate). The samples of AISI 316 stainless steel in solution annealed and 20 % cold worked conditions exposed in a thermal convection loop made o AISI 316 stainless steel for a period of 10500 hours at a maximum temperature of 723 K showed significant changes in the tensile properties. In the second phase of the programme a dynamic sodium loop has been commissioned. In this loop, attempt has been made to simulate the material/environment/temperature conditions in the primary circuit of the reactor. In this paper a brief description of the loop is given and the results obtained on the samples removed from the cold leg (723 K) after 6000 hours of exposure are described.

2. LOOP DESCRIPTION

The schematic of the loop is shown in Fig. 1. The loop consists of two parts: (a) the main loop and (b) the purification loop. The main loop consists of two electromagnetic pumps, three heater units, three sodium to air coolers, six sample holders and an expansion tank.

Design of the sample holder is shown in Fig. 2. The samples of AISI 316 stainless steel are being exposed in these sample holders. The sodium purification loop consists of cold trap economiser, cold trap, plugging indicator economiser, plugging indicator and nickel tube sodium sampler. The sodium has been filled in the loop through a micro filter of 10 µm pore size. The total sodium inventory in the loop is 140 kg. The material used for the construction of the loop is AISI 316 stainless steel. The temperature, level and flow of sodium are being controlled and monitored. The operating conditions of the loop are as follows:

Flow rate in the main loop:	210 litres/hour
Flow rate in cold trap:	75 litres/hour
Flow rate in plugging indicator:	120 litres/hour
Cold trap temperature:	393 K
Number of test sections:	6
Test section temperatures: (a) hot leg	823 K, 723 K, 623 K
(b) cold leg	723 K, 623 K, 523 K
Velocity in test sections:	5 m/s

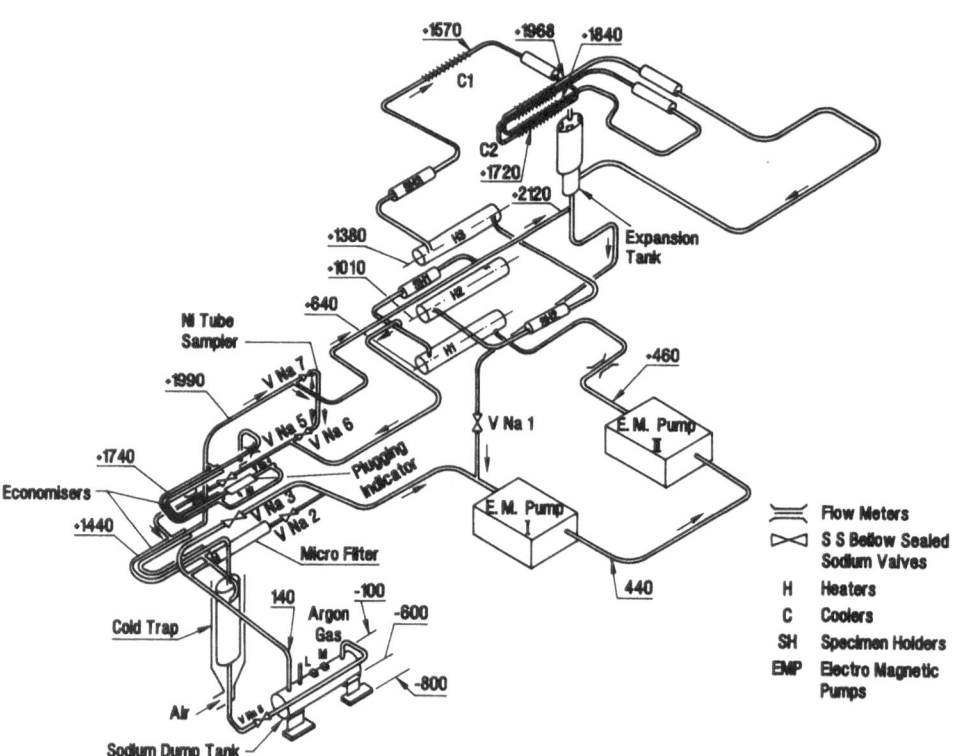

Fig. 1: Layout of the Mass Transfer loop (MTL)

3. MATERIALS, SPECIMENS AND TESTING

Samples from two heats of the stainless steel type AISI 316 were used in all the sample holders. Chemical compositions of the two alloys are shown in Table 1. After 6000 hours of operation one of the sample holders from the cold leg was removed for examinations. This sample holder has been exposed at 723 K. Three types of samples were obtained, (a) 1.6 mm sheets of alloy 1 which were exposed to sodium on both sides, (b) 12 mm thick sheets of alloy 2 which were exposed on one side only and (c) 12 mm thick sheets which were not in contact with sodium. Tensile samples with 20 mm gauge length were machined from these

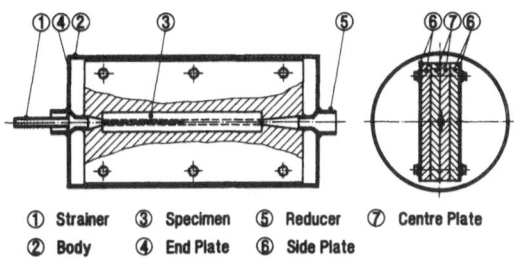

| ① Strainer | ③ Specimen | ⑤ Reducer | ⑦ Centre Plate |
| ② Body | ④ End Plate | ⑥ Side Plate | |

Fig. 2: Mass Transfer Loop Sample Holder

sheets such that the sodium exposed surface formed the gauge length. The samples of size 15 mm x 10 mm were cut for evaluation of microstructure, hardness measurement, surface analysis and elemental depth profiles. Similar samples were machined from unexposed and untreated sheets of the two alloys. Tensile tests were conducted at room temperature at a strain rate of $3 \cdot 10^{-4}$ s^{-1}. Surface analysis and elemental depth profiling was carried out using secondary ion mass spectrometry (SIMS). For revealing the microstructure the polished samples were electrolytically etched in ammonium persulphate solution.

Table 1: Chemical Composition of Steels Used (wt.%)

Alloy	C	Cr	Ni	Mo	Si	Mn	S	P	Fe
1	0,049	16,42	12,63	2,15	0,67	1,77	0,008	0,027	bal.
2	0,054	16,92	11,69	2,19	0,63	1,85	0,015	0,045	bal.

4. RESULTS AND DISCUSSION

The sodium which is being used as coolant in the fast breeder test reactor at Kalpakkam contains about 25 ppm of carbon which is much higher than the carbon contents of sodium used in other countries. However, the carbon content refers to total carbon and not to its soluble form. Most of the published data pertain to the carbon contents of less than one ppm. The ductility of structural materials is likely to be affected adversely if they get carburized by liquid sodium. However, in the present investigation, room temperature tensile properties of the samples removed from the cold leg (723 K) after 6000 hours of exposure did not indicate any significant difference in the case of both the alloys (Table 2).

Table 2: Tensile Test Results, Strain Rate: $3 \cdot 10^{-4}$ s^{-1}

	Unexposed	378	606	66
1	Thermally Aged	357	610	56
	Sodium Exposed(one side)	356	629	53
	Unexposed	276	606	64
2	Sodium Exposed(both sides)	309	619	68

The minor effects observed are related to thermal ageing effects due to stress relieving and carbide precipitation. However, the microstructure did not indicate significant precipitation of carbides (Fig. 3).

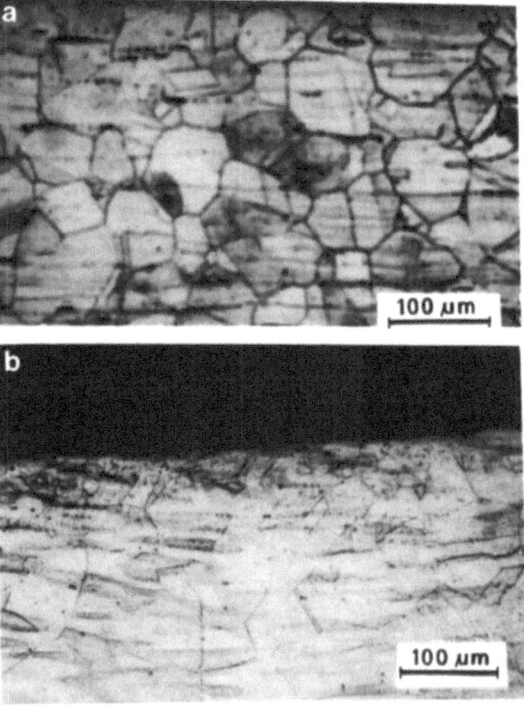

Fig. 3: Microstructure at X-section after 6000 hours of exposure at 723 K, (a) Alloy 1 (b) Alloy 2

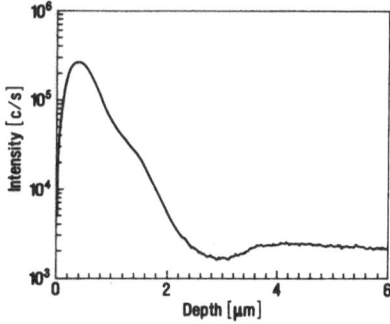

Fig. 4: Depth profile of carbon for a sodium exposed sample after 6000 hours at 723 K (alloy 2)

In the case of Alloy 2 a difference in the etched structure was seen close to the surface (Fig. 3b). This region showed a hardness of 190 VHN as compared to 175 VHN in the centre of the specimen. This is reflected in the increase of yield strength. Surface analysis on sodium exposed samples showed the presence of carbon, calcium, potassium and sodium in addition to the constituents of stainless steel. Also depth profiles (Fig. 4) showed a high level of carbon at the surface. However, the high carbon content was found only in the narrow region of a few microns. This can be explained to be due to the following reasons: (a) carbon in sodium or on the steel surface may be present in the form of stable compounds (not soluble) which have not affected the carbon activity and/or (b) the temperature and duration of exposure were not sufficient to allow diffusion to an appreciable depth. Exposure of the samples at different temperatures for longer durations is in progress and the results from these exposures may help clarify these aspects. Surface analysis with respect to other elements also showed minor variations within the narrow top surface layer, to the extent of less than a micron (Fig. 4).

5. SUMMARY AND CONCLUSIONS

A sodium loop with an on-line purification circuit has been commissioned to evaluate the influence of high carbon sodium on the properties of AISI 316 stainless steel. Primary circuit

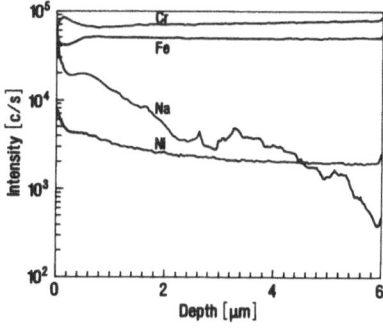

Fig. 5. Depth profile of elements for a sodium exposed sample after 6000 hours of exposure at 723 K (alloy 2)

conditions of fast reactors with respect to material, temperature and velocity have been simulated in this loop. The maximum temperature is 823 K. Based on the analysis of the samples from the two heats of AISI 316 stainless steel removed after 6000 hours from the cold leg of the loop (723 K), the following conclusions are drawn:

1. Exposure at 723 K for 6000 hours did not influence the tensile properties significantly.
2. Microstructure did not indicate any evidence of carburization.
3. SIMS analysis showed a high level of carbon at the surface.
4. Long time exposure is needed to confirm that similar trend is followed.

ACKNOWLEDGEMENT

The authors are thankful to Dr. A.K. Tyagi for carrying out SIMS analysis, Miss Sivai Bharasi for her assistance in conducting tensile tests and Mr. Gurumurthy and Mr. Muthuvel for operating the loop.

References

1. G.J. Lloyd, Atomic Energy Reviews, 16 (1978) 155.
2. H. Huthman, G. Menken, H.U. Borgstedt, and H. Tas, Proc. of Second International Conference on "Liquid Metal Technology in Energy,Production", Richland, Washington, Vol.2 (1980), p. 19/33.
3. P. Marshal, in "The Influence of Environment on Fatigue", Mechanical Engineers Publication Limited, London, New York, (1977) p. 77.
4. K. Natesan, O.K. Chopra and T.F. Kassner, J. Nucl. Mater. 73 (1977) 137.
5. H.U. Borgstedt, G. Drechsler, G. Frees, H.S. Khatak, Z. Peric and B. Seith, Intern. J. of Fracture, 32 (1986) p. R23-R28.
6. H.S. Khatak, B. Seith, G. Frees, G. Drechsler and H.U. Borgstedt, Transaction of Indian Institute of Metals, 42(supplement) (1989) p. S 235.
7. H.U. Borgstedt, G. Drechsler, G. Frees, H.S. Khatak, Z. Peric, C. Phaniraj and B. Seith, in: Failure Analysis, Theory and Practice, Vol.II, Ed. E. Czoboly, (1988) p. 1046.
8. H.S. Khatak, Hasan Shaikh and J.B. Gnanamoorthy, Proc. 4th International Conference on Liquid Metal Engineering and Technology, SFEN, Paris, Vol. 2, 1988, p. 510-1.
9. H.S. Khatak and J.B. Gnanamoorthy, in:"Materials Behaviour and Physical Chemistry in Liquid Metal Systems", Ed. H.U. Borgstedt, Plenum Press, New York. 1982, p. 229.

RESULTS OF MATERIAL TESTING IN THE "CREVONA" AND "FARINA" LOOPS

H.U. Borgstedt and G. Frees

Kernforschungszentrum Karlsruhe GmbH.
Institute of Materials Science
P.O. Box 3640, D-76021 Karlsruhe, Germany

1. INTRODUCTION

The mechanical properties of structural materials of sodium cooled fast reactors have to be evaluated under the influence of the liquid metal environment. The liquid sodium coolant which is circulating through the primary system of a fast neutron reactor is of high purity. Its capacity to reduce passivating layers and to remove carbon from structural steel has to be considered. Thus, a programme to study the influence of very pure flowing sodium on the creep-rupture and fatigue properties of austenitic steels, AISI types 304 and 316L(N), was initiated in 1978, for the performance of which the two sodium circuits CREVONA and FARINA were constructed.. The testing programme was performed in co-operation with Siemens-KWU(Interatom), SCK Mol, AEA Risley, CEGB Leatherhead, CEN de Cadarache and EdF les Renardières. It was related to the German fast reactor SNR 300 and the European Fast Reactor Project.

2. EXPERIMENTAL

2.1 Sodium Loops CREVONA and FARINA

The two sodium loops were constructed in the way that two nearly identical figure-of-eight circuits were connected with different test facilities. The loops and their test sections are described in [1] where also information is given on the data processing unit. The two loops have now successfully been operated for about 80000 hours (CREVONA) and about 30000 hours (FARINA). The characteristics and parameters of the two sodium loops are listed in Table 1.

The carbon activity in the CREVONA loop was measured by means of the Harwell carbon meter [2] (diffusion and gas analysis) and (in 1989-1992) of an electrochemical carbon meter cell [3] constructed by Indira Gandhi Centre for Atomic Research, Kalpakkam, India. The carbon activity at 550 °C was around $2 \cdot 10^{-3}$ for a long period of time.

Table 1. Characteristics and parameters of the sodium loops

	FARINA	CREVONA
Sodium volume	400 l	500 l
Capacity of main heater	72 kW	72 kW
Number of test sections	2	4+1
Maximum temperature of the test sections	550 °C	550/600 °C
Flow velocity in the test sections	3 m/s	3 m/s
Cold leg temperature	400 °C	400 °C
Purification system	cold trap	cold trap
Material of piping	X5 CrNi 18 9	X5 CrNi 18 9
Purity measurements:		
carbon	foil equilibration at 700 °C and carbon meters	foil equilibration at 700 °C and carbon meters
oxygen	emf cell	emf cell
dissolved metals	sampling device	sampling device
Operation experience	> 30000 h	> 80000 h

The oxygen meter cell in FARINA was designed and fabricated by General Electric [4], while in CREVONA cells from Westinghouse [5], Harwell [6] and Rossendorf [7] were used. The chemical analyses for metallic impurities have shown that the concentrations of steel components (Cr, Ni, Mn) were $< 1 \cdot$ µ g/g, while the Fe content in Na did not exceed the value of 4 $\cdot$µ g/g. Contents of heavy metals as Pb, Sn, and Zn, which are soluble in Na to a higher degree, were found to be below the analytical detection limits ($< 1 \cdot$ µ g/g).

The creep-rupture tests in CREVONA were performed with constant load machines, in which the load was directly connected with the lower ends of the specimens. Their upper ends were in fixed position. The load was continuously controlled by means of load cells, the strain was measured with the aid of capacitive strain gages.

Two dynamic material testing machines (MTS, 100 kN) were connected to the hot by-pass tube of the FARINA loop for low-cycle fatigue and fatigue crack growth testing. In all low-cycle fatigue tests, small free cross sections allowed a high flow velocity of the liquid sodium at the tested specimens. Thus, corroding conditions were maintained within the test sections for mechanical testing in flowing sodium. Since sodium corrosion of fatigue specimens which are only exposed to the liquid metal for short periods of time is negligible, the fatigue specimens were pre-exposed for 3000 h at 550 °C to flowing sodium in a particular side loop of CREVONA for pre-corrosion. The measurements of crack-growth rates were performed with CT specimens using the crack-opening displacement method. An encapsulated clip-on gage (Micro-Epsilon) was used directly in the flowing liquid sodium.

2.2. Materials Under Testing

The tested materials were austenitic stainless steels, German no. 1.4948, a specification of AISI type 304 steel for application at high-temperatures, and the structural material of the European Fast Reactor, AISI type 316L(N). The compositions of the materials under test are listed in Table 2. The study of the steel 1.4948 was performed with different batches and variant compositions, details are given in [8]. The materials were tested in the as received condition (solution annealed). Beside specimens of base material, some welded specimens were also be included in the testing programme.

Table 2. Composition of the tested materials

Material	1.4948	316L(N)	1.4909
C	.053	.030	.024
Si	.39	.44	.13
Mn	2.25	1.84	2.01
P	.018	.021	.015
S	.032	.001	<.003
Cr	17.8	17.5	17.44
Ni	10.6	12.3	12.54
Mo	.065	2.47	2.40
Cu	.043	.17	.05
N	.041	.075	.061

2.3 Specimens for Creep-Rupture, Fatigue and Crack-Growth Tests

Cylindrical specimens with diameter of 4 (6) mm were taken perpendicular to the rolling direction for the creep-rupture tests. Their gage lengths were five times larger than the diameters.

Hour-glass shaped specimens of a minimum diameter of 8.8 mm were fabricated for the low cycle fatigue (lcf) tests. The encapsulated specimen for lcf tests is shown in Figure 1. The capsules were also used fore a pre-corrosion of the lcf specimens in flowing liquid sodium under conditions as in the tests. Fatigue crack growth tests were performed using compact tension specimens according to ASTM E 399-83. The specimens were pre-cracked by means of cycling at room temperature according to the recommendations given in the ASTM standard. The pre-cracks had a length of a/W ~ 0.3 to 0.5.

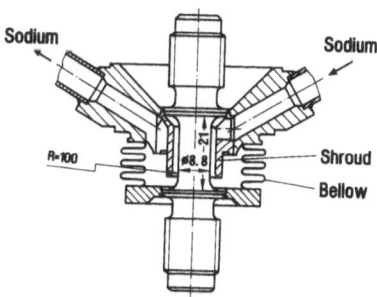

Fig. 1. Encapsulated hour-glass shaped lcf specimen for tests at the FARINA loop.

3. RESULTS OF TESTS

3.1 Creep-Rupture Tests in CREVONA

The creep-rupture tests with steel AISI type 304 were only performed at 550 °C, the material was varied in respect to base metal and weldments, low carbon batch, diffusion heat treated or cold worked material. The time-to-rupture of a number of specimens of the steel in the as specified composition and grain size is shown in Figure 2. The figure compares results of the tests in the CREVONA loop with those of tests in the loop of Interatom [8] and of reference tests in air [8].

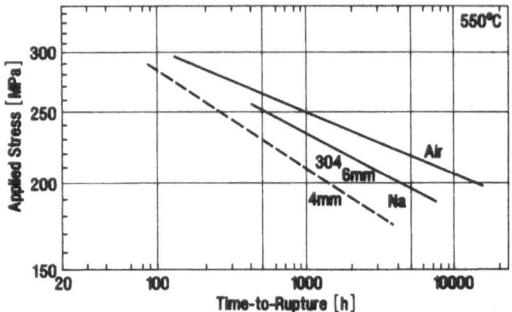

Fig. 2. Time-to-rupture of AISI Type 304 stainless steel at 550 °C in two Na loops and in air [see (8)].

The creep-rupture data of steel AISI type 316L(N) gained in the CREVONA loop were in excellent agreement with data of the co-operating laboratories and results of reference tests [9]. Figure 3 shows the time-to-rupture data for this steel at a temperature of 550 C. Creep life of this material was not influenced by the liquid sodium environment even at 600 °C [9,10], though the surface layers showed a selective corrosion.

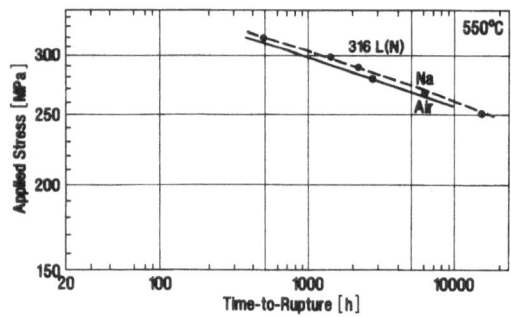

Fig. 3. Time-to-rupture of AISI Type 316L(N) stainless steel at 550 °C in two Na loops and in air (see [9]).

The effect of liquid sodium on the steel Type 304 can be understood as a decarburization of grain boundaries. Figure 4 compares the carbon potentials of the two involved sodium loops with the carbon potentials in stainless steel. It is obvious that the carbon potential in the CREVONA loop is decarburizing for steel type 304, while the carbon potential in the loop of Interatom is in the same level as in the steel. This fact was found to be the main difference in the testing parameters of the two laboratories. A low carbon batch of steel Type 304 did not show different behaviour in the two loops, these findings support the assumption that decarburization might be the factor which causes the reduction of creep strength of the tested material. Specimens with larger diameter (6 mm) did not reveal the reduction of creep strength.

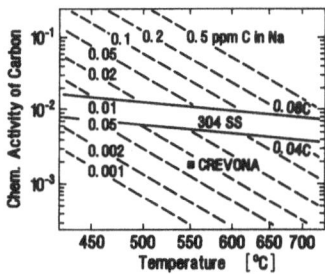

Fig. 4. Carbon potentials in Na of the loops CREVONA and Interatom compared to carbon potentials in steel Type 304 at temperatures of interest.

The low carbon steel Type 316L(N) does not suffer decarburization under the conditions of the CREVONA loop. Thus, the agreement of results of different laboratories is not surprising, since different carbon potentials of the sodium are the only significant differences in parameters. The results gained at 600 °C additionally support this conclusion. Though corrosion of surface layers is characterised by the formation of typical ferrite layers, the mechanical properties in sodium and in air are in excellent agreement.

The effect of the sodium exposure on the nitrogen content of the specimens of Type 316L(N) steel was examined by means of the **nuclear microprobe** (Dr. J. McMillan, AEA Technology, Harwell). The test parameters and matrix concentrations of nitrogen in four specimens are listed in Table 3. Figure 5 shows that, depending on the temperature and duration of exposure to Na, the specimens were denitrided by sodium in more or less thin surface layers, while the reference specimen was nitrided in air at 550 °C.

Table 3. Nitrogen contents of the matrix in four specimens of steel Type 304

Specimen	Medium	Temperature (°C)	Stress (MPa)	Time-to-rupture (h)	Concn. of N (mass %)
1	Na	600	180	10481	0.093
2	Na	550	250	15480	0.081
3	Na	550	265	3537	0.083
4	Air	550	260	10565	0.058

3.2 Fatigue and Crack-Growth Tests in FARINA

The lcf life of specimens of Type 304 steel in liquid sodium at 550 °C after pre-corrosion in a sodium loop in the range of total strains $\Delta\varepsilon = 0.006$ to 0.018 did not differ from results with aged specimens of the same material [11]. The plastic strain was evaluated from the hysteresis curves gained at about the half of the life time. The relations of plastic strain to cycles to fracture fulfil the Coffin relation [12].

$$N_f^{0.5} \cdot \Delta\varepsilon_p = C \qquad (1)$$

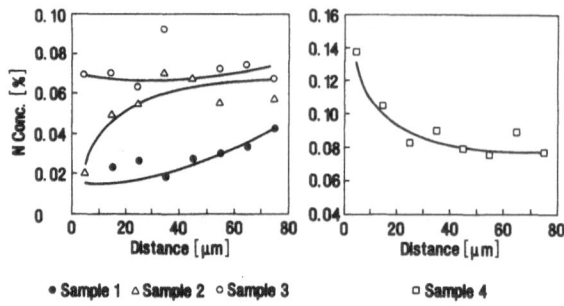

Fig. 5. Profiles of N concentrations in specimens 1-4 of Table 3

with an exponent of 0.507 and a constant C = -0.457. The cycling strain hardening curves did not reveal any significant difference in the lcf behaviour of this steel.

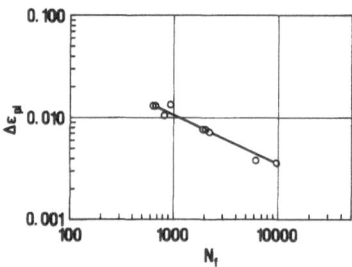

Fig. 6. Coffin relation of specimens of Type 304 steel, batch 325, in lcf tests in FARINA sodium loop

Creep-fatigue tests with hold times up to 30 minutes with the same material showed a small effect of the sodium environment on the creep-fatigue life. The hold times of 10 and 30 minutes were in the strain maximum, the parameters of tests were the same as in the tests without hold time. The effect of sodium might be ascribed to the creep damage, since the specimens showed typical creep damage.

The test arrangement for fatigue crack growth tests in liquid sodium was described in an earlier publication [3]. The encapsulated clip-on-gage, type Micro-Epsilon, allowed the direct measurement of the crack opening displacement inside the sodium test section. The measurements were limited to a maximum temperature of 550 °C.. The fatigue crack growth was calculated as a function of the stress intensity factor by use of the 7-point polynomial method directly by the processor. The Paris law for a typical specimen of the steel 1.4948 was also gained from the processor:

$$da/dN = 4.763 \cdot 10^{-10} \cdot \Delta K^{3.826} \qquad (2)$$

A comparison with results which were published on crack-growth tests with similar steels [14,15] indicated that our results were in good agreement with the scatter band of data.

4. DISCUSSION AND CONCLUSIONS

The creep-rupture tests of different steels in liquid sodium at up to 600 °C have shown that the liquid metal environment does not deteriorate the matrix material. A leaching of non-metallic elements as carbon and nitrogen may occur in surface layers. This effect may reduce the strength of thin layers, thus it may be of importance for thin walled components. The leaching of carbon seems to be supported by the loss of chromium out of the same surface layers. A weakening of grain boundaries seems to be the result. The effect does not occur, if the carbon potential in sodium is comparably high to that of the steel. Thus, the low-carbon steel type 316L(N) has the same mechanical strength in sodium even of low carbon potentials, and the slight losses of nitrogen do not influence its mechanical properties.

Sodium has a beneficial influence on the fatigue properties of the tested group of steels. This fact is ascribed to the reducing properties of the liquid metal. The reduction of oxides layers on crack surfaces may be the reason for self healing of cracks during the compressional phases of the cycles. Creep-fatigue is more or less related to creep damage of the materials. Crack growth under cycling stress is not influenced by the presence of the liquid alkali metal. There is practically no liquid metal corrosion at the tip of cracks, since the transport of corrosion products through the liquid phase is characterised by long ways. The flowing sodium has no exchange with sodium inside of cracks. Thus, the higher concentration of dissolved material in the residential sodium minimises the further dissolution and corrosion.

The typical effects of sodium on the surfaces of steels are limited to the surface grains of the steels. Large numbers of very small cracks are the worst results of sodium corrosion on stressed austenitic steels. The formation of such small cracks is not relevant for fracture mechanical considerations. The small dimensions and the large number of such micro cracks do not cause any stress concentrations on individual crack tips. Thus, the micro cracks are not forced to grow into dimensions, under which they would proceed according to the Paris law. The mechanisms and growth rates of crack formation under cycling stress are, therefore, the same as in air.

Acknowledgements

We gratefully acknowledge the contributions of Mrs. Z. Peric and Messrs. Drechsler, Kleiber, Seith, and Wollensack to the experimental work, we are thankful to Mrs. Dr. Ch. Adelhelm for the chemical analytical work, to Mrs. B. Bennek-Kammerichs for the metallographic studies and to Mr. M. Schirra for the reference tests. The former Fast Breeder Project of the Kernforschungszentrum Karlsruhe has generously supported this work, we are indebted to Dr. W. Marth, Dr. G. Heusener and Mr. H. Plitz. Co-operation and fruitful discussions on the results of our studies with Dr. C.A.P. Horton, formerly with the CEGB, UK, and Dr. H. Huthmann, SIEMENS-KWU, during the course of this work were very helpful.

REFERENCES

1. H.U. Borgstedt, G. Drechsler, G. Frees and E. Wollensack, in: "Material Behavior and Physical Chemistry in Liquid Metal Systems", Plenum Press, New York, 1982, p. 185.
2. R.C. Asher, L. Bradshaw, T.B.A. Kirstein, T.H. Nixon and A.C. Tolchard, in: "Liquid Alkali Metals", The British Nucl. Energy Soc., London, 1973, p. 133.
3. S. Rajendran Pillai, R. Ranganathan and C.K. Mathews, in: "Liquid Metal Engineering and Technpology, Soc. Franc. d'Energie Atomique, Vol. 2, Paris 1988, paper 504.

4. G.J. Licina, P. Roy and C.A. Smith, in: "Material Behavior and Physical Chemistry in Liquid Metal Systems", Plenum Press, New York, 1982, p. 297.

5. B.R. Grundy, E. Berkey, E.T. Weber and W.A. Ross, Trans.Am.Nucl.Soc. 14 (1) (1971) 186.

6. R.C. Asher, R. Dawson, D.C. Harper, T.B.A. Kirstein, F. Leach, S.Y. Moss, A.N. Moul, R.G. Taylor, R. Thompson, and C.C.H. Wheatley, Liquid Metal Engineering and Technology, Vol. 1, British Nucl. Energy Soc., London 1984, p.393.

7. H. Ullmann, H.-J. Lang, W. Richter, F.A. Kozlov, T.A. Vorob'eva, Yu.A. Tsoj and S.A. Davydov, Liquid Metal Engineering and Technology, Vol. 1, British Nucl. Energy Soc., London 1984, p.387.

8. H.U. Borgstedt and H. Huthmann, J. Nucl. Mater. 183 (1991) 127.

9. H.U. Borgstedt, P. Debergh, C.A.P. Horton, H. Huthmann and D.S. Wood, Internat. Conf. on FAST REACTORS AND RELATED FUEL CYCLES FR'91; Proceedings, Atomic Energy Soc. of Japan, Tokyo, 1991, Vol. IV, P4.10-1.

10. M.P. Mishra, H.U. Borgstedt, G. Frees, B. Seith, S.L. Mannan and P. Rodriguez, J. Nucl. Mater. 200 (1993) 244.

11. H.U. Borgstedt and G. Frees, in: Low Cycle Fatigue and Elasto-Plastic Behaviour of Materials, Ed. K.-T. Rie, Elsevier, London 1987, p. 378.

12. L.F Coffin, Jr., Applied Materials Research (1962) 129.

13. H.U. Borgstedt, G. Drechsler, G. Frees, H.S. Khatak, Z. Peric, B. Seith, Intern. J. Fracture 32 (1986) R23.

14. H. Huthmann, O. Gossmann, in: Liquid Metal Engineering and Technology, British Nucl. Energy Soc., London (1984), Vol. 2, p. 453.

15. L.A. James, R.L. Knecht, Metal. Trans. 6A (1975) 109.

COMPARISON OF CREEP RUPTURE PROPERTIES OF TYPE 316(N) STEEL IN AIR AND FLOWING SODIUM

H. Huthmann,* C.A.P. Horton** and H.U. Borgstedt[+]

* Siemens AG-KWU, Bergisch Gladbach, Germany
** Consultant, Great Bookham, England
[+] Kernforschungszentrum Karlsruhe GmbH., Karlsruhe, Germany

1. INTRODUCTION

The behaviour of austenitic stainless steel Types 304 and 316 in liquid sodium has been the subject of extensive investigations [1-4]. For Type 304 steel it has been demonstrated that under decarburizing conditions the stress rupture properties can be adversely affected by sodium [5].

On 316L(N) steel which is the structural material of the European Fast Reactor (EFR) uniaxial creep rupture tests have been performed in the non-isothermal sodium loops at EdF, France, Interatom and KfK, Germany, CEGB and UKAEA, England, with running times up to 27,000 h. By comparison with the results from parallel tests in air on the same heats it should be decided whether any sodium factor for creep rupture properties has to be regarded in design.

In evaluating the 316L(N) data consideration should also be given to the possible mechanisms of the effect of sodium on mechanical properties so that the data can be extrapolated to longer times with greater confidence.

2. EXPERIMENTAL

2.1 Materials Characterization

For the in-sodium and the in-air creep rupture tests two heats of 316L(N) steel have been used, one heat supplied by Creusot-Loire (Tole SQ) the second by Krupp SW (Krupp-528). A characterization including chemical composition and tensile properties of these heats is given in Table 1.

Liquid Metal Systems, Edited by H.U. Borgstedt
and G. Frees, Plenum Press, New York, 1995

Table 1:Characterization of 316L(N) heats used for creep rupture tests in flowing sodium

	Tole SQ	Krupp-528
Type	ICL 167 SPH	316L mod
Supplier	Creusot-Loire	Krupp SW
Heat no.	T9075	013824(Interatom.528)
Plate no.	T81/Tole SQ	5
Plate dimensions	30-2000-5000 mm	30-1000-1000 mm
Melting process	Electro furnace + A.S.V.	ESR
Solution annealing	90 min 1100 °C	30 min 1080 °C/H_2O
Grain size (ASTM)	2 - 4	2 - 4
Delta ferrite (Fischer)	0 - 4%	0
Non-metallic inclusions *	SS: 1 OA: 3	SS: 0 OA: 0
acc. to DIN 50602	OS: 8 OG: 2	OS: 0 OG: 2-3
max. values		

* SS: long sulphides; OA: dissolved oxides; OS: long oxides; OG: globular oxides.

Chemical Composition (wt %)

Element	C	S	P	Si	Mn	Ni	Cr	Mo	N	Co	Cu	B(ppm)
Tole SQ	.030	.001	.021	.44	1.84	12.3	17.5	2.47	.075	.15	.17	11
Krupp 528	.024	<.015	.015	.13	2.01	12.54	17.44	2.40	0.61	.04	.05	7

Tensile Properties at Room Temperature (transverse)

	$R_{p0.2}$ (MPa)	RT (MPa)	A(5d) (%)	Z (%)
Tole SQ	277 - 280	567 - 578	58 - 57	82 - 80
Krupp 528	273 - 280	592 - 598	46 - 47	58 - 59

Both heats fulfil the specification of RCC-MR [6]. In Tole SQ a ferrite content of 0 to 0.4 % has been found, whereas in the Krupp cast no ferrite was detectable. Concerning the maximum values of non-metallic inclusions the Krupp heat appears to have a very high purity, whereas Tole SQ has more impurities, especially long oxides.

The sensitization behaviour of Tole SQ is given in Fig. 1   showing that sensitization will occur after 500 h at 600 °C and after 4000 h at 550 °C . No sensitization diagram is available for the Krupp cast, but a very similar behaviour is expected because of the similar contents of C, N, Cr and Mo.

The creep rupture tests have been performed in the as-received (solution annealed) condition and for Tole SQ, additionally in a sensitized condition, given by a heat treatment of 1000 h at 600 °C to a piece of material from the plate before manufacturing the specimens.

2.2 Specimens

All laboratories used cylindrical specimens with 4 to 9 mm diameter, taken perpendicular to the rolling direction of the plate.

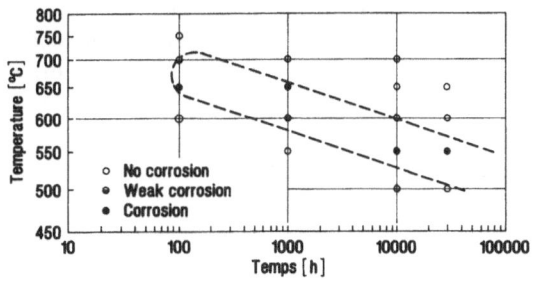

Fig. 1: Sensitization diagram for Tole SQ, from occurrence of intercrystalline corrosion after long time heat treatments (from Le Huede, EdF).

2.3 Sodium Loops

All in-sodium tests were performed in non-isothermal pumped sodium loops simulating the conditions of a primary sodium loop of a FBR. The characterizing properties of the loops at EdF, KfK and Interatom are given in Table 2. Detailed descriptions of the sodium loops used at CEGB, EdF, Interatom, KfK and UKAEA are given in the references [7-10].

Table 2 Characterization of sodium loops used for creep rupture tests

	CINATRA/EdF	CREVONA/KfK	AWN/Interatom
Status	Dismantled	Operational	Dismantled
Material of pipes	316 L SPH	X5CrNi18 9	X6CrNi18 11
Inventory of Na		500 kg	600 kg
Flow of Na		0,38/0,55 m^3 h^{-1}	0,08/0,18 m^3 h^{-1}
Flow rate at specimens		3 m s^{-1}	3 m s^{-1}
Temperatures (°C)			
Testsection		550 / 600	500 / 550
Main Loop		500	520
Cold leg		400	380
Cold trap		125 ±5	125 - 130
Chemistry			
Foil equilibration		Carburizing slight	Carburizing slight
for C and N		denitriding	denitriding
Carbon activity			
(Harwell meter)		2·10^{-3}	6·10^{-3}

2.4 Test Facilities

For the performance of creep rupture tests in flowing sodium special test chambers have been used, where the elongation of the specimens during testing is enabled by bellows. The elastic forces from the bellows and additionally the dynamic pressure of sodium have to be taken into account for the determination of the applied load for the specimens. Because of experience of many years and numerous tests for comparison and validation from the participating laboratories the necessary corrections to the applied loads, which are in the order of about 5 % of the total load, are not called in question. The detailed descriptions of the test facilities are given in Refs. [8-10].

3. AVAILABLE DATA AND COMMON EVALUATION

3.1 Overview of Compiled Test Data

In total, results from 96 creep rupture tests including 81 creep curves have been transferred to an IBM mainframe computer for display and common analyses. Most tests have been performed on the SQ-cast in the as-received condition at 550 °C (23 in-air, 25 in-sodium) with running times up to 27,165 h in sodium. 11 tests (7 in-sodium) have been performed on the SQ-cast in the previously sensitized condition. On the Krupp-cast in the as-received condition, 26 tests (11 in sodium) at 550 °C and 11 tests (4 in sodium) at 600 °C are available.

3.2 Common Evaluation of Compiled Creep Curves

The detailed data and results of the compiled creep rupture tests are given in Tables 3, 4, 5 and 6.

Beside the usual parameters for creep rupture tests as load, rupture time (RUPTIME) or test time (TESTIME), elongation after loading (EPSO), elongation after rupture (ELONG) and reduction of area (R of A) two values for time to tertiary creep (T1 & T2), two minimum creep rates (MCR1 & MCR2) <and two times to minimum creep rates (TMIN1 & TMIN2) are given.

This was necessary because in the long running tests on the as-received material sensitization took part during the creep tests and a second area of minimum creep rate occurred in the creep curves. This behaviour and the meaning of the given parameters is demonstrated in Figs. 2a and 2b giving the creep curve and creep rates versus time for the test IA in Na at 275MPa(13820 h).

In short time tests (less than about 7000 h at 550 °C) and in the tests on previously sensitized material only one minimum creep rate occurred and consequently only the first values are given (T1, MCR1 and TMIN1).

In the analyses of the creep curves the creep rate at each data point along the creep curve has been calculated, averaging the rate at each point over a chosen time span. The chosen time span must be sufficiently large to smooth out experimental fluctuations in strain measurements, but small enough to accurately reveal a true minimum creep rate.

After storing the creep rates the computer programme identified the minimum values, constructed a tangent to the creep curve at the appropriate point (time TMIN1 and TMIN2) and additionally a 0.2 % strain offset line for the determination of T1 and T2. Clearly when values for both T1 and T2 were obtained from a single tests, only T2 could represent actual time to tertiary creep. To avoid confusion the values obtained for T1 and T2 were analysed separately as "times to transition". It is likely that they were related to separate mechanisms, one before and one after sensitization.

Table 3: Results from creep rupture tests on SQ cast tested at 550 °C (first 4 groups tested in air)

Load	Ruptime	Epso	Elong	R of A	T1	T2	MCR1	MCR2	T min1	T min2
360	1092	13.4	23.0	26	-				-	-
360	811	13.6	25.0	35	550		1.34E-08		281	-
330	1945	10.6	19.0	19	-				-	-
310	3193	8.4	18.0	21	-				-	-
290	5959	5.5	14.0	21	-		8.81E-09		-	-
290	5150	8.3	15.0	16	3000		4.03E-10		1221	-
275	7743	6.1	14.0	22	-				-	-
265	13348	4.9	15.5	19	4900	10400	3.00E-10	1.03E-09	1801	-
260	6300*	1.3	-	-	-		2.39E-10		-	6500
240	18204	3.3	12.0	13	-			2.83E-10	-	-
240	20221*	4.9	-	-	8800			3.25E-10	2960	≥16416
240	34216	4.9	-	-	6100				1934	≥25082
380	392	17.2	23.8	37.0	269	-	3.33E-08	-	183	-
340	1142	13.4	17.0	29.1	815	-	7.94E-09	-	490	-
300	6704	8.0	11.1	22.4	4450	-	9.22E-10	-	1286	-
275	12107	6.0	10.7	17.8	5275	9400	4.78E-10	1.09E-09	2219	6750
240	27001	2.4	-	-	-	-	3.72E-10	2.66E-10	4145	≥21671
380	353	14.4	32.0	43.5	315		3.92E-08		202	
360	779	14.25	24.5	30.9	522		1.42E-08		337	
330	2204	11.4	20.0	23.0	1350		2.21E-09		481	
320	2158	13.0	19.0	23.0	1525		2.92E-09		400	
320	2752	11.7	18.9	21.0	2000		2.63E-09		1050	
300	4652	9.7	16.5	18.7	3050		1.43E-09		1448	
360	786	14.7	25.0	38	488		1.34E-09		252	
310	2535	7.6	23.0	34	-		-		-	
265	9410	5.6	13.4	19	-		-		-	
265	10400	9.0	-	-	-		-		-	
380	726	17.2	23.8	37.0	269		1.1E-08	-	183	-
360	1150	15.7	26.3	30.0	760		6.06E-08	-	373	-
340	1748	10.3	18.7	26.2	1330		6.72E-09	-	857	-
300	3412	5.3	15.5	18.3	2660		2.61E-09	-	1364	-
275	12801	7.0	-	-	6260		2.43E-10	6.33E-10	2530	6750
275	1138*	-	-	-	-		8.89E-10	-	-	-
275	13820	5.6	-	-	4900		4.86E-10	5.64E-10	1665	10244
275	6329	5.4	13.3	12.9	5600		7.50E-10	-	2832	-
240	14258*	4.1	-	-	-		-	3.26E-10	-	-
240	27165*	4.4	-	-	-		4.03E-10	2.00E-10	7373	16319
380	570	17.85	27.5	-	350		1.69E-08		228	
380	824	13.67	26.3	27.5	490		1.40E-08		284	
360	1328	19.36	22.2	28.0	920		5.22E-09		482	
360	1107	15.42	25.5	-	650		4.75E-09		300	
330	2875	12.90	21.3	23.0	2010		1.06E-09		1120	
330	2487	10.68	18.4	22.5	1360		2.29E-09		516	
330	2273	13.00	18.4	21.5	1310		2.38E-09		723	
320	3547	11.53	20.2	18.5	2325		1.81E-09		1085	
320	3755	11.83	17.4	19.0	2260		2.38E-09		610	
340	2312	13.5	-	-	1750		1.56E-09		726	
320	2697	11.6	19.1	27.6	1950		2.64E-09		1012	

* Test was terminated before rupture of specimen

Part 1: EdF, in air;φ=5 mm   Part 2: Interatom, in air;φ=4 mm
Part 3: KfK, in air;φ=4 mm   Part 4: PA, in air;φ=3,8 mm
Part 5: EdF, in Na;φ=4 mm   Part 6: Interatom in Na;φ=4 and 6mm
Part 7: KfK, in Na;φ=4 mm   PN/UK, in Na.φ=3,8 and 5 mm

Table 4: Results from creep rupture tests on Krupp cast tested at 550 °C

Load	Ruptime	Epso	Elong	R of A	T1	T2	MCR1	MCR2	T min1	T min2
310	758	13,2	29,8	33,6	335		2.72E-08		116	
280	2946	11,0	30,1	33,4	1945		8.89E-09		1281	
260	12002	8,2	35,4	48,3	5800		1.62E-09		2680	
220	19139*	6,6	-	-	-		-	2.83E-10	-	7064
360	166	20,3	40,3	45,2	42	-	5.61E-08	-	18	-
340	481	7,3	30,0	35,5	110	-	2.49E-08	-	32	-
320	670	24,0	30,3	55,5	310	-	2.89E-08	-	104	-
300	1691	11,7	25,7	29,2	350	9400	4.06E-09	-	117	-
260	10565	7,0	38,0	56,3	4100	-	2.36E-09	-	1847	
220	32528*	5,5	-	-	-	32300	-	1.86E-10		16385
360	417	16,5	18,7	40,1	352		1.58E-08		265	
340	718	-	31,3	34,6	-		-		-	
310	1066	-	18,6	23,3	-		-		-	
280	2899	8,2	21,6	23,0	1760		4.83E-09		308	
260	26709	7,4	23,0	27,5	-	14000	-	5.08E-10	-	7603
310	918	13,2	27,5	26,6	365		2.19E-08		154	
260	3018	-	17,8	19,5	2980		2.15E-09		1453	
220	18958*	5,8	-	-	-	17400		2,31E-10		1106
220	15976*	0,6	-	-0	-			8.61E-11		68675
320	476	19,8	28,6	32,0	220		3.78E-08		98	
300	1449	15,5	27,5	43,5	350		1.00E-08		215	
290	1636	11,5	33,0	40,0	1000		1.82E-08		726	
280	2763	11,3	35,0	39,5	1500		5.06E-09		937	
270	6313	10,3	35,5	40,5	3300		4.06E-09		1991	
255	13562*	8,7	-	-	6000	9200	1.86E-09	2.83E-09	3827	7314
250	15480	8,6	33,0	40,0	7150		1.92E-09		3597	

* Test was terminated before rupture of specimen

Part 1: Interatom, in air;ϕ=4 mm

Part 2: KfK, in air;ϕ=4 mm

Part 3: UKAEA, in air;ϕ=5 and 9 mm

Part 4: Interatom, in Na;ϕ=4 mm

Part 5: KfK, in Na;ϕ=4 mm

Table 5. Results from creep rupture tests on sensitized SQ cast tested at 550 °C (in air upper part)

Load	Ruptime	Epso	Elong	R of A	T1	T2	MCR1	MCR2	T min1	T min2
360	223	14,6	33,0	33,0	138		1.12E-07		66	
290	1874	8,2	-	-	1820		3.19E-09		1106	
265	7605	6,8	17,0	19,0	4770		1.34E-09		2423	
240	38098	5,6	15,0	19,0	23000		4.06E-10		8715	
360	159	13,8	26,6	36,0	119		8.42E-08		81	-
360	94	13,1	30,2	39,7	77		2.32E-07		55	-
330	353	11,6	25,8	29,0	-		-		-	-
310	732	9,0	19,0	23,4	580		1.79E-08		327	-
265	2782	6,0	13,9	21,1	2650		2.74E-09		1112	
265	3537	6,0	13,8	14,4	3250		1.41E-09		1882	16385
240	11447	-	16,2	23,4	-		--		-	

All tests at EdF, Les Renardieres, specimens of 5 mm diameter in air, 4 mm diameter in Na.

4. STATISTICAL ANALYSES OF CREEP DATA

4.1 Methods of Statistical Analyses

The aim of the statistical analyses was to compare the in-air and in-sodium behaviour for each combination of cast, material state and test temperature. This was accomplished by using a graphical procedure, (SAS/GRAPH) to decide wether the mean curves were significantly different. The theory behind is illustrated in Fig. 3.

Table 6: Results from creep rupture tests on Krupp cast (solution annealed), tested at 600 °C (in Na upper part)

Load	Ruptime	Epso	Elong	R of A	T1	T2	MCR1	MCR2	T min1	T min2
240	799	8,4	41,0	37,8	510	-	6.00E-08	-	404	-
230	1560	6,0	34,0	39,5	640	1135	2.89E-08	4.81E-08	374	900
200	6690	4,2	48,5	52,8	1600	3900	6.06E-09	7.92E-09	976	3291
180	10480	3,6	40,3	43,8	-	4300	-	4.08-09	-	2737
280	80	10,7	52,0	53,9	14	-	4.28E-07	-	5	-
260	184	4,0	49,7	59,0	36	-	1.69E-07	-	15	-
240	500	7,5	53,7	64,0	241	-	9,75E-08	-	100	-
220	1940	4,5	63,0	75,1	770	-	2.25E-08	-	386	-
200	2650	3,6	59,0	61,4	1150	-	1.76E-08-	-	757	-
180	11234	1,7	44,7	48,0	-	3800	-	3.00E-09	-	1737
130	27515*	-	-	-	-	19200	-	2.56E-10	-	7751

All tests at KfK with specimens of 4 mm diameter

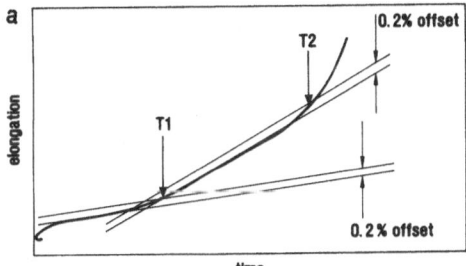

Fig. 2a: Determination of transition times T1 and T2 from creep curve.

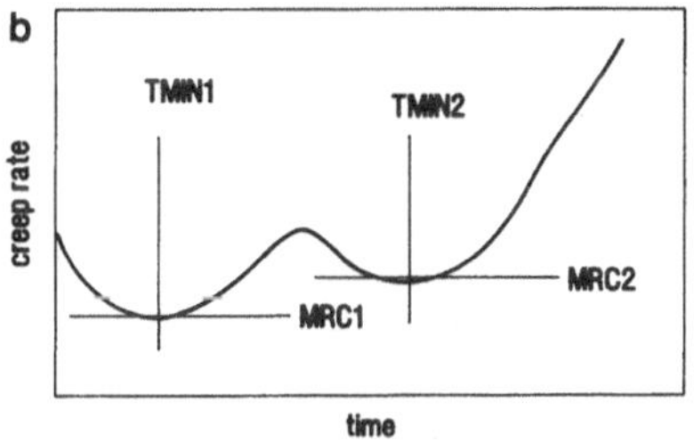

Fig. 2b: Determination of minimum creep rates MRC1 and MRC2 and correlated times TMIN1 and TMIN2 from plot of creep rate versus time.

The applied stresses σ were plotted against time to rupture RUPTIME, transition times T1 and T2, and minimum creep rates MCR1 and MCR2 in the form of log-log plots.

For the evaluation of the creep ductility the rupture times were plotted against creep strain and reduction of area (R of A).

The plots involving times to rupture (RUPTIME) generally used only those values where failure had occurred.

The plotted data were best described by linear or quadratic relationships between the pairs of variables and regression analyses was used to produce the best fit line, together with the 95 % confidence limits on the mean lines in each case. The SAS/GRAPH procedure, used for this, automatically plotted the mean and limit lines when plotting the chosen data. The regression was carried out on "Y", the vertical axis, as a function of "X", the horizontal axis, and so it was necessary to plot the dependent variable (e. g. rupture time) on the vertical axis and the independent variable (e. g. stress) on the horizontal axis.

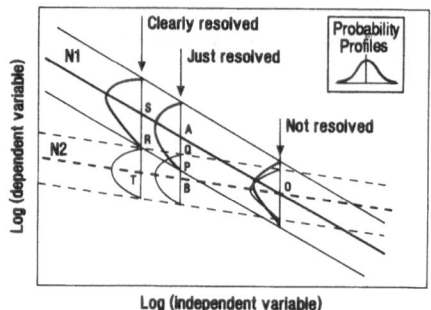

Fig. 3: Schematic representation of two mean curves and their 95 % confidence limits. The mean curves intersect at O.

Table 7: Results and statisitcal comparisons between in-air and in-sodium behaviour: creep strengthsin the case of the SQ and KRUPP casts in the as-received state, a number of the comparisons showed small but significant differences (i.e. 95 % confidence, see Fig. 4) between in-air and in-sodium behaviour at high stresses, or shorter times, only. However, these differences disappeared with decreasing stress/increasing time.

Cast [°C]	log σ vs:-				
	log t_r	log T1	log T2	log MCR1	log MCR2
SQAR/550	Na > air at σ > 330 MPa Fig.4	Na > air at σ > 330 MPa	Insufficient data	Na > air at σ > 320 MPa	Na > air at σ > 270 MPa
KRAR/550	Na > air 275 <σ < 330 MPa Fig.5	No sign. diff. in data range	Insufficient data	Na > air at σ > 300 MPa	Insufficient data
SQSENS/550	Na > air at σ < 310 MPa Fig.6	Na > air at σ > 320 MPa	No data	No signif. difference	No data
KRAR/600	No sign. diff. in data range Fig. 7	Na > air at σ > 230 MPa	Insufficient data	No signif. difference	Insufficient data

4.2 Results from Statistical Analyses

Table 7 shows the results from log-log plots of stress versus time to rupture, times to transition and minimum creep rates examined for each of the four cast/state/temperature combinations. An inequality shown in the table, e. g. Na > Air, signifies that there is at least 95 % confidence in the mean values for Na and Air being different over at least part of the range of test conditions.

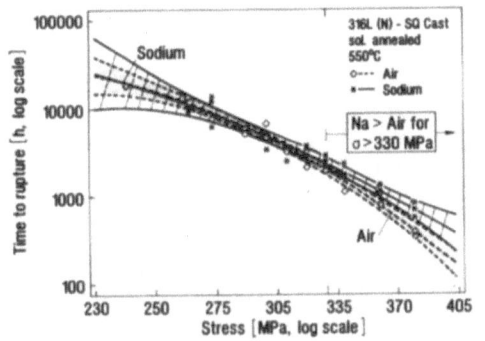

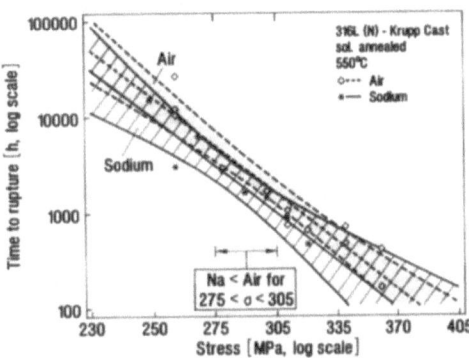

Fig. 4. Time to rupture versus stress for 316L(N), SQ-cast, solution annealed, tested at 550 °C in air and sodium.

Fig. 5. Time to rupture versus stress for 316L(N), Krupp-cast, solution annealed, tested at 550 °C in air and sodium. All tests at EdF, Les Renardieres, specimens of 5 mm diameter in air, 4 mm diameter in Na.

In contrast to this behaviour, cast SQ in the sensitized state showed lower in-sodium rupture values (Fig. 6) and times to tertiary creep (Fig. 8) over the whole test range. These differences became significant for stresses below about 310 MPa and increased with decreasing stress or longer test times.

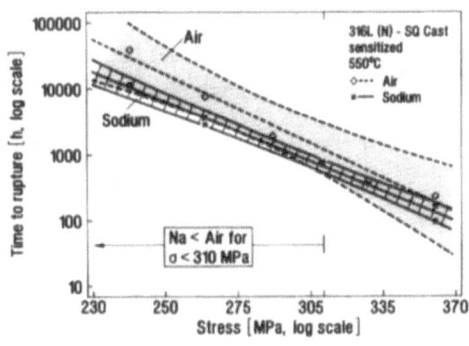

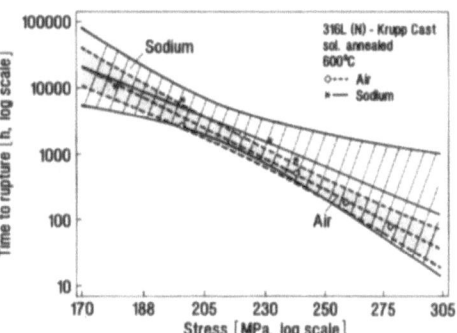

Fig. 6. Time to rupture versus stress for 316L(N), SQ-cast, sensitized, tested at 550 °C in air and sodium.

Fig. 7. Time to rupture versus stress for 316L(N), Krupp-cast, solution annealed, tested at 600 °C in air and sodium.

143

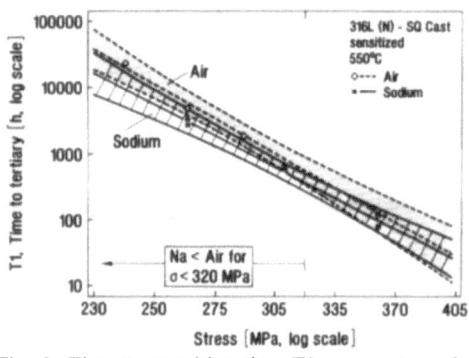

Fig. 8. Time to transition time T1 versus stress for 316L(N), SQ-cast, sensitized, tested at 550 °C in air and sodium.

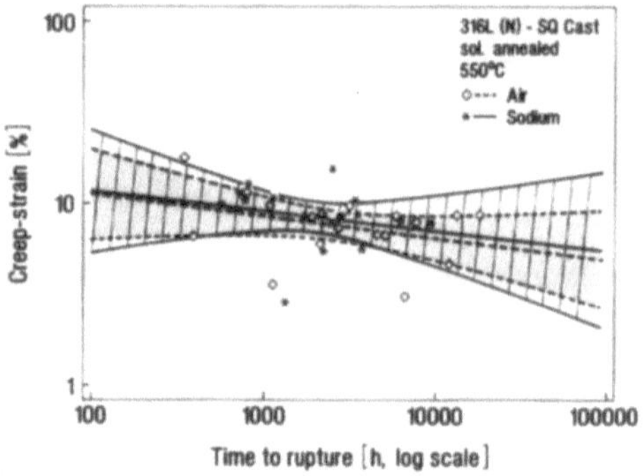

Fig. 9. Creep strain versus time to rupture for 316L(N), SQ-cast, solution annealed, tested at 550 °C in air and sodium.

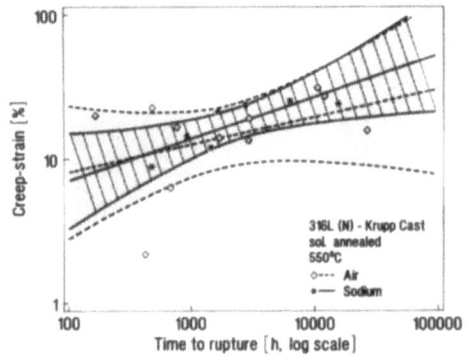

Fig 10. Creep strain versus time to rupture for 316L(N), Krupp-cast, solution annealed, tested at 550 °C in air and sodium.

Table 8.Results of statistical comparisons between in-air and in-sodium behaviours: creep ductilities

Cast [°C]	log t_r vs:-	
	log(Creep strain)	log(R of A)
SQAR/550	No significant difference Fig. 9	No significant difference
KRAR/550	No significant difference Fig. 10	No significant difference
SQSENS/550	Na < air at rupture times> 1000 h, Fig. 11	No significant difference
KRAR/600	Na < air at rupture times < 2000 h, Fig. 12	Na < air at rupture times < 2000 h, Fig. 13

SQAR...Cast SQ, as received   KRAR...Krupp cast, as received   NB: as received = solution annealed
SQSENS...Cast SQ, sensitized   creep strain = ELONG - EPSO

Table 8 shows the results from log-log plots of time versus rupture strain and reduction in area, also examined for each of the four cast/state/temperature combinations. In this examination only the KRUPP cast, tested at 600 °C, showed a significant difference, with slightly lower ductilities in sodium for test times less than about 2000 h (see Fig. 12 and 13). In the case of the sensitized SQ cast there were only three in-air tests with values for creep strain or reduction in area (see Fig. 11) thus giving wide 95 % confidence limes. Nevertheless the creep strain in sodium is significantly lower for rupture times larger than 1000 h (see Fig. 11). But this reduced ductility is not confirmed by the comparison of the R of A values.

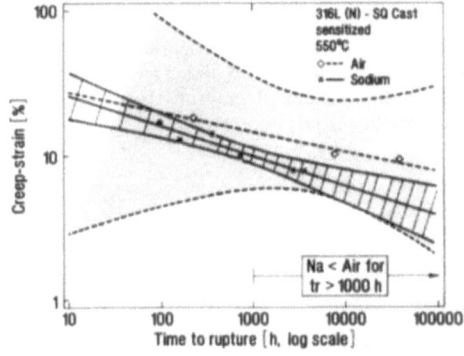 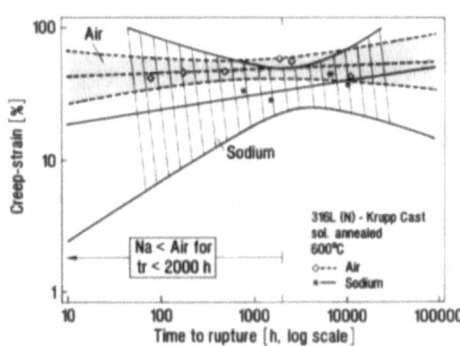

Fig. 11. Creep strain versus time to rupture for 316L(N), SQ-cast, sensitized, tested at 550 °C in air and sodium.

Fig. 12. Creep strain versus time to rupture for 316L(N), Krupp-cast, solution annealed, tested at 600 °C in air and sodium.

Conventional plots were produced for log(STRESS) versus log (TIME TO RUPTURE) showing the correct regression lines - i. e. produced with log (TIME TO RUPTURE) as the dependent variable. The plots are given as Fig. 14 (SQ cast, sol. annealed), 15 (SQ cast, sensitized) and 16 (Krupp cast, 550 °C and 600 °C).

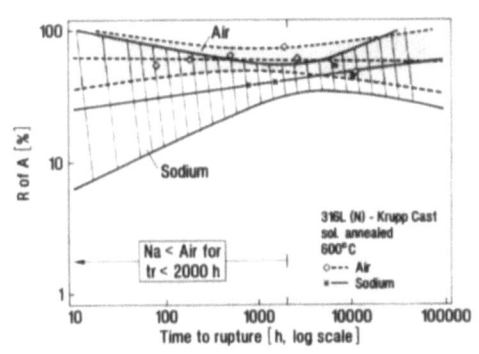 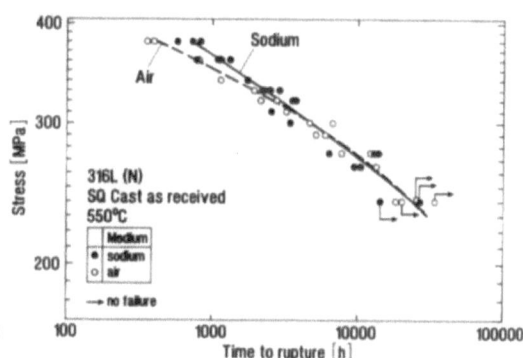

Fig. 13. Reduction of area versus time to rupture for 316L(N), Krupp-cast, solution annealed, tested at 600 °C in air and sodium.

Fig. 14. Stress versus time to rupture for 316L(N), SQ-cast, solution annealed, tested at 550 ° in air and sodium. Regression curves are only based on rupture data.

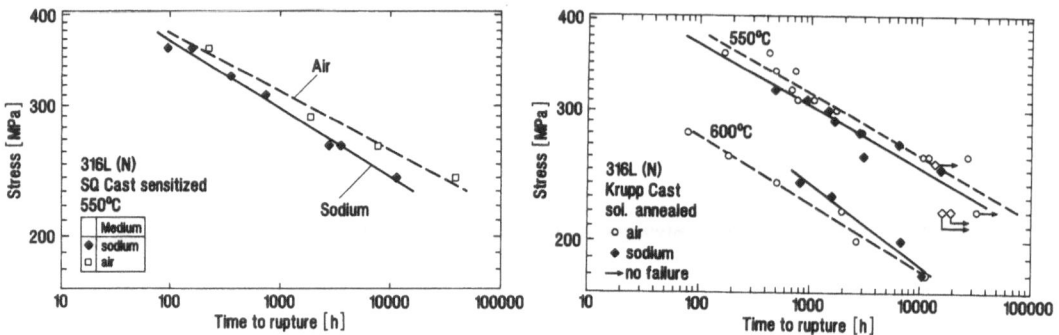

Fig. 15. Stress versus time to rupture for 316L(N), SQ-cast, sensitized, tested at 550 °C in air and sodium.

Fig. 16. Stress versus time to rupture for 316L(N), Krupp-cast, solution annealed, tested at 550 and 600 °C in air and sodium. Regression curves are only based on rupture data.

4.3 Discussion of Results from Statistical Analyses

The small differences observed between in-sodium and in-air behaviours of the as-received SQ and Krupp casts do not indicate serious consequences for design. Thus, in the case of SQ, the rupture strength was apparently increased by sodium at the shorter test times, Fig. 14. In the case of the Krupp cast, the mid-range decrease in in-sodium strength was caused by a single, outlying test result, Fig. 16. For both casts there were no significant differences between the in-air and in-sodium behaviours at longer times.

The only major effect of sodium, which may have consequences for the design was observed on the sensitized SQ cast. For the discussion of this effect it has to be taken into account that the rupture stress values for the SQ cast in the sensitized condition and in the as-received condition show a convergence at longer times (see Fig. 17).

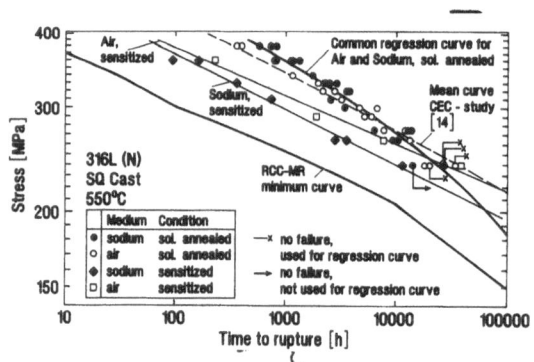

Fig. 17. Comparison of creep rupture strength of SQ-cast in sensitized and solution annealed condition, tested at 550 °C and comparison with mean curve from CEC study contract [14] and RCC-MR minimum curve.

This tendency is confirmed when additionally TESTIME data for unfailed tests are included. A criterion for doing this is to include TESTIME data as RUPTIME data if the TESTIME value is equal or greater than the mean value at that stress level. This is the case for three tests on SQ cast in as-received condition tested at 240 MPa in air (EDF and IA) and one test in sodium at 240 MPa (IA). Including these TESTIMES as RUPTIMES produced the modified

146

regression lines for SQAR/Air and SQAR/Na as presented below. These curves do not differ very much from those shown in Fig. 14.

Because SQAR/Air and SQAR/Na modified rupture data coincide at long times they can be treated as one data set (SQAR/All) for extrapolation purposes. The regression line for this is given by the equation:

$$\text{SQAR/All} \quad \log t_r = -30.012 + 35.283 \log \sigma - 8.753 \, (\log \sigma)^2$$

and it is plotted in Fig. 17 together with the lines for sensitized SQ cast.

Additionally in Fig. 17 the mean curve for in-air data of different heats of 316L(N), determined within a CEC-study contract [11] and the minimum curve from RCC-MR [6] are plotted.

In the CEC-study contract the mean curve was given by

$$\log t_r = [25875 + 4233.3 \log \sigma - 2558 (\log \sigma)^2]/T$$

with the Temperature T in Kelvin. This curve is less bent than the curves for the SQ-cast in the solution annealed condition. But because the CEC mean curve is determined from many different casts and even not confirmed by times to rupture larger than 40 000 h, the curves for the SQ cast will be used for the further comparison. Additionally the strong decrease in the RCC-MR minimum curve for rupture times larger than 10 000 h confirms the bending of the curves for the SQ-cast.

Fig. 18 shows the same data plotted together with the 95 % confidence limits on the position of mean lines for SQAR/All and SQSENS/Na. The extrapolations indicate that the difference between SQAR/All and SQSENS/Na becomes insignificant (< 95 % conf.) at stresses below about 210 MPa, but for higher stresses the stress rupture strength is significantly reduced by the sodium environment.

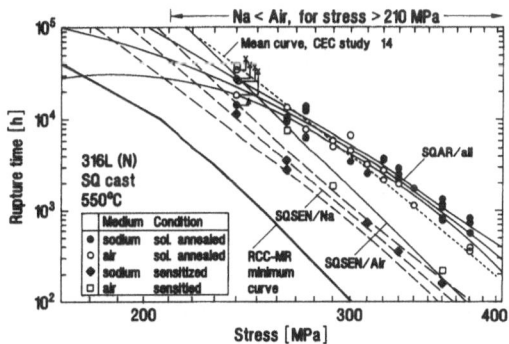

Fig. 18. Rupture time versus stress for SQ-cast (solution annealed and sensitized) with 95 % confidence limits tested in air and sodium. Regression and 95 % confidence limits for solution annealed SQ-cast are calculated for common data set of in-air and in-sodium data. Additionally the mean curve from CEC study contract [14] and MCC-MR minimum curve are given.

In comparison to the minimum curve from RCC-MR the band of the 95 % confidence limits of the SQSENS/Na data is well above (i. e. At higher rupture times) and no indication for convergence is given. However, it must be remembered that the test data in Fig. 18 represent one cast only and so, no account has been taken of cast to cast variations. The data are best taken as an indicator of the magnitude of the effect of sodium on pre-sensitised material and how this compares with the behaviour of initially solution treated material in the long term.

5. METALLOGRAPHICAL EXAMINATIONS

5.1 Fracture Surfaces and Side Surfaces

A comparison of fracture surfaces of specimens from the SQ-cast tested in the sol. annealed condition at 275 MPa in air and in sodium by SEM photographs shows larger continuous areas of intercrystalline fracture for the specimens tested in sodium than the air-specimen. In the specimens tested in sodium these intercrystalline cracks start from the surface of the specimen. This was clearly found in SEM-photographs of the side surface of the specimens tested at 550 °C in sodium.

5.2 Statistics of Cracks

A quantitative measure of cracks can be given by counting the number of cracks with a certain crack depth observed on longitudinal sections of the gauge length of the specimens. For the SQ-cast tested in the sensitized condition it is shown (Fig. 19) that the in-sodium specimen has a higher number of deep cracks than the in-air specimens. The latter have the highest number of cracks in the category with 0 - 50 μm depth. The SQ-cast in the solution annealed condition (Fig. 20) also shows for the in-sodium specimens the highest numbers for deep cracks, but for the in-air specimen (IA) more cracks were observed for 50 - 100 μm depth than for 0 - 50 μm depth.

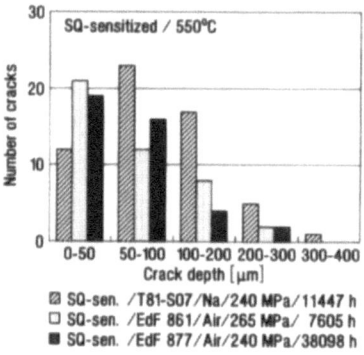

Fig. 19. Comparison of number of cracks (within a range of 10 mm from fracture surface) for specimens from SQ-cast, tested in sensitized condition at 550 °C.

Contrary to the behaviour of the SQ-cast, sensitized condition, the Krupp-cast tested in the sol. annealed condition shows significantly more short cracks (0 - 50 μm) for the in-sodium test than for the in-air test (Fig. 21).

5.3 Longitudinal Sections

The reason for the higher number of deep cracks, found in the SQ-cast tested in sodium after sensitization, is given in the optical micrographs of longitudinal sections of specimens tested in air and in sodium (Fig. 22). The cracks in the in-air specimen are wider, filled with oxides and blunted at the crack tip, whereas the cracks in the in-sodium specimen are deeper and less blunted, giving the impression that a solution process may occur along the grain boundaries. Specimens tested in sodium in the solution annealed condition do not develop comparable sharp and deep surface cracks (Fig. 23, specimen KfK 255 MPa, $t_r > 13\ 562$ h).

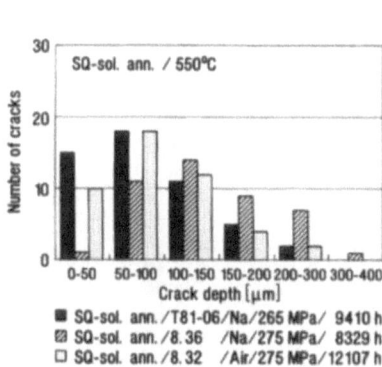

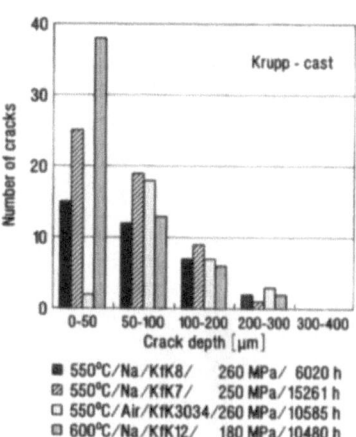

Fig. 20. Comparison of number of cracks (within a range of 10 mm from fracture surface) for specimens from SQ-cast, tested in solution annealed condition at 550 °C.

Fig. 21. Comparison of number of cracks (within a range of 10 mm from fracture surface) for specimens from Krupp-cast, tested in solution annealed condition at 550 and 600 °C.

An additional difference between in-air and in-sodium specimens is the formation of ferrite layers. These ferrite layers have been clearly identified on creep rupture specimens (Krupp cast) tested at 600 °C in sodium for more than 6 000 h (Fig. 24) and on specimen KfK tested at 550 °C for 13 562 h without rupture (Fig. 25). Because a certain time is necessary for its formation, the ferrite layers are not protecting the pre-sensitized material.

5.4 Exchange of Carbon and Nitrogen

An influence on the mechanical properties in a liquid metal environment may occur from an exchange of interstitial elements, if a substantial depth is affected.

The exchange of carbon is determined by the carbon potentials available in sodium and in the steel. The chemical activity of carbon in the two casts have been calculated in dependence of temperature and the concentrations of C, Ni, Mn, and Cr by the formula:

$$\ln a_C = \ln(0.0048.\%C) + (0.525 - 300/T).\%C - 1.845 + 5100/T - 0.021 - 72.4/T).(\%Ni + \%Mn) + (0.248 - 404/T).\%Cr - (0.012 - 9.422/T).(\%Cr)^2 + 0.033.\%Cr$$

For 550 °C (T = 823 K) and the concentrations given in Tab. 1 the following values were calculated:

$$a_C = 6.2 \cdot 10^{-4} \text{ for SQ-cast} \quad \text{and}$$

$$a_C = 5.17 \cdot 10^{-4} \text{ for Krupp-cast}$$

These values are lower than the carbon activities measured [5] in the sodium loops of KfK (a_C (CREVONA) = $6 \cdot 10^{-3}$) and Interatom a_C (AWN) = $2 \cdot 10^{-3}$. Therefore specimens from both casts will be carburized in these loops which will not have a harmful effect on the creep properties.

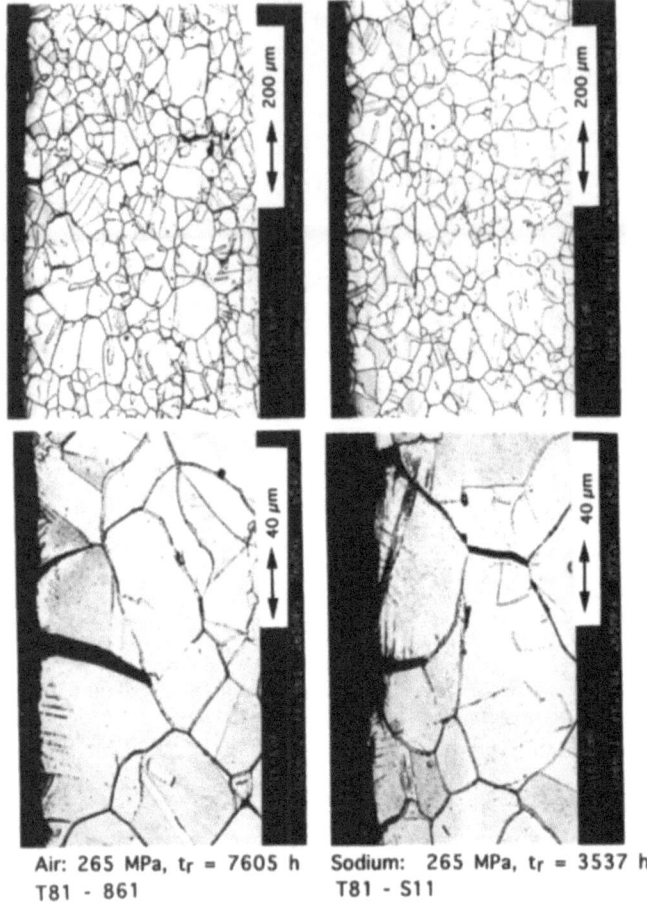

Air: 265 MPa, t_r = 7605 h   Sodium: 265 MPa, t_r = 3537 h
T81 - 861   T81 - S11

Fig. 22: Comparison of longitudinal sections of specimens from SQ-cast, tested in sensitized condition in air and in sodium at 550 °C

The carbon activity of the CINATRA loop at EdF, Les Renardieres, has not been measured, but because 316L(N) steel has been used as structural material for this loop, no decarburizing conditions, which may reduce the creep rupture strength [5], are expected for the tested casts.

Therefore the exchange of carbon will not explain the reduction in rupture strength occurring in the sensitized SQ-cast.

For the control of the exchange of nitrogen measurements of N-profiles have been performed with the method of Nuclear Microprobe [12,13] on some selected specimens. The results, given in Fig. 26, show a dependence on the environment, temperature and time: In the surface area of the Krupp cast tested in sodium nitrogen losses are observed. For the specimen tested at 600 °C the N-content is reduced to below 0.03 % for a depth of 50 μm from the surface and even in a depth of 180 μm the level of the matrix content is not reached. For the specimen tested at 550 °C the reduction in N-content is less pronounced but still significant to a depth of 30 μm below the surface.

Fig 23: Longitudinal section of specimen KfK from Krupp-cast, tested at 550 °C in sodium at 255 MPa for 13562 h (no rupture)

In contrast, the specimen of the Krupp cast tested in air shows a significant increase in the N content for up to about 50 μm from the surface. The measured matrix N content of this specimen is in agreement with the chemical analysis, Table 1, whilst that measured in the two in-sodium Krupp specimens is higher.

Finally, the EdF specimen (SQ-cast, sensitized) tested in sodium at 550 °C does not show a significant loss in the N-concentration (perhaps a small reduction up to 25 μm depth) and the matrix content measured by Nuclear Microprobe is in agreement with the chemical analysis.

Therefore it can be concluded that the exchange in nitrogen is not responsible for the reduced times to rupture measured for the sensitized SQ-cast.

6. DISCUSSION OF MECHANISMS

In the statistical analyses of rupture times it was found that a significant effect of sodium occurred for the previously sensitized material. The rupture stress and ductility were decreased in sodium, Table 7.

The metallographical examinations show that the effect of sodium is based on the development of surface cracks along grain boundaries which seem to grow more easily in sensitized material under sodium environment. In air the crack tips are blunted and oxidized and further growth is inhibited.

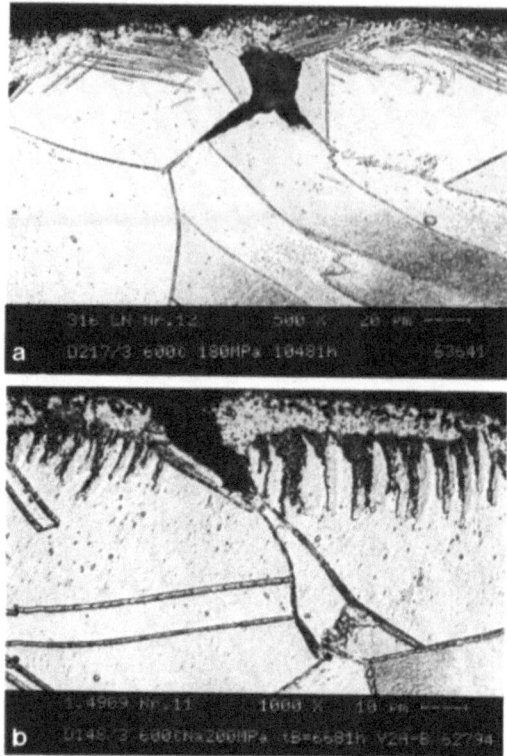

Fig. 24. Micrographs of 316L(N) specimens (Krupp-cast), tested in sodium at 600 °C, showing the formation of ferrite layers at the specimen-sodium interface;

a) specimen KfK 11, Na, 600 °C, 200 MPa, t_r = 6681 h

b) specimen KfK 12, Na, 600 °C, 180 MPa, t_r = 10480 h

The obvious microstructural feature of sensitized material is the presence of grain boundary precipitated in the form of carbo-nitrides $\{M_{23}(C,N)_6$ and $M_6(C,N)\}$. It is possible that these suffer a dissolution process when in contact with sodium and this will aid grain boundary cracking in the presence of the applied stress. It is not possible to identify any further detail of this process from the present results but it is likely that decarburization or denitrification are involved.

This mechanism appears to have an important influence on sensitized material under high loading conditions in sodium environment. For sol. annealed material in long running tests with low load levels in which the material becomes sensitized during the test it may be postulated that a similar sodium effect will occur. But this presumption is disproved by four in-sodium tests with the SQ-cast (sol. ann.) with rupture times between 9 410 h and 13 820 h and by one in-sodium test with 27 165 h running time at 240 MPa (see Fig. 17). Although sensitization will occur already after about 4 000 h (see Fig. 1), these rupture (running) times are as long as in air at the same stress level. That means that the material sensitized under stress conditions in sodium environment does not show a harmful effect of sodium and the normal creep damage by void formation over the total bulk of the specimens leads to fracture.

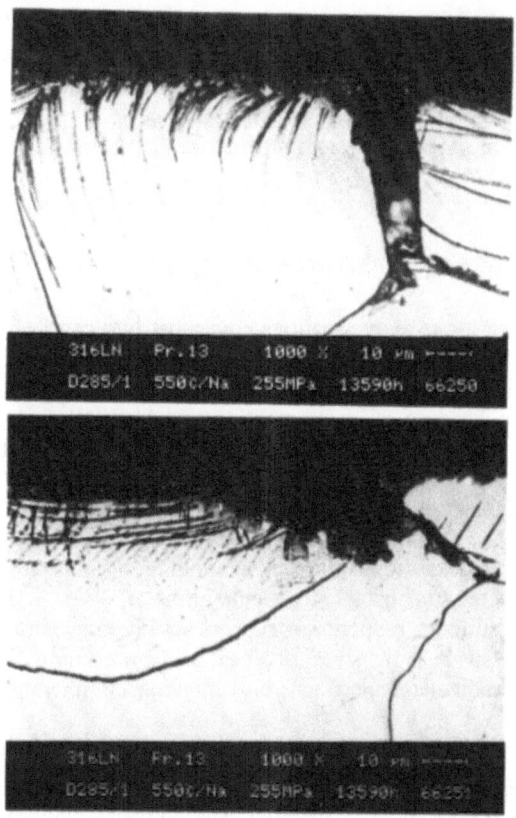

Fig. 25. Micrographs of specimen KfK (Krupp-cast), tested in sodium at 550 °C for 13562 h (no rupture), showing the formation of ferrite layers at the specimen-sodium interface.

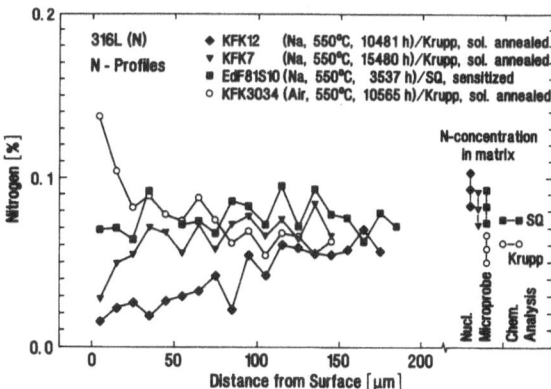

Fig. 26. Nitrogen profiles measured with Nuclear Microprobe on 316L(N) specimens after creep rupture tests in air and sodium.

153

A reason for this may be that the precipitations occurring under stress conditions in sodium environment are different from those occurring by previous sensitization. The sodium itself may change the kind of grain boundary precipitations near the surface, e. g. only carbo-nitrides will form, which are stable in the sodium environment.

In addition the formation of ferrite layers on the surface of the austenitic material may act as a protective layer (see Fig. 24).

7. ASSESSMENT OF EFFECT OF SODIUM FOR DESIGN

The test results reported in the above sections cover just two casts of 316L(N) steel and so cannot be used to guarantee an absolute lower limit for use in design. However, the results can be used to indicate the likely magnitude of any effect of sodium on the rupture strength of 316L(N) steel. This is so because all casts meeting the 316L(N) specification will develop similar microstructures which will be similarly involved in any grain boundary dissolution process in sodium.

The main indications of the rupture strengths in the presence of air or sodium, as shown in Figs. 17 and 18 are:

i)   There is no evidence of a deleterious influence of sodium on the rupture strength of solution annealed material subjected to constant stress conditions

ii) Pre-sensitized material suffers a reduction in rupture strength in sodium by an amount which increases progressively as the stress decreases. The maximum decrease is about 10 %. This applies to 4 mm diameter specimens fully immersed in flowing sodium of reactor grade purity.

iii) The rupture strengths of the solution annealed and pre-sensitized materials converge at longer imes/lower stresses. Extrapolation indicates that there is no significant difference between the strengths of pre-sensitized material in sodium and solution annealed material at stresses less than about 210 MPa.

It is apparent in Fig. 17 that the RCC-MR minimum curve is more closely aligned to the sensitized data at short and intermediate times, but follows the downturn of the solution annealed data at long times. In view of this and the above indication of in sodium behaviour, the following considerations for design are proposed.

Components fabricated from solution annealed 316L(N) and subjected to stress in flowing sodium at temperatures around 550 °C are expected to exhibit rupture strengths equivalent to those in air. This is also the case for pre-sensitized material loaded in 550 °C sodium at stress levels below 210 MPa.

On the other hand, pre-sensitized material subjected to stress values above 210 MPa will exhibit a reduced rupture strength in sodium. This reduction in rupture strength is likely to be up to 10 % for 4 mm thick sections surrounded by flowing sodium at 550 °C.

Loads above 210 MPa, with high strain ranges, could occur in the case of creep-fatigue. However, it is less easy to appreciate what situation would result in pre-sensitized material being loaded in sodium. This requires careful consideration of fabrication and operating conditions of the plant in question.

8. CONCLUSIONS AND RECOMMENDATIONS

Within the present study contract an extensive compilation and analysis of creep curves and rupture data on 2 heats of 316 L(N) were performed including

-   43 tests at 550 °C in sodium up to 27 000 h

- 4 tests at 600 °C in sodium up to 10 500 h
- 49 parallel tests in air.

The statistical analysis of data with 95 % confidence limits shows:

i) for the solution annealed condition no deleterious effect of sodium on
  stress rupture strength
  time to onset of tertiary creep
  minimum creep rates
  ductility (elongation and reduction of area) at 50 °C;
  but at 600 °C a small reduction in ductility for $t_r < 2000$ h occurs;

ii) for pre-sensitized condition:
  sodium reduces stress rupture strength
  and onset of tertiary creep
  and creep strain for $t_r > 1000$ h

iii) Differences in rupture strength for solution annealed material and pre-sensitized conditions
  tested in sodium become insignificant for stresses below about 210 MPa.

From the metallographical examinations it is suggested that the effect of sodium is based on the development of surface cracks along grain boundaries growing more easily in pre-sensitized material in a sodium environment. Although solution annealed material is expected to sensitize during test, grain boundary cracking does not appear to have been encouraged by sodium. This may be due to the kind of grain boundary precipitation and/or the formation of protective ferrite layers at the specimen surfaces. In air the crack tips are oxidised and blunted, thus reducing further crack growth.

Although the observations have involved only two heats of 316L(N) steel the results can be used to indicate the likely magnitude of sodium effects relevant to design stresses in components of 316L(N) steel operating in high purity sodium.

i) Design can be based on in-air rupture data for components starting operation in solution annealed condition and subjected to constant stress conditions.

ii) For material sensitized before loading in sodium, design may be based on in-air rupture data for stresses below about 210 MPa at 550 °C. On the other hand stresses above 210 MPa may lead to a reduction in rupture strength and this may be as much as 10 % in the case of a 4 mm thick component immersed in flowing sodium at 550 °C.

The authors of this contribution have carried out this work to the best of their professional ability, taking account of the information and techniques available at the time, but cannot take personal responsibility for the accuracy of the test results, particularly those from laboratories outside their control. The influence of sodium indicated in this document applies only to high purity sodium. Any introduction of oxygen or water vapour to the sodium can seriously increase its deleterious influence on material properties. Designers wishing to use these results should satisfy themselves as to the accuracy of the original data and of the relevance of the operating conditions of the intended application.

Acknowledgement

The authors greatfully acknowledge the willingness of the following organizations for providing the data basis of this study contract:

AEA-Technology Atomic Energy Authority, Risley, England

EdF - DEM Electricité, de France - Département Etude des Matériaux, Les Renardières, France

National Power National Power, PLC, Technology and Environmental Centre, Leatherhead, England

KfK Kernforschungszentrum Karlsruhe, Institut für Materialforschung III, Karlsruhe, Germany

Siemens AG - KWU Siemens AG - Power Generation Group (KWU), Bergisch Gladbach1, Germany

Special thanks are given to the colleagues of these organizations for the performance of creep rupture tests and metallographical investigations.

References

1. R. S.Fidler and M. J.Collins , Atomic Energy Rev. 13, (1975) 3.
2. G. J.Lloyd , Atomic Energy Rev. 16, (1978) 155 - 208
3. K.Natesan, O.K.Chopra and T.F. Kassner,. Sec. Int. Conf. on Liquid Metal Technology in Energy Production, Richland, Washington, (1980)19.41 - 19.49
4. C.A.P. Horton, R.S. Fidler, and B. H. Targett, Conf. on Liquid Metal Engineering and Technology, BNES, London, (1984) Vol.2 pp. 439 - 444
5. H.U. Borgstedt,and H. Huthmann, J. Nucl. Mater. 183 (1991) 127-136
6. RCC-MR, Section II-Materials Design and Construction Rules for Mechanical Components of FBR Nuclear Islands AFCEN, June 1985
7. H.Huthmann,and G. Jenner Conf. on Liquid Metal Engineering and Technology, BNES, London, (1984) Vol. 2, pp. 473 - 477
8. H.U. Borgstedt,G. Drechsler ,G. Frees,. and E. Wollensack, in: Materials Behaviour and Physical Chemistry in Liquid Metal Systems, Ed. H. U. Borgstedt, Plenum Press, New York, (1987) pp. 185-191
9. Ph. Debergh,. and M. Bethmont, EdF report: HT-46/NEQ 1274-A, 27 Fevrier 1992
10. H. Huthmann,H.U. Borgstedt, H. Tas, in: Behaviour and Physical Chemistry in Liquid Metal Systems, Ed. H.U. Borgstedt, Plenum Press, New York (1987), pp. 141-151
11. D.Lehmann ,V.B. Livesey ,E. te Heesen ,H. Breitling , CEC Study Contract RA1-0182-F Evaluation of the Stress to Rupture and Creep Properties of Type 316L(N) Steel for Design Use - Final Report - CEA-CEREM N. T.   SRMA 92-1992, Decembre 1992
12. J.W. McMillan.,in "Analysis of Non-Metals in Metals", 1981, Walter de Gruyter & Co., Berlin, New York pp. 173-192
13. J.W. McMillan ,Report AEA-Technology (Harwell Laboratories) to Dr. H. U. Borgstedt, 3rd November 1992

HIGH-CYCLE FATIGUE BEHAVIOUR OF ALLOY 718
IN LIQUID SODIUM

Eiichi Yoshida, Ryuji Komine, Fumiyoshi Ueno and Yusaku Wada

Materials Development Section, O-arai Engineering Center
Power Reactor and Nuclear Fuel Development Corporation
4002 Narita, O-arai-machi, Ibaraki-ken 311-13, Japan

1. INTRODUCTION

In the vicinity of the core outlet of a fast breeder reactor, thermal striping arises from the coolant temperature difference between the outlet of the fuel assemblies and the control rod assemblies, and causes the structural material at the upper part of the core to suffer high-cycle thermal stress. A structural design standard is prescribed to prevent the failure by such thermal stress in the structural materials, and the Alloy 718 which is superior in the high-cycle fatigue strength might be used at this upper core structure for future reactors. However, there are few studies conducted on the fatigue properties of Alloy 718 at the high-cycle region in flowing sodium. Further Alloy 718 is a Ni-base alloy containing more than 50% Ni, so it is pointed out that the selective dissolution of Ni in alloy into the sodium [1,2] may exert influence on the fatigue failure life of Alloy 718.

In this paper the effects of sodium on the fatigue behaviour was evaluated based on the results of high-cycle fatigue test up to 10^8 cycles in elevated temperature sodium on Alloy 718 after pre-exposure to flowing sodium for 10,000 hours.

2. EXPERIMENTAL PROCEDURE

2.1 Test Material

The test material was solution treated Alloy 718 plate of 25 mm thick. The chemical composition is shown in Table 1 and the mechanical properties in Table 2. The configuration of test specimens used in the in-air and in-sodium fatigue tests are shown in Figure 1. Both test specimens are a solid bar type with a parallel section of 20 mm in length and 6 mm in diameter. The axis of the specimen is parallel with the rolling direction of the material, and the surface of the specimen was polished in the axial direction with #400 emery paper to remove machining flaws.

After the pre-exposed of specimens, bellows was jointed to the shoulders of the specimen by electron beam (EB) welding to minimise heat affected zone formation. The load imposed on the bellows is designed to be less than 1/100 that on the parallel section of test specimen

Liquid Metal Systems, Edited by H.U. Borgstedt
and G. Frees, Plenum Press, New York, 1995

and is neglected in the analysis. An inner spacer served to give a flow rate to the sodium on the surface of parallel section of the test specimen.

Table 1 Chemical composition (mass %)

C	Si	Mn	P	S	Ni	Cr	Mo	Cu	Al	Ti	Nb
0.06	0.16	0.09	0.002	0.002	52.56	18.77	3.03	0.04	0.61	0.99	5.38

Solution heat treatment: 955 °C x 1 h, AC
Ageing heat treatment: 720 °C x 8.0 h, FC /625 °C x 8.0 h, AC

Table 2 Mechanical properties

Tensile properties			Creep-rupture properties			Grain size
0.2% YS kg/mm^2	UTS kg/mm^2	Elong. %	Stress kg/mm^2	Rupture time h	Elong. %	ASTM no.
R.T. 114.6	139.0	15.2	-	-		< 6.0
(649 °C) 90.2	108.6	16.4	(649 °C) 70.3	68.5	42.5	

2.2 Pre-test treatment

The specimens were pre-exposed to flowing sodium for 10,000 hours (hereinafter "the sodium exposed material") under non-stress conditions. The pre-exposure to sodium was conducted using a sodium test loop constituted by austenitic stainless steels (Types 304 and 316). The flow sheet of the sodium exposure test section is shown in Figure 2. For the conditions of sodium exposure test, the sodium temperature was 600 °C and 650 °C, the oxygen concentration in sodium was about 1 ppm and sodium flow rate on the surface of parallel section of the specimen was 1m/sec. Further the test specimens with the thermal ageing were prepared, which was treated at the same temperature for the same time as the sodium exposed materials (hereinafter "the thermal aged material") in order to separately evaluate the effects of sodium environment.

Table 3 Test conditions of high-cycle fatigue test

Pre-test treatment of specimen	None (as received)	Sodium exposed 600 & 650 °C, 10^4 h 1 m/s, ~ 1 ppm O$_2$	Thermally aged 600 & 650 °C, 10^4 h
Type of loading	Uniaxial push-pull	←	←
Control	Load control	←	←
Loading wave form	Triangular	←	←
Frequency	10 Hz	←	←
Test temperature	550,600,650 °C	600, 650 °C	←
Test environment	In air In sodium	In-sodium	←
Sodium velocity	- 1 m/s	1 m/s	←
Oxygen content	- ~ 1 ppm	~ 1 ppm	←

2.3 Test Conditions

The outline of the fatigue test section in air and in sodium is shown in Figure 3. The test machine used in this study was a push-pull type with an electro-servo-hydraulic system having a maximum load capacity of ± 10 tons. This machine was installed in the sodium loop, and was designed to be capable of testing in air too.

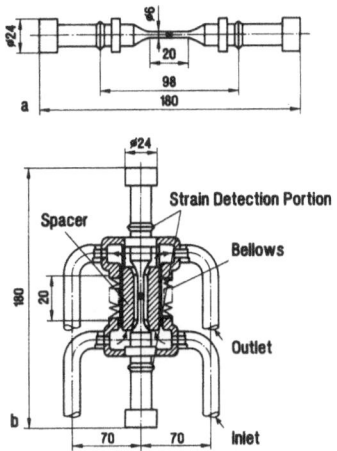

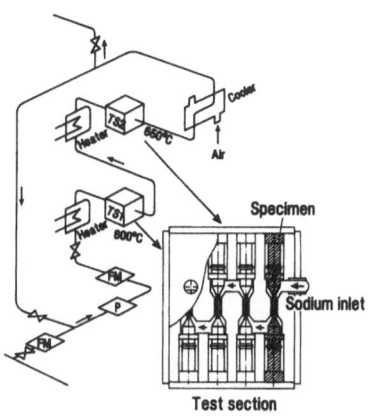

Fig. 1. High-cycle fatigue specimen in air and sodium
(a) specimen in air
(b) specimen in sodium (unit: mm)

Fig. 2. Flow sheet of the sodium exposure test loop and detail of the test section

The conditions of the high-cycle fatigue tests are shown in Table 3. The high-cycle fatigue tests were conducted at 550, 600, and 650 °C under load control mode with frequency set at 10 Hz. The test temperatures were at the same level as the pre-exposure temperature, and controlled within ± 2 °C during the test by the thermocouples set in the upper and lower plenums of the specimen. The sodium flow rate was 1.0 m/s on the surface of the specimen. The purity of sodium was controlled by maintaining the cold-trap temperature at 120 °C, and the oxygen concentration in sodium was about 1 ppm as obtained from Eichelberger's formula [3].

3. EXPERIMENTAL RESULTS

3.1 High-Cycle Fatigue Life of As-received Material

The comparison between the fatigue lives in sodium and in air of the as-received material at 550, 600, and 650 °C is shown in Figure 4. The total strain range $\Delta\varepsilon_t$ was obtained by the following equation for the stress range $\Delta\sigma$ and the Young's modules E, because the stress is in elastic region.

$$\Delta\varepsilon_t = \Delta\sigma / E \tag{1}$$

In the case of the as-received material, the high-cycle fatigue life in sodium was nearly equal to that in air and no significant effect of sodium environment on fatigue life could be found out. The fatigue life of Alloy 718 did not depend on the test temperature within the range of 550-650°C in this study.

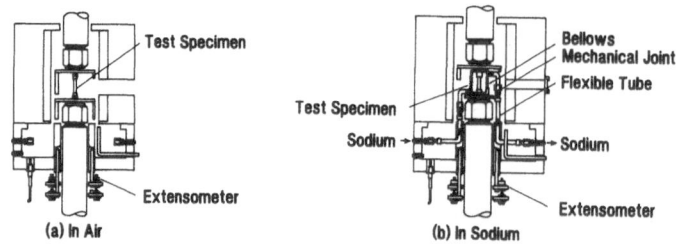

(a) In Air   (b) In Sodium

Fig. 3. Outline of the fatigue test section in air and sodium

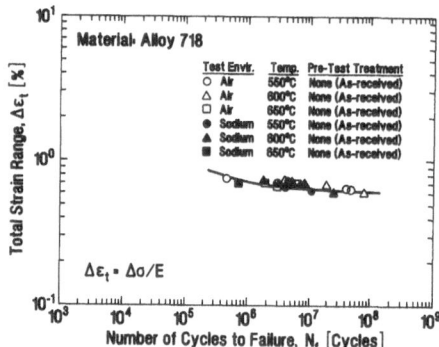

Fig. 4. Comparison of high-cycle fatigue life for Alloy 718 in sodium and air

3.2 High-Cycle Fatigue Life of Sodium Exposed and Thermal Aged Materials

The in-sodium high-cycle fatigue lives of the sodium exposed material at 600 and 650 °C are shown in Figures 5 and 6, respectively. In the figures, the average fatigue failure curve of Type 304 stainless steel is shown. At 600 °C, the fatigue life of the sodium exposed material was slightly shorter than that of the as-received material, but it remains equal to that of the thermal aged material. Therefore the effects of sodium environment on the fatigue life was not significant at 600 °C. While in 650 °C sodium, the fatigue life of the sodium exposed material was lower than that of the thermal aged material, and the effects of 10,000 hours exposure to sodium was recognised. However, the high-cycle fatigue life of Alloy 718 was fairly longer than the average fatigue life of Type 304 stainless steel, even after the material was pre-exposed to sodium or thermal aged for 10,000 hours.

3.3 Fracture Surface

The typical SEM micro graphs of fatigue fracture surfaces are shown in Figure 7. It was found that cracks of in-air tests for the as-received material initiated from the specimen surface and propagated transgranularly, and the striations were clearly observed on the fracture surface. In the fatigue fracture surface tested in sodium for the sodium exposed and thermal aged materials, the striations were also observed at inner site and the starting point of cracks was the specimen surface. But the striations were rather ill-defined compared with those in air. This was similar to the results of fatigue test for Type 304 stainless steel [4,5]. For the material sodium exposed to sodium at 650 °C, the fracture surface affected by sodium corrosion was observed at crack initiation area.

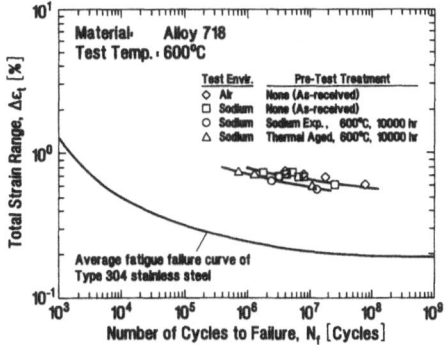

Fig. 5. Comparison of high-cycle fatigue life for Alloy 718 tested at 600 °C

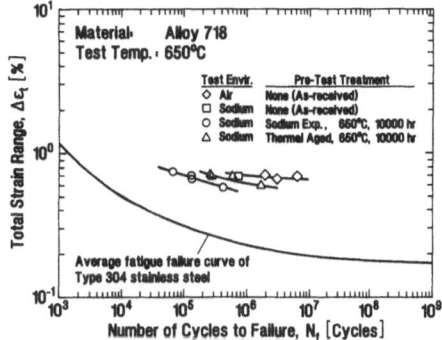

Fig. 6. Comparison of high-cycle fatigue life for Alloy 718 tested at 650 °C

4. DISCUSSION

In the case of sodium exposed material at 650 °C, the fatigue life showed a trend of becoming shorter than that of the thermal aged materials, and the effect of sodium environment on fatigue life was recognised. The reduction of fatigue life of the sodium exposed materials at 650 °C was considered to be mainly affected by the long time pre-exposure to sodium rather than the environmental effect during the fatigue tests.

The change of surface morphology, chemical composition and micro structure by exposure to flowing sodium for 10,000 hours before the fatigue tests are shown in Figures 8 and 9, respectively. In the case of sodium exposed material at 600 °C for 10,000 hours, the machining flaws on the specimen surface did not disappear and the extent of corrosion caused by the sodium exposure is small (Figure 8-a). Also no remarkable composition change was recognised near the surface region. However, in 650 °C sodium, the selective dissolution of Ni was particularly observed and the intergranular corrosion was observed through a depth of 15-20 μm from the sodium exposed surface (Figures 8-b, 9-a). Near the surface of specimen, the traces of the precipitants fallen-off and the massive precipitations (Ni3Nb) at grain boundaries were also observed (Figure 9). It is considered that those changes in composition and micro structure of the grain boundaries are connected with the initiation of fatigue cracks.

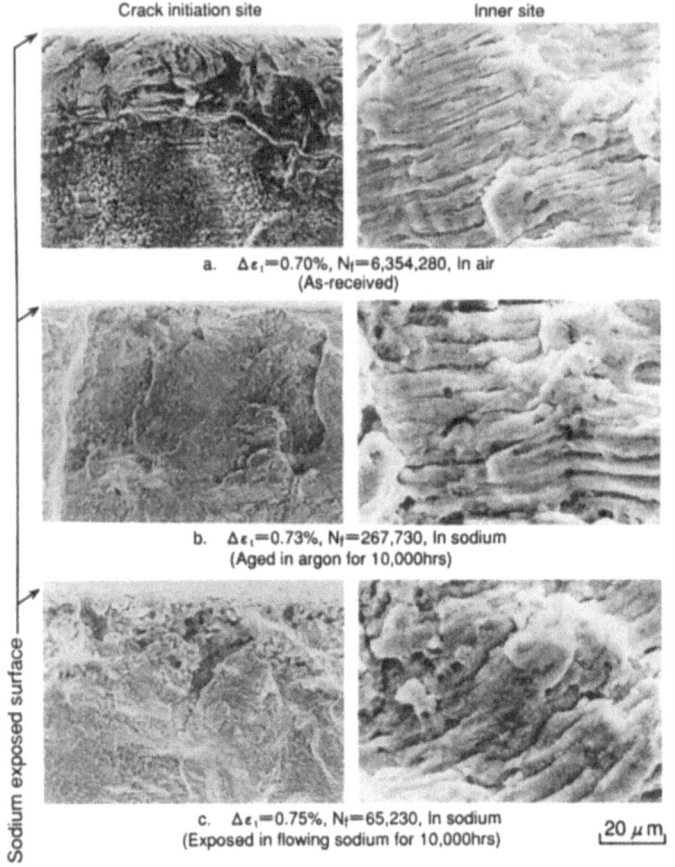

Crack initiation site   Inner site

a. $\Delta\varepsilon_t$=0.70%, N_f=6,354,280, In air
(As-received)

b. $\Delta\varepsilon_t$=0.73%, N_f=267,730, In sodium
(Aged in argon for 10,000hrs)

c. $\Delta\varepsilon_t$=0.75%, N_f=65,230, In sodium
(Exposed in flowing sodium for 10,000hrs)   20 μm

Sodium exposed surface

Fig. 7. SEM micro graphs of the fracture surface after high-cycle fatigue tests at 650 °C

The typical examples of the surface cracks after fatigue tests are shown in Figure 10. The cracks were propagated from these weakened grain boundaries and precipitants fallen-off in the sodium exposed material at 650 °C. It was found that cracks had propagated over 2 to 3 crystalline particles along the grain boundaries, and then progressed transgranularly. In the thermal aged Alloy 718 tested in 650 °C sodium, most cracks propagated transgranularly, but their depth was shallow compared with those of the sodium exposed material. In the sodium exposed and thermal aged materials tested at 600 °C, cracks also propagated transgranularly.

Based on the above, the results of the analysis of the high-cycle fatigue behaviour of Alloy 718 are summarised in Table 4. It is considered that the reduction of high-cycle fatigue life of the sodium exposed material at 650 °C was caused by earlier intergranular crack initiation near the surface due to the intergranular corrosion and coagulation of precipitated particles along the grain boundary during long term sodium exposure. A conceptual view of this process is shown in Figure 11.

The intergranular corrosion of this material in sodium was observed under the high temperature conditions above 700 °C by previous examinations [2,6,7]. The influence of intergranular corrosion on the fatigue behaviour must be considered, if Alloy 718 is applied under such high temperature sodium environment.

Generally, the rate of sodium corrosion with dissolution of alloying elements in the steel shows an Arrhenius type relationship to temperature, which can be expressed as follows.

$$A = A_0 \times \exp(-E/RT) \tag{2}$$

Where A is corrosion rate, E is the activation energy for the corrosion process, T is the Kelvin temperature, R is gas constant and A_0 is a constant.

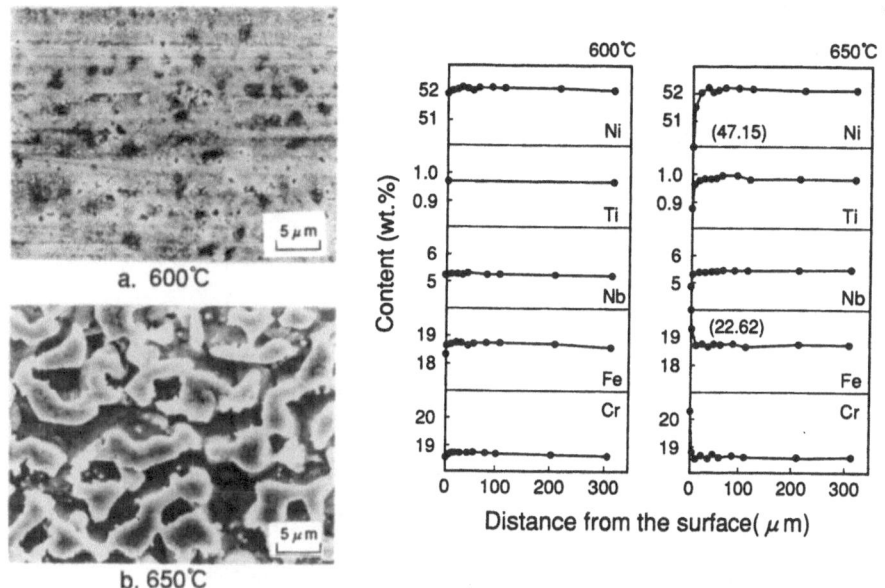

Fig. 8. Change of surface morphology and chemical compositions by exposure to flowing sodium for 10,000 hours

The rate of general and intergranular corrosion of Alloy 718 in sodium obtained from this study and relevant data [6-8] are shown in Figure 12 in comparison with the data for austenitic stainless steels. The intergranular corrosion rate is higher than that of general corrosion at 650 °C and shows a remarkable temperature dependence, because the diffusion of alloy elements from the grain boundary is dominant at higher temperature. But the intergranular corrosion rate under the condition of temperatures below 600 °C becomes lower than the data scatter band of general corrosion, and the general corrosion becomes the dominant corrosion process. Thus the general corrosion rates of Alloy 718 are almost the same as those of austenitic stainless steels below 600 °C, and good compatibility with sodium is shown.

The corrosion rate increases proportional to the Ni content in the steel [8, 9], but this Ni dependency is small in a precipitation hardening alloy like Alloy 718 [2], and not sensitive to the oxygen concentration in sodium[10]. Further, no intergranular corrosion of Alloy 718 has ever been observed at temperatures below 600 °C. If the operating temperature for

Alloy 718 is less than 600 °C for the use to prevent the thermal striping, it can be expected that the effect of the intergranular corrosion on the fatigue life would be as small as shown in Figure 5. Consequently, Alloy 718 has good compatibility with sodium below 600 °C, and the effect of sodium environment is negligible. And then, the thermal strain arising from the thermal striping is fairly smaller than the allowable values given by the fatigue curve and the endurance limit suggested by P. Marshall et al [11]. As the result, the design fatigue curve for sodium exposed conditions below 600 °C can be determined as the best fit curve in air with a margin 2-20 (a factor of 2 on the total strain range or a factor of 20 on the fatigue life) [12].

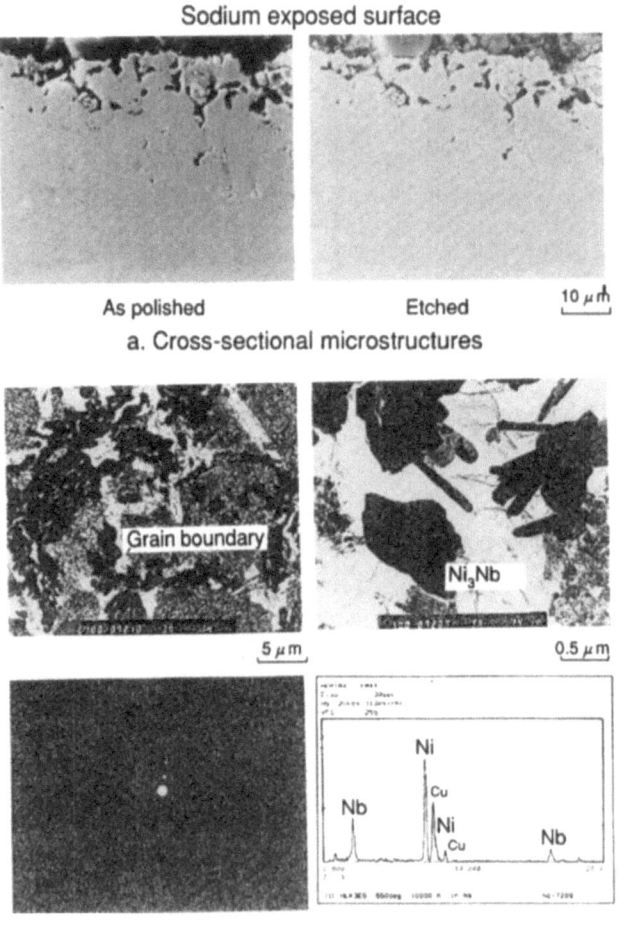

Fig. 9. Micro structures of Alloy 718 after exposure to flowing sodium
for 10000 hours at 650 °C by optical and electron microscopy

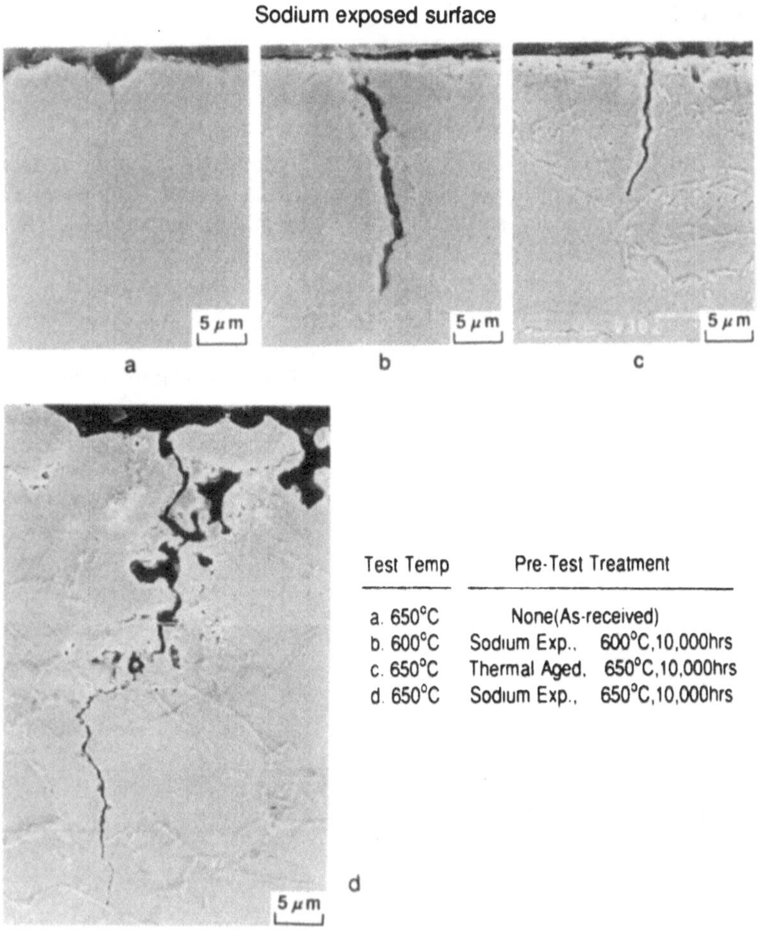

Fig. 10. Typical examples of the surface crack of Alloy 718 after high-cycle fatigue tests in flowing sodium

Table 4. Summary of fatigue behaviour for Alloy 718

	600 °C	650 °C
As-received material (RA)	- Fatigue life: in air + in-sodium - Crack initiation; transgranular	- Fatigue life: in air + in-sodium - Crack initiation; transgranular
Thermal aged material (Aged) 10000 h	- Fatigue life; aged < AR (in-sodium) - Crack initiation; transgranular	- Fatigue life; aged < AR (in-sodium) - Remarkable precipitants
Sodium exposed material (Na) 10000 h	- Fatigue life: Na < AR(in-sodium) Na + aged (in-sodium) - Crack initiation; transgranular - No effective corrosion	- Fatigue life: Na < AR (in-sodium) Na< aged (in-sodium) - Crack initiation intergranular - Weakened grain boundary - Remarkable precipitants

5. CONCLUSIONS

(1) The high-cycle fatigue life of Alloy 718 exposed to flowing sodium at 650 °C for 10,000 hours was slightly less than that of as-received and thermal aged materials. At 600 °C, no sodium environmental effect could be detected.

(2) The transgranular fatigue crack propagation was predominant in the as-received and the sodium exposed materials at or below 600 °C, while it was observed at 650 °C, that cracks initiated from the grain boundary, and propagated by the transgranular mode. The reduction of the high-cycle fatigue life was due to the intergranular corrosion during long-term sodium exposure and coagulation of precipitated particles along the grain boundary.

(3) In comparison with Type 304 stainless steel, Alloy 718 showed superior high-cycle fatigue strength. The effect of sodium environment was negligible below 600 °C. It was concluded that Alloy 718 had superior resistance to thermal striping by sodium mixing.

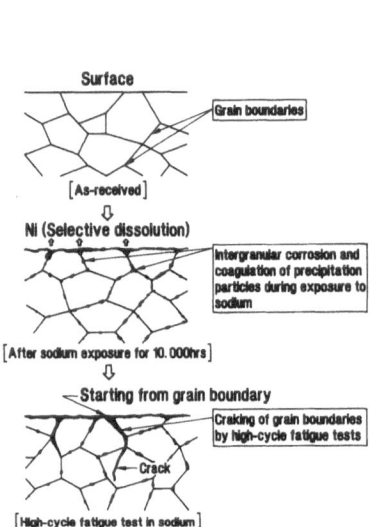

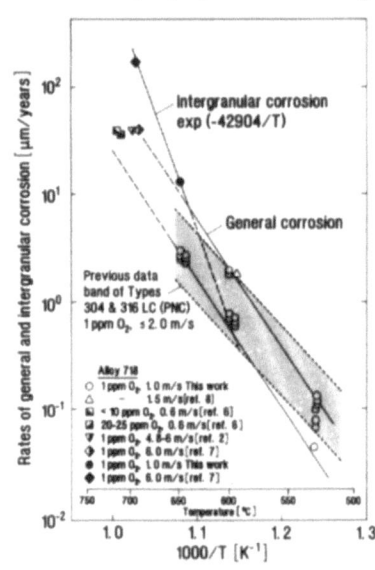

Fig. 11. Concept of the development of intercrystalline corrosion in Ni base alloys due to dissolution of Ni

Fig. 12. Rates of general and intercrystalline corrosion of Alloy 718 in Na

REFERENCES

1. H.U. Borgstedt and C.K. Mathews, Applied Chemistry of the Alkali Metals, Plenum Press, New York and London, 1987.
2. S.A. Shiels, C. Bagnall, et al, Int. Conf. on Liquid Metal Technology in Energy Production, Richland, VII-2, 1980.
3. R.L.Eichelberger, USAEC report, AI-AEC-12685, 1968.
4. T. Maruyama, S. Kato, R. Komine, et al, Proc. Int. Conf. on Liquid Metal Engineering and Technology, LIMET'88, 512, 1988.
5. W.J. Mills and L.A. James, ASME 79-PVP-83, 1979.
6. G.A. Whitlow, J.C. Cwynar, R.L. Miller, et al, Chemical Aspects of Corrosion and Mass Transfer in Liquid Sodium, Proc. of the Symposium, Detroit, 1971.
7. W.F. Brehm and R.P. Anantatmula, Int. Conf. on Liquid Metal Technology in Energy Production, Champion, VIIb-8, 1976.
8. A.R. Keeton and C. Bagnall, Int. Conf. on Liquid Metal Technology in Energy Production, Richland, VII-3, 1980.
9. T. Itaki, S. Yuhara, et al, Materials for Nuclear Reactor Core Applications BNES, London, 38, 1987.
10. A.W. Thorley and C. Tyzack, Liquid Alkali Metals, BNES, London, 41, 1973.
11. P. Marshall and C.R. Brinkman, Nucl. Energy, 20 (1981).
12. Criteria for Design of Elevated Temperature Class 1 Components in Section III, Division 1, of the ASME Boiler and Pressure Vessel Code, 1976

SODIUM FOR FAST BREEDERS
- PRODUCTION, PURIFICATION, AND QUALITY

Michel Salmon

Métaux Spéciaux, S.A.
F-73600 Moutiers, France

1. INTRODUCTION

1.1 History of Métaux Spéciaux in Sodium Production

Since 1920, Métaux Spéciaux, under different names, has been involved in the manufacture of metallic sodium. The first process used is known as Castner process, in which sodium is produced by electrolysis of melted caustic soda, giving hydrogen as a by-product.

In 1950, a new process was implemented, using chloride as raw material and producing sodium and chlorine. The metallic sodium has been produced continuously from the fifties up to now, with a thirteen thousand metric tons per year capacity.

The standard quality of the sodium getting out of the facility is known as technical grade sodium, the major impurities are calcium and metallic oxides, the amount of potassium depends on the purity of the salt.

1.2 The Nuclear Grade Sodium Speciality

Beside the standard grade, we had to face a demand from nuclear industry in the sixties when the fast breeders were being developed in France by CEA (Commissariat a l'Energie Atomique).

At this time a co-operation started with CEA in order to produce small quantities of nuclear grade sodium for experimental loops and for the first prototype RAPSODIE, experimental fast breeder of 25 MW thermal power, upgraded to 40 MW some years after.

In order to increase the production and make it more reliable, we developed a strongly improved process and built facilities which were able to produce 10 tons a day.

From these facilities, around 8 000 tons of nuclear grade sodium were delivered, for experimental loops and for French breeders, 1 200 tons for PHENIX in the 70's and close to 6 000 tons for SUPERPHENIX in the 80's. Then came the 90's ...

Liquid Metal Systems, Edited by H.U. Borgstedt
and G. Frees, Plenum Press, New York, 1995

1.3 Japanese Development Programme (MONJU)

After running of experimental fast breeder JOYO, PNC (Power Reactor and Nuclear Fuel Development Corporation), the Japanese state Company, decided to build a new reactor called MONJU, supplying an electrical power of 280 MW, i.e. a thermal power of 714-MW, containing about 1 700 tons of sodium. Métaux Spéciaux was chosen as sodium supplier. The steps of production, purification, transportation, storage and delivery were completed during the year 1991 under a severe quality control. These operations are described hereafter.

2. PRODUCTION OF TECHNICAL GRADE SODIUM

2.1 Electrolysis Process

The sodium chloride is prepared to meet grain size and purity specifications. It is melted in a liquid bath of other chlorides at 600°C and electrolysed between a graphite anode and a steel cathode. Sodium is collected at the cathode and chlorine at the anode. This process is known as Downs process, from the inventor's name.

2.2 Characteristics of Technical Grade Sodium

Owing to the calcium chloride in the bath and the high solubility of calcium, the sodium contains high amounts of calcium and a low temperature filtration is necessary to lower calcium content, but the result is limited to about 300 - 400 ppm. A typical analysis of standard grade sodium is listed in Table 1.

Table 1. Composition of standard grade sodium

Element	typical
Na	99,9 %
Ca	350 ppm
K	200 ppm
Ba	< 5 ppm
Fe	< 5 ppm
Cl	< 10 ppm
O	unknown

3. PURIFICATION TO NUCLEAR GRADE

3.1 Specifications of Nuclear Grade Sodium

The Japanese specifications are more severe than those given for the French programme. The main difference concerns oxygen and hydrogen which are limited to 10 ppm and 1 ppm on average respectively at the moment of filling up the MONJU facility; peak values of 30 and 5 ppm respectively are accepted provided the specification is respected. The reason is that the lower is the amount of oxygen, the shorter is the purification in the MONJU facility; nevertheless the surface of vessels, of piping and of fuel elements can release a much larger oxygen amount. The specifications of sodium for MONJU are listed in Table 2.
The contents of Al, Cd, Cr, Cu, Fe, Mg, Mn, Mo, Ni and Si were also controlled.

3.2 Purification Process

The purification process is based on the preferential reaction of oxygen with calcium , which gives calcium oxide totally insoluble in sodium. The next step is the separation.

Table 2. Specification of sodium for MONJU

Element	ppm maximum	ppm
O	< 30	< 10
H	< 5	< 1
C	< 30	
Cl	< 30	
B	< 4	
K	< 300	
Ca	< 10	
Li	< 10	
U	< 0,01	

The oxide formed is very fine and separation is carried out through different steps of settling and filtration. The final filtration is done using micro filters made of sintered stainless steel.

The sodium at this stage is pure and oxide free, but before shipment, it is put into a storage tank and sampled for quality control. As we want to take the utmost care about quality, we have installed an additional purification control system including a cold trap and a plugging indicator. These devices developed by CEA have been used for many years and are considered very accurate.

The quality control before shipment is performed by analysis of all specified elements, of a sample taken from the bulk. It will be described later.

4. TRANSPORTATION - STORAGE - DELIVERY

4.1 Sodium Transportation

The organisation of delivery of bulk sodium to Japan, although already performed since four years by Métaux Spéciaux, has to take account of heavy constraints. The trip takes 4 full weeks by sea, the shipping companies accept only 4 containers per ship, exceptionally 8, at Kobe port in Japan.

4.2 Tank Containers

Métaux Spéciaux has developed a special container for overseas sodium transportation. Sodium is transported cold and in solid state in a tank equipped with oil heater for sodium melting and all necessary fittings to allow safe unloading in good quality conditions. The container has been developed by Métaux Spéciaux in co-operation with French partner BSLT, tank containers constructor, and approved by PNC.

4.3 Sodium Unloading Facilities

Before delivery to the MONJU facility, the sodium is unloaded into storage tanks, in order to keep permanently at hand the requested quantity for transfers of sodium to intermediate vessels and also to make possible an additional purification in the storage tanks if needed.

Fig. 1. Container for overseas transport of sodium
It has been designed for a working pressure of 4 bar - 60 psig
working temperature 130°C (266 °F); capacity 18,5 Mt - 40 800 lbs
transportation agreements world-wide: road, train, sea.

4.3 Sodium Unloading Facilities

Before delivery to the MONJU facility, the sodium is unloaded into storage tanks, in order to keep permanently at hand the requested quantity for transfers of sodium to intermediate vessels and also to make possible an additional purification in the storage tanks if needed.

The facilities were built and operated by a consortium of Japanese partners and Métaux Spéciaux. The consortium, co-ordinated by NIC (Nissho Iwai Corporation), included HZC (Hitachi Zosen Corporation), SDK (Showa Denko K.K.) and PJ/MSSA (Pechiney Japan/Métaux Spéciaux S.A.).

The quality control was performed through a plugging temperature measurement; the plugging indicator was operated in a special way so as to measure the saturation temperature and know the amount of oxygen and hydrogen. With sodium unloaded at 120°C (250°F) then heated at 160°C (320°F) and circulating at this temperature, the saturation temperature could never be measured because it was lower than the limit of 110°C (230°F).

Under these conditions the cold traps fitted on the tanks have never been used except for tests during the operation of the facilities.

4.4 Oxygen and Hydrogen Content vs Temperature

The amount of dissolved oxygen and hydrogen in liquid sodium are given by the following experimental laws, where C is the mass concentration in ppm and T the absolute temperature in K:

OXYGEN CONCENTRATION IN LIQUID SODIUM [1]

$$\log_{10}(C_{ppm}) = 6.250 - 2444.5 / T_K$$

HYDROGEN CONCENTRATION IN LIQUID SODIUM [2]

$$\log_{10}(C_{ppm}) = 6.467 - 3023 / T_K$$

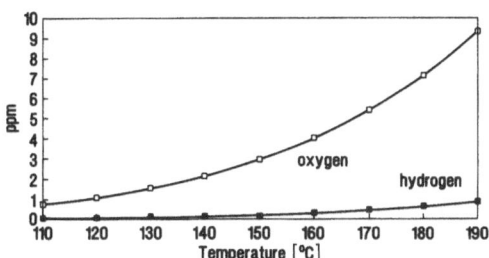

Fig. 2. Saturation concentrations of oxygen and hydrogen in sodium after eqs. (1) and (2)

4.5 Sampling and Analyses

Beside the plugging measurement, sampling was carried out before every transfer to the MONJU facility and samples were analysed.

5. SAMPLING METHOD

The samples are contained in stainless steel U-shaped tubes about 1 cm diameter. The tubes are flushed with sodium during 24 to 48 hours in order to clean the surface as much as possible. Then the flow is stopped and the sample is quenched by air blower to solidify and cool the sodium in 2 minutes. This sampling method was recommended by PNC.

The sampling equipment is designed as shown in Fig.3.

One sample is made of two tubes containing about 100 g of sodium each. After cooling, the tubes are removed quickly while metallic plugs are screwed on at both ends to prevent oxidation. Such samples can be kept for a long time and easily transported to any laboratory.

Before analysis a "secondary sampling" is carried out in a glove box to get pieces of sodium to be analysed. In the glove box, the requested quantity of sodium is extruded by pressing the tubes as one does for tooth paste but with a hydraulic tool. The first centimetre is rejected then the extruded rod is cut into pieces for analyses. Another way of recovering the sodium is to melt it by external resistance heating; this is the case for oxygen analysis, because it is necessary not to leave oxides of sodium, if any, on the inner surface of the tube.

The preparation of the tubes before sampling is decisive because the analytical result depends also on the residual impurities remaining on the surface. From our experience it is obvious a flushing temperature of 140°C (285°F) is not enough to ensure that oxygen is desorbed and oxides are dragged along.

We noticed that a strong pickling followed by humid air exposure generates iron oxides and more generally metal oxides which remain on the inner surface of the tubes.

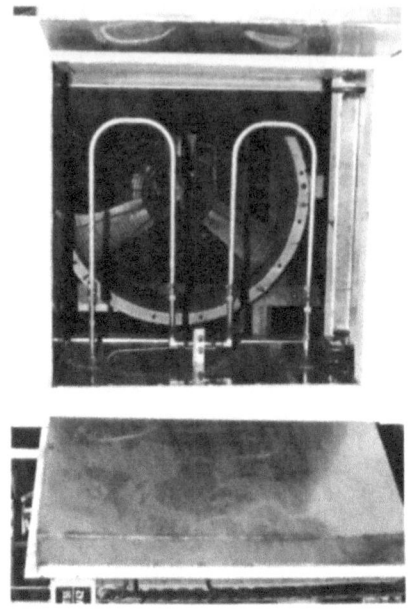

Fig. 3. Sampling device for taking samples from containers

6. ANALYTICAL PROCEDURES

The analytical procedures used for quality control have been decided jointly by PNC and Métaux Spéciaux. The general flow sheet is as follows:

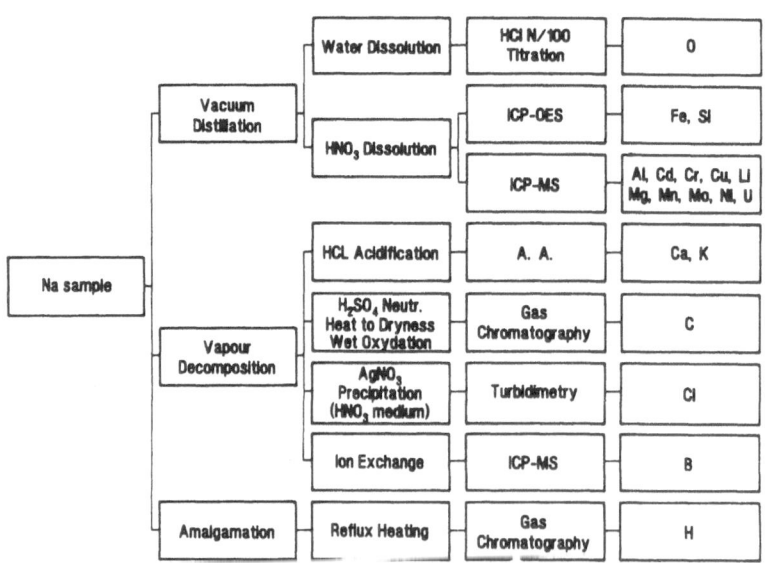

Fig. 4. Analytical procedures used for MONJU sodium

The sensitivity of analyses of O and metallic elements has been improved by concentration of impurities. The distillation of the sample under vacuum evaporates metallic sodium, and the remaining residue contains non volatile elements or compounds such as oxides and metallic elements. The detection limit can be lowered to 0.05 to 0.1 ppm depending on the elements.

In order to prevent contamination of the samples, the glove box used for sample handling is supplied with high quality argon purified through an efficient purification unit able to guarantee about 1 ppm of oxygen and 1 ppm of moisture.

The residue is analysed by Induction Coupled Plasma Spectrography, Optical Emission Spectrography for Fe and Si, Mass Spectrography for others.

Amalgamation has been used for hydrogen; this method has a detection limit of 0.05 ppm.

7. ANALYTICAL RESULTS

The results of chemical analyses of 92 samples of the sodium shipped for MONJU are listed in Table 3.

Table 3.   Mean values of contents of impurities in sodium, global results from the analyses of 92 samples before shipment

Element	O	H	C	Cl	B	K	Ca	Li	U
Specification	< 10	< 1	< 20	< 30	< 4	< 300	< 10	< 10	< 0,01
Average	2,8	0,10	7	3	0,5	163	1,7	0,01	0,001
STD DEV	1,5	0,04	3	1	0,06	27	0,8	0,001	0,000
Maximum	7	0,30	17	7	0,9	215	4,9	0,02	0,001
Minimum	1	0,05	5	1	0,2	118	0,2	0,01	0,001

Figs. 5-7 present the results of all the samples as bargraphs for oxygen, hydrogen and calcium which are the major impurities to be removed.

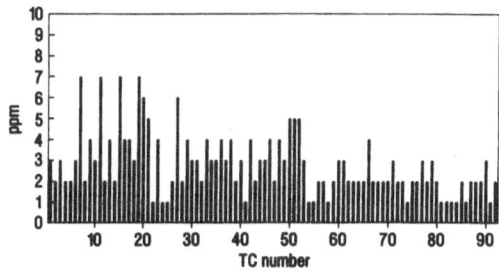

Fig. 5. Oxygen contents of 92 samples of MONJU sodium

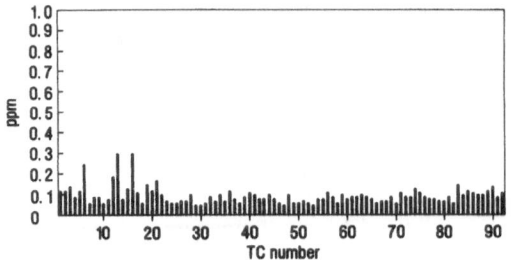

Fig. 6. Hydrogen contents of 92 samples of MONJU sodium

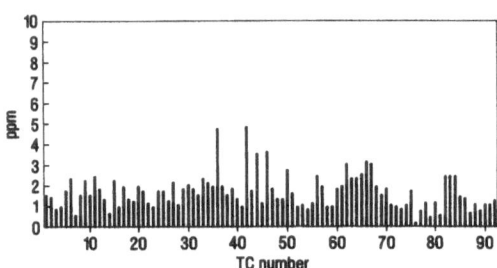

Fig. 7. Calcium contents of 92 samples of MONJU sodium

References

1. J.D. Noden, J. Brit. Nucl. Energy Soc. 12 (1973) 57-62 and 329-331.
2. A.C. Whittingham, J. Nucl. Mater. 60 (1976) 119-131.

EXPERIMENTAL INVESTIGATION OF AN AUTOMATIC PLUGGING METER FOR MEASURING IMPURITIES IN SODIUM

Hong Shunzhang, Zhang Zhong*,
Zhang Jianmin, Sang Weiliang, Qian Chengyao,
Zhou Xuezhi, Li Chongxiang, Xu Weinan**

* China Institute of Atomic Energy
P.O.Box 275(53), Beijing, China
** Xi`an Jiaotong University

1. INTRODUCTION

Plugging meters can be operated manually and automatically. The manual plugging meter was reported before [1]. Many automatic plugging meters have been reported in previous papers [2-6]. In the past few years an automatic plugging meter has been developed and operated in the mode of both programme-controlled and vibration using an IBM computer.

Impurities deposited in the plugging orifices mainly are oxygen and hydrogen, the precipitating amount of the impurities depends on species of the impurity, difference of concentration of impurity (c), temperature at the plugging orifices and deposition time.

In order to reduce the deviation between plugging temperature and saturation temperature of impurities, we have tried to diminish the cooling rate at the orifices, thus deposition of the same appreciable amount of impurities can be obtained under conditions of a longer deposition time. A programme which takes advantage of the temperatures at the orifices to control the cooling rate of the blower is compiled. The lower the temperature at the orifices is, the lower is the cooling rate. When the temperature at the orifices is 383 K, the cooling rate is 0.5 K/min.

2. EXPERIMENTAL DEVICE AND PRINCIPLE

The automatic plugging meter was designed by ourselves and tested in the Multipurpose Sodium Purification Loop [7]. It can be operated in the mode of both programme- and vibration-controlling. The former one is as follows: the plugging meter consists of the main body of plugging meter, flow meter, filter, economizer, blower, electric and steering computer etc..

Fig. 1 shows its flowsheet. The schematic diagram of the steering computer for the automatic plugging meter is shown in Fig. 2. Heating of the main body of the plugging meter and lifting of the central rod are controlled by on-off output unit. The blower of the plugging meter is controlled by module output unit. The rotating speed of the blower can be regulated from 500 cycl/min to 300 cycl/min.

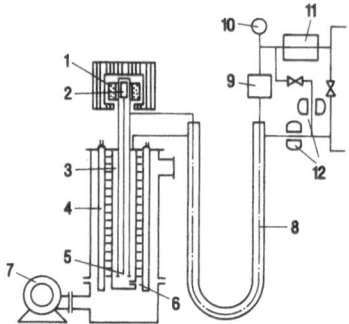

Fig. 1. Schematic of the automatic plugging meter

1.Electromagnetic inductive coil;
2.Iron core;
3.Cooler;
4. Electrical heating element;
5.Plugging orifice;
6.Thermo-couple;

7.Blower;
8.Heat Exchanger;
9. Filter;
10.Sodium pressure gauge;
11.Electromagnetic pump;
12.Electromagnetic flowmeter;

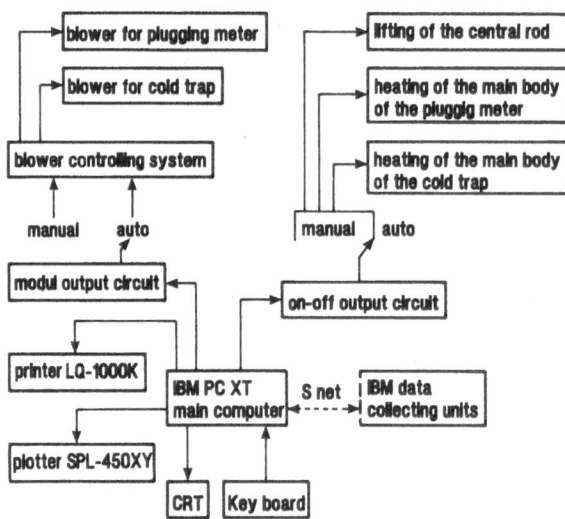

Fig. 2. Schematic diagram of the steering computer for the automatic plugging meter

The software of the steering computer for the plugging meter was compiled by means of an advanced IBM BASIC language and the IBM MACRO-ASSEMBLY language. It can be found that the blower will start to work as soon as the flow through the orifices falls inside the limits of 5.96 0.05 · 10^{-2} m^3/h after the operation mode is selected, the bypass flow of the plugging meter is 0.210 0.02 m^3/h, and the rotating speed of the blower changes with the temperature of the plugging orifices. The temperature of the plugging orifices is low, so is the rotating speed. This makes it possible to measure low plugging temperatures precisely and to avoid the solidifying of sodium in the plugging orifices. When the flow through the orifices is reduced to 0.98 · 5.95 · 10^{-2} m^3/h, the screen will show the plugging temperature immediately. According to the formula of the impurity solubility, the concentration of O and H can be calculated and printed, it is assumed that O and H have same plugging temperatures and the plugging temperatures are very close to the saturation temperatures.

In the meantime, the plugging orifices are kept in cooling condition until a plug of 10 % in the orifices is reached. Up to this point, a cycle of operation of the plugging meter is finished, the unplugging will begin. In order to enhance the dissolution of impurities in the orifices, the blower ceased to work, the heating elements begin to work and the rod is lifted. After 10 minutes, the impurities in the orifices are entirely dissolved, the next measuring cycle may then be started. The results measured can be both printed and drawn.

3. EXPERIMENTAL RESULTS AND DISCUSSION

The sodium in the main loop of Multipurpose Sodium Purification Loop is circulated at flow rate of 0.8 m^3/h at the temperature of 650 K. The cold trap temperature is automatically controlled at preset point by the computer and the temperature controlling deviation is ñ 1 K, the flow through the cold trap is 7 ·10^{-2} m^3/h.

3.1 Programme-Controlled Automatic Plugging Meter

In order to comprehend the conditions of impurity measurement at different impurity levels by means of the plugging meter, five impurity levels, namely five different cold trap temperatures have been tested. In a sense, the plugging meter is calibrated by means of controlling cold trap temperature. The constant impurity concentration is obtained by keeping the cold trap temperature on the preset value for 5 hours. Figs. 3-7 indicate flow-temperature curves of the programme-controlled automatic plugging meter at different cold trap temperatures. The plugging meter works very well, even if the plugging temperature is as low as 318 K. Nevertheless the temperature at the orifices is low, the rotating speed is low also, thus the plugging meter has enough time to cause orifices 2 % and even 10 % plug.

Fig. 8 shows plugging temperatures at different cold temperatures measured by both manual plugging meter and automatic plugging meter. It can be seen that the results from the automatic plugging meter are close to the higher plugging temperatures obtained from manual plugging meter. This is due to the fact that the plugging temperature measured by automatic plugging meter is obtained under the conditions of 2 % plug in the orifices. These results are lower than those under the conditions of just starting plugging in the orifices. The average deviation listed in Table 1 is 3.5 0.8 K, this deviation is equivalent to a difference of 1 ppm oxygen impurity at the plugging temperature of 473 K.

Fig. 8 also shows that the cold trap temperature is about 20 K higher than the corresponding plugging temperature. The reason has been found from the experiment. It is known that the distribution of the temperatures is not the same in radial direction. The cold trap is usually cooled by the blower and the temperature near the inner wall of the lower part of the trap cylinder is lower than that near the centre where the measuring point of cold trap temperature is located. This difference which is shown in Table 2 is about 11 K. In order to verify it, another test has been done. When the cold trap works, and the blower keeps in static condition to make the difference of the temperatures at radial direction smaller, then higher plugging temperatures were measured. The results are shown in Table 3.

In general the plugging temperatures are close to the coldest temperatures in the cold trap. It proves the plugging meter is an accurate impurity on-line instrument.

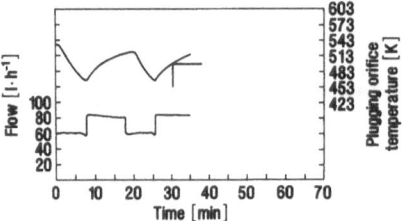

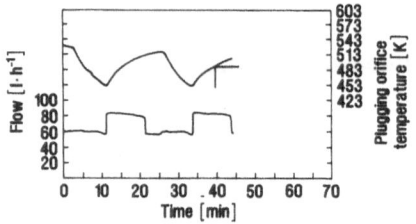

Fig. 3. Temperature and flow curves of the plugging meter at 498 K at the temperature of the cold trap

Fig. 4. Temperature and flow curves of the plugging meter at 473 K at the temperature of the cold trap

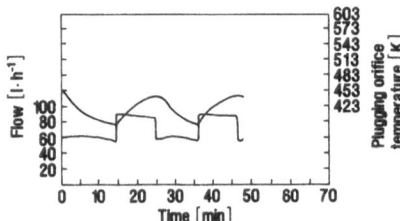

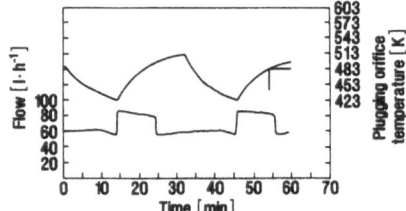

Fig. 5. Temperature and flow curves of the plugging meter at 448 K at the temperature of the cold trap

Fig. 6. Temperature and flow curves of the plugging meter at 423 K at the temperature of the cold trap

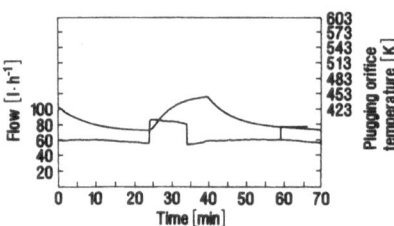

Fig. 7. Temperature and flow curves of the plugging meter at 413 K of the cold trap

3.2 Vibration Automatic Plugging Meter

In the programme-controlled automatic plugging meter, it is necessary that lifting of the central rod in the plugging meter by means of an electromagnetic induction device enhances the dissolution of the impurities in the orifices. However, during the work of this device, vibration noise occurs which may affect the life time of the plugging meter. Therefore, we tried to find another operating mode, namely, the "vibration mode". In this mode, it is unnecessary to lift the central rod, while the orifices is unplugged. The unplugging is mainly caused by stopping the blower and increase the heating power. Fig. 9 shows the flow-temperature curve measured by the vibration automatic plugging meter at the cold trap

178

temperature of 413 K. This curve allowed to observe the course of impurity concentration change during impurity adding from the cold trap. The plugging meter took continuous measurement of plugging temperatures. The result is shown in Fig. 10. It is obvious that this operation mode is perfect.

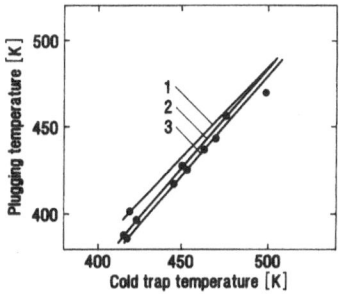

Fig. 8. Plugging temperatures at different cold trap temperatures

Table 1. Effect of different plugging levels at the plugging orifices on determination of plugging temperatures

cold trap temperature /K	plugging temperature of 2% plug in the orifices / K[1]	plugging temperature of 2% plug in of starting plugging in the orifices /K[1]	difference / K
498	467.5	471	3.9
473	455.4	458	2.6
448	427.3	430	2.7
423	392.5	399	6.5
413	385.4	388	2.6

[1] average value of two measurements

Table 2. Radial temperature distribution at bottom of the cold trap

central temperature of cold trap bottom / K	inner wall temperature of cold trap / K	difference / K
130.7	119.9	10.3
143.1	131.0	12.1
148.1	137.1	11.0

Table 3. The plugging temperature measured at different temperatures of the cold trap without cooling by the fan

cold trap temperature / K	high plugging temperature / K
435	431
431	425
430	423

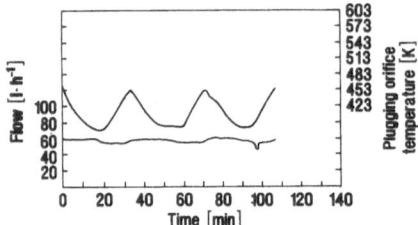 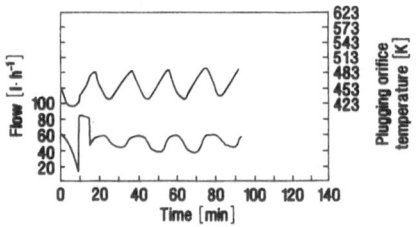

Fig. 9. Temperature and flow curves of the vibration mode automatic plugging meter at 413 K of the temperature of the cold trap

Fig. 10. Temperature and flow curves of the vibration mode automatic plugging meter in the course of change of the cold trap temperature

Acknowledgements

The authors wish to thank Mr. Xing Chaoqing and Mr. Jia Zaozhong for assemblying the plugging meter and Miss Chao Yaping, Messrs. Shen Fenhyang, Cao Zeng and Yu Chunli for operating the sodium loop.

REFERENCES

[1] Hong Shunzhang, "Investigation of A Manual Plugging Meter for Measuring Impurities in Sodium", Proceeding of the Fourth International Conference on Liquid Metal Engineering and Technology, Vol.3,17-21 October 1988.

[2] R. Hans and H.-J. Weiss, "Sodium Contamination Measurement with Plugging Meter and Hydrogen Leak Detection System", Siemens Review XL 11,No.5, 1975.

[3] Hoschour, W.M. and Gasey L.F., "Design and Thermal Analysis of Sodium Specialities",Components for HNPF, USAEC Report, NAA-SR-5445 Atomic International, Feb. 15, 1961.

[4] Davis, K., "Development of a Rapid Operation Plugging Meter", RLAA-SR-4573, 1959.

[5] Davison, D. F., Roch, P. F., "An Experimental Continous Indication Plugging Meter For Impurity Monitor in Liquid Metal", UKAEA, TRG Report 1640(R),1968.

[6] Delisle, J.P., "Automatic Measurement of Plugging Temperature of Na and NaK", The European Atomic Energy Society Colloquim on Liquid Metal, Aix-en-provence, 30 Sep. to 2 Oct., 1963.

[7] Xin Chaoqing and Hong Shunzhang et al., "Design of Multipurpose Sodium Purification Loop", China Fast Reactor Special Issue (part 1), p.37, 1991.

HIGH PURITY ALKALI METAL (SODIUM)
AS CHEMICAL-THERMAL MEDIUM

O.V. Starkov

Institute of Physics and Power Engineering
249020 Obninsk, Russia

The properties of sodium make the metal interesting for a number of remarkable applications. Liquid sodium of reactor grade purity does not show any adsorption activity. Furthermore, it does not cause any degradation of structural materials in respect to their mechanical properties. This statement was proved for steels in the course of a large number of investigations. The statement is also valid for refractory alloys as for instance molybdenum-based alloys.

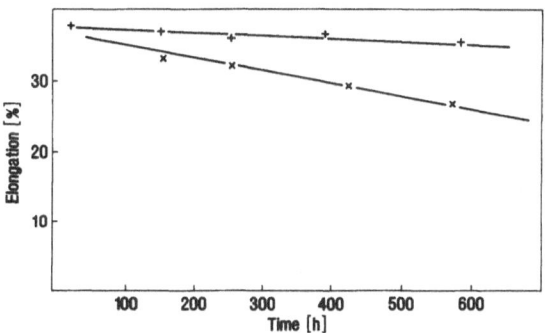

Fig. 1. Change of length of a molybdenum alloy aged at 900 °C in sodium (1) and in vacuo (2)

Sodium has good reducing properties as well as low partial pressure. The reduction of oxide films on steel starts at 300 - 350 °C. Sodium reduces fluorides and chlorides, a fact which is already used for the production of several metals (e.g. production of rare earth metals).

Sodium has a very low capacity to dissolve the main alloying components of steels as chromium, iron, tungsten, nickel. Corresponding to the low solubilities, concentrations of these transition elements do not exceed 0.1 - 10 ppm. Thus, steels and alloys do not corrode due to thermal-chemical treatment in liquid sodium.

Liquid Metal Systems, Edited by H.U. Borgstedt
and G. Frees, Plenum Press, New York, 1995

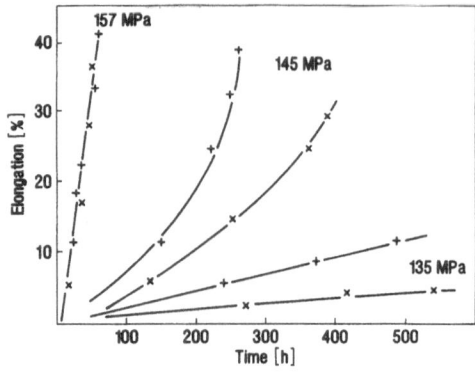

Fig. 2. Creep curves of molybdenum alloy at 900 °C in sodium (+) and in vacuo (x)

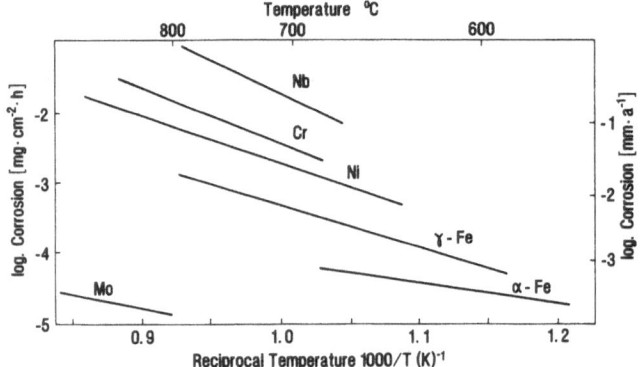

Fig. 3. Corrosion rates of transition metals in flowing sodium:
γ-Fe - austenitic steel
α-Fe - ferritic steel

Uniform heating may be obtained in systems with liquid sodium, since the thermal conductivity of the metal is high. This fact is important for the application in thermal-chemical treatment processes.

Vapour and liquid of sodium or lithium promote the nitriding of metals, since the presence of the liquid metals significantly reduces the activation temperature of the nitrogen molecule.

Intensive carbon transfer occurs in liquid metals as sodium or lithium, the rate of transfer might be controlled and adjusted by the parameter maintained in the liquid metals.

Liquid sodium is a medium in which intensive hydrogen transfer may take place. This is important for the technology of refractory metals as niobium, zirconium or titanium.

The chemical technology of the production and application of high-purity sodium has reached commercial level.

The problem of fire hazards connected with the use of alkali metals can be minimised by means of principally approved measures. It is experimentally established that the sodium-potassium-caesium alloy oxidises without the formation of aerosols, i.e. fire without smoke.

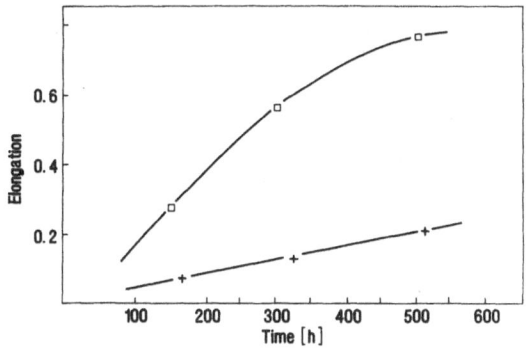

Fig. 4. Change of ductility of steel 10X16H15M3 during nitriding at 700 °C
with p_{N_2} =0.032 MPa in sodium (1) and in the vapour phase (2)

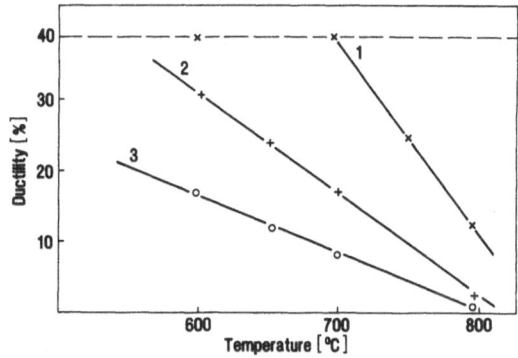

Fig. 5. Ductility of steel 10X16H15M3 after nitriding with p_{N_2} = 2.1 MPa in 200 h at 700 °C
1 - in nitrogen atmosphere
2 - in nitrogen over sodium
3 - in sodium under nitrogen

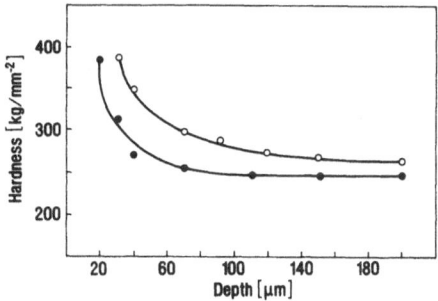

Fig. 6. Micro structure and micro hardness of austenitic steel 10X16H15M3 (700 °C, 4 h,)

183

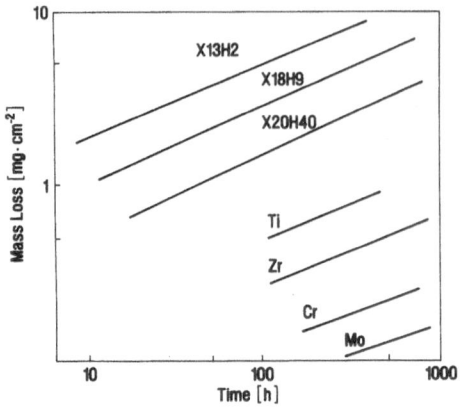

Fig. 7. Absorption of carbon by steels and several transition metals at 700 °

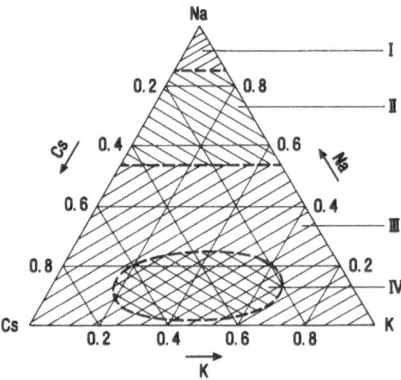

Fig. 8. Aerosol content of the atmosphere above burning Na-K-Cs alloy

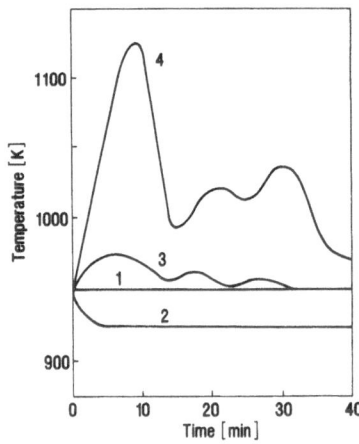

Fig. 9. Comparison of aerosol formation above the Na-K-Cs eutectic alloy and Na:
1-temperature of alloy   3-temperature of sodium
2-temperature of atmosphere above the alloy   4-temperature of atmosphere above sodium

CONCLUSIONS

The properties of liquid sodium are favourable for the application as a chemical-thermal medium

1. The technology of handling and application of sodium metal is developed.
2. The excellent thermal conductivity guaranties uniform heating of volumes of molten sodium.
3. Sodium does not cause a degradation of mechanical properties of several metallic materials.
4. The corrosivity of sodium against transition metals is negligible up to high temperatures.
5. Sodium reduces compounds of transition metals as oxides, chlorides or fluorides and may be applied in metallurgical processes.
6. Sodium enables nitrogen to form metal nitrides.
7. Sodium has the capacity to transfer carbon between different materials.
8. Relative stable hydride is formed between sodium and hydrogen.
9. A Na-K-Cs alloy of certain composition does not develop aerosol at heating to high temperatures ("fire without smoke").

AN ELECTROCHEMICAL HYDROGEN METER FOR USE IN SODIUM - BASIC STUDIES AND OPERATIONAL EXPERIENCE

R. Sridharan, K.H. Mahendran, T. Gnanasekaran, G. Periaswami
and C.K. Mathews

Materials Chemistry Division,
Chemical Group, Indira Gandhi Centre for Atomic Research,
Kalpakkam - 603 102, India.

1. INTRODUCTION

In the fast reactor steam generator, high-pressure steam and high-temperature sodium are separated by a single wall. Any manufacturing defect in this can lead to contact between sodium and water and the consequent sodium-water reaction. Caustic produced in the exothermic reaction can attack the tube wall and cause rapid expansion of the leak, resulting in sodium-water explosions [1]. Thus it is necessary to detect these leaks at the very inception for the safe operation of fast reactors. This is best done by monitoring the sodium at the steam generator outlet for its hydrogen concentration. The conventional hydrogen leak detection systems are based on hydrogen diffusion through nickel membrane. This system is complex involving many components and is also expensive. On the other hand, electrochemical hydrogen meter originally proposed by Smith [2,3] appears inherently simple and truly on-line. Development and testing of such a meter at Kalpakkam has already been reported [4]. This paper describes further work done at Kalpakkam on the meter testing and on basic properties of the electrolyte.

2. PRINCIPLES OF ELECTROCHEMICAL HYDROGEN METER

The electrochemical hydrogen meter developed in our laboratory, is shown in Fig. 1 and is represented below :

$$[H]_{Na} \parallel CaCl_2\text{-}CaH_2 \parallel Li\text{-}LiH \text{ (ref)} \tag{1}$$

The EMF of the cell is given by

$$E = (RT/2F) \ln pH_2(ref)/pH_2(sample) \tag{2}$$

where : E = EMF in volts
R = Universal gas constant, Joule mol^{-1} K^{-1}
F = Faraday's constant, 96487 C mol^{-1}

Liquid Metal Systems, Edited by H.U. Borgstedt
and G. Frees, Plenum Press, New York, 1995

pH_2 (ref) = Partial pressure of hydrogen in reference Li-LiH electrode

pH_2 (sample) = Partial pressure of hydrogen in sample electrode

The hydrogen concentration in sodium is obtained from the emf of the cell and using Sievert's law for the solubility of hydrogen in sodium:

$$1/2H_2 \Leftrightarrow [H]_{Na} \qquad (3)$$

$$[C]_{Na}^H = K \ pH_2^{1/2} \qquad (4)$$

where : $[C]_{Na}^H$ = concentration of hydrogen in sodium, in ppm

pH_2 = equilibrium pressure of hydrogen in sodium, in Pascals, Pa

K   Sievert's constant, in ppm Pa$^{-1/2}$

Smith used $CaCl_2$ with CaH_2 dissolved in it as electrolyte [2,3] and found it to yield >95% theoretical response in the temperature range of 653 to 823 K when used with Li/LiH reference electrode. However, the electrical conductivities of this biphasic electrolyte as well as the transport number for hydride ion in it have not been measured so far. A detailed knowledge of these properties would be required for complete exploitation of this electrolyte for measurement of low levels of hydrogen in liquid metals as well as in gas streams. A detailed phase diagram study of this system have also not been established, although the schematics are available from an earlier work [5,6]. This paper reports results of our ongoing work in these measurements as well as our experience with such meters in operating sodium systems.

3. EXPERIMENTAL

3.1 Phase diagram studies

$CaCl_2$ (Pro Analysi grade, E Merck, Germany) used in this work was purified by treating it with HCl as described earlier [4]. CaH_2 was prepared by hydriding calcium metal contained in an iron thimble welded to a sealed stainless steel tube. The thimble was heated in a flowing hydrogen gas stream at 923K for about 50 h. Hydrogen diffusing through iron thimble reacts with calcium metal and forms calcium hydride. Calcium metal for this purpose was prepared by distilling 99.95% metal under vacuum at 1173 K. Desired compositions of $CaCl_2$ and CaH_2 were mixed inside an argon atmosphere glove box, approximately 50 mg of the mixture is taken in a completely sealed stainless steel crucible and used for the DSC studies. DSC-111 of Setaram, France was used for this purpose. Argon containing 0.11% H_2 was used as the protection gas during the run. CaHCl is reported to have a decomposition pressure of 10^{-3} torr at 903 K [5,6] and use of this protective gas would not lead to decomposition of CaHCl during the DSC experiments. Heating rate of 1K per minute was employed for these experiments.

4. CONDUCTIVITY MEASUREMENT

The composition of the electrolyte chosen for conductivity measurements was $CaCl_2$-5.88 mol % CaH_2. The electrolyte was prepared from appropriate quantities of CaH_2 and purified $CaCl_2$. Magnesium was found to be the major impurity in the electrolyte prepared and analysis by atomic absorption spectrophotometry showed its concentration to be 575 ppm.

The geometry of the electrolyte assembly used for conductivity measurements is shown in Fig. 2. The electrolyte is contained in a thin walled iron cylinder (9 mm O.D., 60 mm long and 0.5 mm thickness). An iron rod of 4 mm dia and 60 mm long is positioned at the centre of the electrolyte and this served as another electrode. Total conductivity of the biphasic electrolyte in this assembly was measured by a Solartron Frequency Response Analyser model no. SI-1255. Measurements were made at temperatures of 673 to 748 K in steps of 25 K. The gas environment was argon containing 95 ppm and 4.85 % H_2 in the experiments.

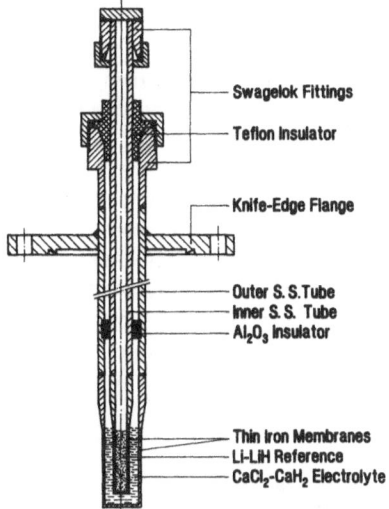

Fig. 1. Schematics of the electrochemical hydrogen meter

5. TESTING OF METERS WITH GAS REFERENCE ELECTRODES

The meters with Li-LiH have relatively high temperature coefficient. Theoretical value of this coefficient at a meter operating temperature of 723 K and at a cold trap temperature of 393 K is 1.47 mV/K. This calls for a very good control of the sodium temperature in an operating plant. As gas reference electrodes are expected to have lower temperature coefficients, meters were tested in a bench top sodium loop with argon gas having 1100 and 95 ppm of hydrogen as reference. The meter used for this purpose had a configuration identical to the one shown in Fig. 1, except that it did not have the Li-LiH reference. The reference gas, after passing through molecular sieves and reduced form of BASF-R3/11 catalyst with a view to remove oxygen and moisture, was passed over inner iron membrane. The meters were operated at 723 and 673 K and their output measured as a function of cold trap temperature. The bench top loop used for this purpose is described in reference 4.

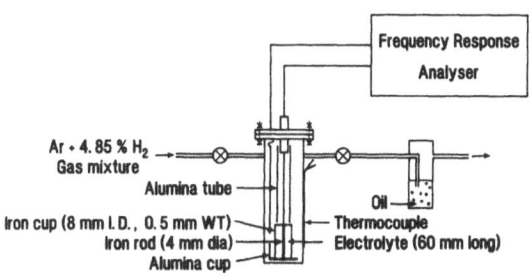

Fig. 2. Experimental set-up used for conductivity measurements

6. TESTING OF METERS IN REACTOR LOOPS

Four electrochemical hydrogen meters were incorporated in the secondary sodium circuits of the fast breeder test reactor at Kalpakkam. The meters were calibrated in bench top sodium loops prior to installation in the reactor. Fig. 3 shows the calibration graphs for these meters as a function of reciprocal cold trap temperature. Of these four meters, performance of two meters had deteriorated with time after a few cycles of sodium dumping and refilling. Electrolyte cracking is suspected to be main reason for this behaviour. The other two meters have been working very satisfactorily for more than two years in the sodium circuit. Qualification of these meters for detection of a steam leak as well as their in-loop calibrations were performed by injecting hydrogen into sodium maintained at temperatures ranging from 453 to 673 K [7]. Injection of hydrogen gas was effected using an injection facility available in the steam generator circuit. These experiments involved injection of an appropriate amount of hydrogen gas needed to increase its concentration in sodium by 50 ppb. The meter response as a function of time after injection of hydrogen was recorded.

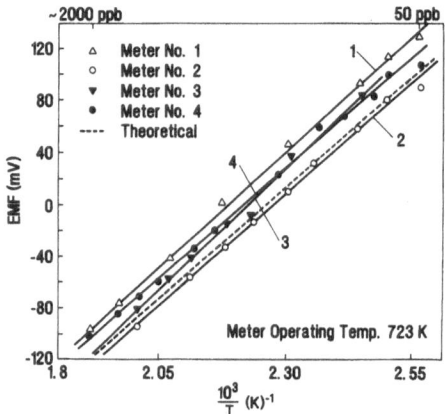

Fig. 3. Meter output as a function of reciprocal cold trap temperature, T = 723 K

7. RESULTS AND DISCUSSION

The phase diagram elucidated from the present DSC experiments is shown in Fig. 4. The eutectic temperature deduced from the experimental results is 893 K and this differs by 20 K from the value reported by Ehrlich et al [5,6].

The eutectic composition, 21 mol % CaH_2, shown in the above figure has been obtained by extrapolation of the liquidus line. Ehrlich et al reported 973 K as the melting point of CaHCl whereas our results indicate a higher temperature. More experiments are needed to confirm these observations as well to construct the other part of the phase diagram.

A typical impedance plot, imaginary Z_{im} (capacitive) against real Z_{re} (resistive) impedances, for measurement at 723 K in argon-4.85% hydrogen environment is shown in Fig. 5. Impedance of the electrolyte is obtained from the intersect of the arc of the circle with Z_{re} axis.

This was used to calculate the bulk conductivity of the electrolyte and the effective area of the electrolyte in the geometry used was calculated as described in reference [8]. Conductivity values obtained from the complex impedance plots as a function of temperature were used to calculate the activation energy from Arrhenius equation. The plot of log (σ T) against 1/T is shown in Fig. 6 and is expressed as below:

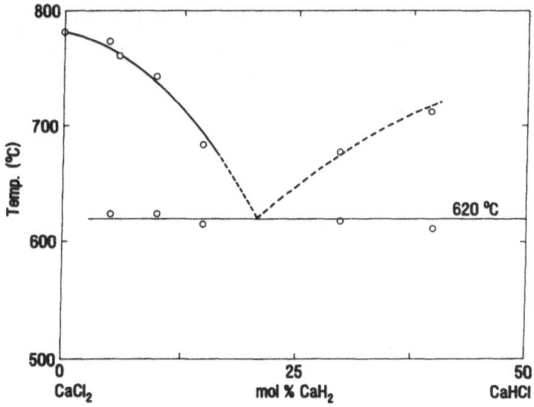

Fig. 4. Phase Diagram of CaCl$_2$-CaH$_2$ system

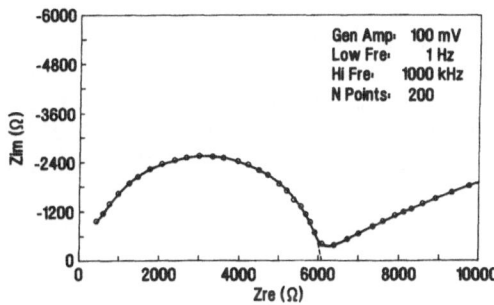

Fig. 5. Impedance plot for CaCl$_2$-CaH$_2$ electrolyte; T = 723 K; Ar - 4.85 % H$_2$ environment

$$\log (\sigma T) = -4.533 \cdot 10\text{-}3/T + 3.729 \qquad (5)$$

where : σ = the total conductivity (Siemens cm^{-1}) and
T, = the temperature in Kelvin.

The activation energy derived from this result is 0.17 eV. Similar measurements in argon with 95 ppm hydrogen environment did not show any significant change in conductivities.

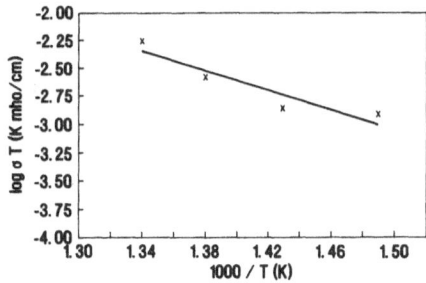

Fig. 6. Arrhenius plot for conductivity of CaCl$_2$-CaH$_2$ electrolyte

The meter output as a function of hydrogen in sodium for two different gas references is shown in Fig. 7. The meter out-put were much lower than the theoretical values. The reasons for this observations are not clear, though theoretical behaviour under these conditions are expected.

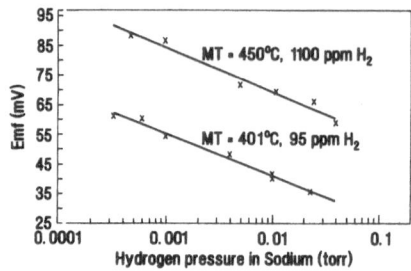

Fig. 7. Output of the electrochemical hydrogen meter with gas phase reference electrodes

The response of the electrochemical hydrogen meters for injections of hydrogen in sodium loop are shown in Fig. 8. The meters responded well to hydrogen injections and the decrease in EMF after each injection was quantitative. The time after which the meter responded could be correlated to the time taken by sodium from the injection port to reach the meter module in the circuit. However, the meters did not respond to injection of hydrogen into sodium at 453 K. Kinetics of hydrogen dissolution is very low at these temperatures and the injected gas would escape to the argon cover gas.

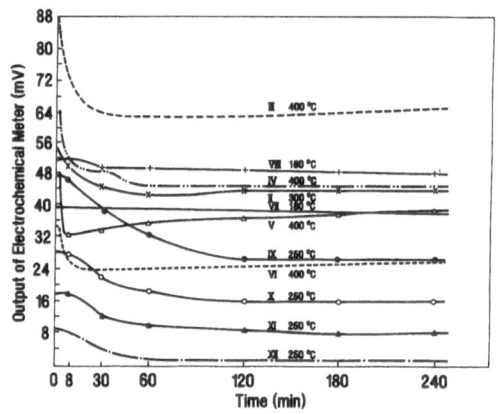

Fig. 8. Response of the meter for in-loop hydrogen injection

References

1. R. Hans and K. Dumm , Atomic Energy Review, 15 (1977) 611.
2. C. A. Smith, in Proc. of the Int. Conf., on Liquid Alkali Metals, British Nuclear Energy Society, Nottingham (1974) p.101.
3. C. A. Smith, Central Electricity Generating Board, U.K., Report, R/D/BN-2331 (1972).
4. T. Gnanasekaran, K. H. Mahendran, R. Sridhran, V. Ganesan,G. Periaswami, and C. K. Mathews, Nucl. Technol.,90 (1990) 408.
5. P. Ehrlich and H. Gortz, Z. Anorg. Allgem. Chem., P 288 (1956) 148.
6. P.Ehrlich and H. Kulka, ibid, 288 (1956) 156; L. Gentoch, ibid, 283 (1956) 58.

7. Chande, S. K., and Rajendran, B., Soviet-Indian Seminar On Steam Generat.ors For Fast Reactors, USSR, Obninsk, October 7-14,1990

8. Friedman, L.M., Oberg, K.E., Boorstein, W.M., and Rapp, R.A., Met. Trans., $\underline{4}$ (1973) 69.

MONITORING THE COMPOSITION OF LIQUID Pb-17Li: A STATE - OF - THE - ART APPRAISAL

P. Hubberstey

Chemistry Department
University of Nottingham
Nottingham, NG7 2RD,
England, United Kingdom

1. INTRODUCTION

The lithium-lead eutectic alloy, Pb-17Li, has been proposed as liquid breeder material in both water-cooled [1-5] and self cooled [3, 6, 7] blanket designs for future fusion reactors because of its relatively low freezing point (508 K; 235 °C) [8], its relatively low reactivity towards water [9, 10] and air [9] and its intrinsic neutron multiplication [11].

The chemical and physical properties of Pb-17Li have been extensively studied during the past decade and are now well documented. [12-15]

Tritium generation {scheme 1; [13]} and accidental ingress of air or water {scheme 2 [9]}:

$$^6Li_3 + {}^1n_0 \text{ (slow)} \rightarrow {}^3H_1 + {}^4He_2$$
$$^7Li_3 + {}^1n_0 \text{ (fast)} \rightarrow {}^3H_1 + {}^4He_2 + {}^1n_0 \text{ (slow)}$$

scheme 1

$$Li_{(Pb-17Li)} + 1/4\ O_2(g) \rightarrow 1/2\ Li_2O(s)$$
$$Li_{(Pb-17Li)} \text{ (excess)} + 1/2\ H_2O(g) \rightarrow 1/2\ Li_2O(s) + 1/2\ H_2(g)$$
$$Li_{(Pb-17Li)} + H_2O(g) \text{ (excess)} \rightarrow LiOH(s) + 1/2\ H_2(g)$$
$$Li_{(Pb-17Li)} + 1/4\ CO_2(g) \rightarrow 1/2\ Li_2O(s) + C(s)$$

scheme 2

will result in lithium depletion from the liquid eutectic and modification of its physical and chemical properties. A method of detecting change in the lithium content of the breeder is thus essential for the safe and efficient operation of fusion reactors containing this breeder material. In this paper, the methods which have been proposed for monitoring the composition of Pb-17Li are critically appraised. They fall into two distinct types; batch and continuous methods. The former, which are attractive owing to their inherent simplicity, split into destructive and non-destructive methods; the latter, which despite their greater complexity are favoured owing to their in-line application, are necessarily non-destructive.

Liquid Metal Systems, Edited by H.U. Borgstedt
and G. Frees, Plenum Press, New York, 1995

2. BATCH METHODS

2.1 Destructive Methods

Adelhelm et al [16] have described an analytical method for the determination of lithium in Pb-17Li; it is similar to techniques used by the commercial manufacturers of Pb-17Li {Pechiney (France), Metallgesellschaft (Germany) and Nuova Samin (Italy)} in their quality assessment and control and by chemists studying the solubilities of metals in Pb-17Li [17]. Samples are taken from loops / rigs, product batches or static vessels and destroyed by digestion in nitric acid. Owing to the possibility of segregation, taking samples has to be done carefully; it is best to take small samples from stirred material and analyse them in their entirety [34].

The lithium content of the solution thus prepared is measured by atomic absorption spectroscopy using an air-acetylene flame. At the resonance line (λ = 670.8 nm) neither spectral nor chemical interference is observed, and the absorption is linear in the range 0.05 - 3.00 mg dm^{-3}, with an extinction of 0.380(1) at 2.00 mg dm^{-3}. The method measures the total amount of lithium in the sample; it is not discriminatory between that dissolved in the eutectic alloy and that present in precipitated impurities (e.g., oxides, hydroxides). A standard deviation of less than 0.3 at % Li has been achieved in the composition range Pb-10.7Li to Pb-23.2Li [16]. Adelhelm et al [16] used the technique to determine the composition of the alloy in the PICOLO loop after 3000 hours of operation at 823 K; a composition of 17.20 ± 0.12 at % Li was found.

2.2 Non-destructive Methods

All the methods described depend on the correlation of the liquidus temperature of the alloy with its composition. The quality of the liquidus data is fundamental to the accuracy of the technique. The liquidus of the lead-rich section of the Pb-Li phase diagram has recently been redetermined to a high standard. [8] These data (Figure 1) are such that composition can be determined to an accuracy of 0.17 at % Li per 1 K.

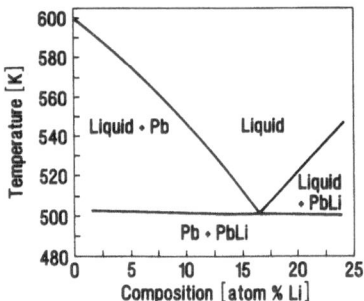

Fig. 1. The lead-rich section of the Pb-Li phase diagram

Liquidus temperatures can be determined either by thermal analytical methods or by detection of the abrupt change in a physicochemical property which occurs as the system passes through the phase boundary. Thermal analysis methods are notoriously unreliable [18, 19]; they are non-equilibrium methods being used to determine equilibrium data. Furthermore, although it is possible to incorporate a thermal analytical facility in a loop system, unless a system can be designed in which it is possible to isolate a small sample from the loop, the efficacy of the technique will be compromised owing to the non-isothermal conditions set up during the necessarily spontaneous cooling process. The measurement of a physicochemical property (e.g. density, surface tension, electrical resistance) as a function of temperature overcomes all these problems as the data are accrued under equilibrium conditions. Electrical resistance is normally chosen owing to its ease and accuracy of measurement.

The superiority of the electrical resistance - temperature (R-T) method over thermal analytical methods is depicted in Figure 2 in which R-T data are compared to temperature - time (TA) and differential temperature - time (DTA) plots for a Pb-Li alloy containing ~7 mol % Li. Section AB represents the resistance of the liquid alloy, section BC the resistance of the two phase {Pb-Li(l) + Pb(s)} region, section CD the change in resistance during the invariant eutectic reaction and section DE the resistance of the two phase {Pb(s) + LiPb(s)} region. The most significant feature of the R-T plot is the clarity of the discontinuity (point B) representing the liquidus. The corresponding discontinuity on the TA curve is almost non-existent, while that on the DTA curve is much more diffuse. The discontinuity (point C) representing the start of the eutectic reaction is also clearly defined; in this case, however, the corresponding inflections on the TA and DTA curves are equally distinct.

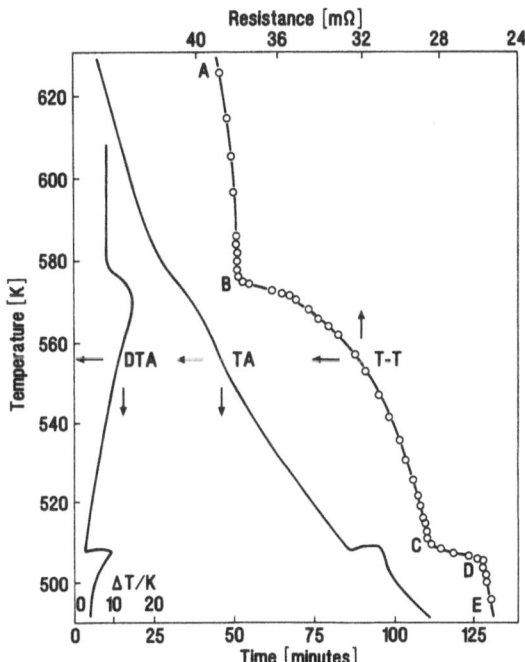

Fig. 2. Comparison of resistance-temperature (R-T), temperature-time (TA) and differential temperature-time (DTA) plots for Pb-7Li

Hubberstey et al [20] have developed a monitor based on the R-T method. A schematic represtation of a facility which can be incorporated into a loop / rig system is shown in Figure 3; detailed dimensions have been reported [20]. It comprises an insulating narrow bore capillary (A) fitted with a small reservoir (B) lowered into a pool of liquid alloy (C). Resistance data are determined by the four terminal method, the electrical connections (E_1-E_4) being attached to the pool and the small reservoir. A constant current (3A) is passed through the monitor and a standard resistance (0.01Ω), arranged in series. The potential differences across the monitor, V_x and standard, V_s, are measured using a digital multimeter. The resistance of the monitor, R_x Ω, is calculated from the expression:

$$R_x = 0.01V_x/V_s \qquad (1)$$

The method of temperature control depends on the specific application. Temperature fluctuations and gradients, however, must be restricted to ±1 K. The temperature of the monitor is measured using chromel-alumel thermocouples (T_1-T_4) located in pockets adjacent the capillary.

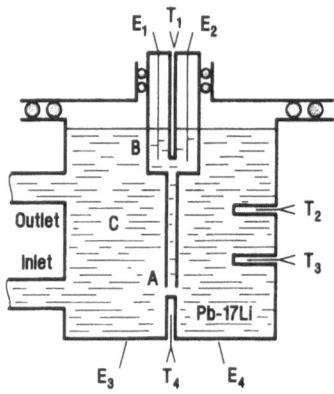

Fig. 3. Schematic representation of a resistance-temperature (R-T) monitor

As noted earlier, the accuracy of the technique depends not only on the accuracy to which the liquidus temperature can be determined but also the quality of the liquidus data. Given an accuracy of 0.17 at % Li per 1 K for the hypoeutectic liquidus of the Pb-Li phase diagram [8], a standard deviation of ± 2 K in the measurement of the liquidus temperature gives an accuracy of ± 0.35 at % Li in the alloy composition.

As yet, no trials of this technique have been carried out on large scale applications. It has been used successfully, however, in small scale experiments to study the reaction of Pb-Li alloys with water vapour, carbon monoxide and carbon dioxide. [9,20] The results of a typical experiment are depicted in Figure 4, which is a composite diagram containing four measured resistance-temperature plots [before (ABC) and after three additions of carbon dioxide (DEF; GHI; JKL)] and the hypoeutectic liquidus (temperature-composition plot) drawn with a common temperature (ordinate) axis. The liquidus phase boundaries are readily identified by the sharp inflections (points B, E, H, K) which separate the single phase {Pb-Li(l)} and the biphasic {Pb(s) + Pb-Li(l)} regions. Correlation of the temperatures of the inflections with the hypoeutectic liquidus (points B′, E′, H′, K′) gives the composition of the alloy before and after the carbon dioxide additions. The results are consistent with abstraction of oxygen from the gases to form Li_2O.

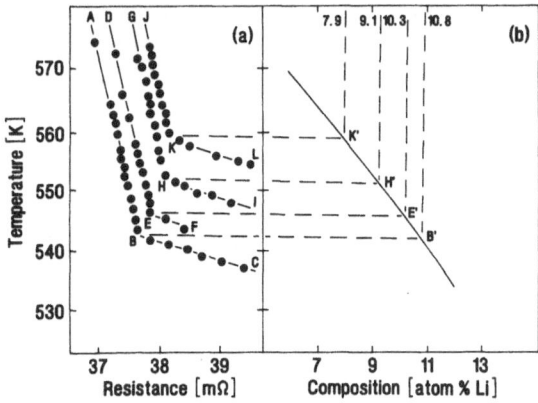

Fig. 4. Derivation of composition data from resistance-temperature data (a) using the hypoeutectic liquidus of the Pb-Li phase diagram (b).

3. CONTINUOUS METHODS

Three concepts have been proposed for the on-line determination of the lithium content of Pb-17Li in fusion reactor systems. Two are electrochemical sensors, one of which monitors directly lithium activity in the alloy, the other of which determines lithium activity from measured oxygen activities. The latter method is restricted to systems saturated with Li_2O and hence would not be applicable to loop/rig systems containing temperature differentials, as saturation will only occur in the coldest regions of such systems. The third is an electrical resistivity monitor which is based on the fact that the electrical resistivity of liquid metal binary alloys is dependent on composition.

Both electrochemical sensors rely on the relationship between lithium activity and composition, which is depicted in Figure 5 as a series of isotherms ($623 < T$ (K) < 923). The isotherms were derived [11] from emf data reported by Saboungi et al. [21]; these are the most extensive and self-consistent of the four published sets of data on Pb-Li thermodynamics [21-24]. Saboungi's experimental data at 770 K and 869 K are included in Figure 5 for comparison. The isotherms fall into two sections: a linear region at $x_{Li} < 17$ at % Li is followed by a curved region at $x_{Li} > 17$ at. % Li. Over the linear region, the composition dependence of a_{Li} in Pb-17Li at 723 K is given by the expression [11]:

$$\ln a_{Li} = -7.606 + 1.05 \ln x_{Li} \qquad (2)$$

Similarly, the temperature dependence of a_{Li} in Pb-17Li is given by the expression [11]:

$$\ln a_{Li} = -16.324 + 9.48.10^{-3} (T/K) \qquad (3)$$

An analogous expression quoted by Smith et al. [25]:

$$\ln a_{Li} = -6960 (T/K)^{-1} + 0.0245 \qquad (4)$$

gives very similar values (at 750 K, $a_{Li} = 0.996 \cdot 10^{-4}$ {eq (3)}; $a_{Li} = 0.956 \cdot 10^{-4}$ {eq (4)}).

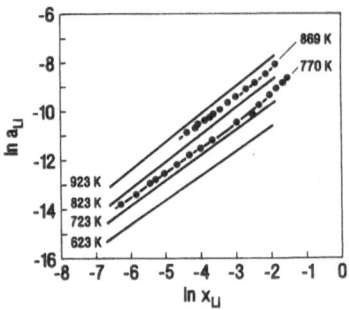

Fig. 5. Activity-composition data for the Pb-Li system

3.1 Electrochemical Lithium Sensor

Tas et al. [26-29], working at the SCK/CEN Laboratories at Mol (Belgium) have developed an electrochemical lithium sensor for use in Pb-17Li. The sensor compares the lithium activity in the liquid alloy and in a biphasic {Li$_3$Bi(s) + Li-Bi(l)} mixture (overall composition 40 at % Bi, 60 at % Bi) using a sodium β-alumina electrolyte:

$$Pb-17Li(l) \parallel \beta-Al_2O_3 \parallel Li_3Bi(s) + Li-Bi(l)$$

The Li-Bi mixture was chosen rather than pure lithium for two reasons. Firstly, it is much less chemically reactive than lithium, which is not compatible with oxide ceramics. Secondly, its lithium activity ($a_{Li} = 0.102.10^{-4}$ at 750 K) is much lower than that of pure lithium ($a_{Li} = 1$) and more comparable with that of Pb-17Li {$a_{Li} = 0.956.10^{-4}$ at 750 K [25]};the sensitivity of the technique is maximised by minimising the difference in the lithium activities between reference and sample electrodes. An advantage of this reference electrode is the fact that its lithium activity is essentially independent of composition over a wide temperature and composition range (688-900 K and 30-50 at % Bi). At higher temperatures the reference electrode moves into a single {Li-Bi (l)} phase regime and its emf is no longer independent of composition. At lower temperatures, a different two phase regime {Li$_3$Bi(s) + LiBi(l)} is encountered. Under these conditions the lithium activity is independent of composition but its temperature dependence is different from that of the original two phase region {Li$_3$Bi(s) + LiBi(l)}. [28].

The emf E [V] generated by the sensor when inserted in Pb-Li alloys

$$E = R T \{\ln[a_{Li(Pb-Li)}/a_{Li(Bi-Li)}]\}/F \tag{5}$$

is dependent not only on the temperature and composition dependence of the lithium activity of Pb-Li alloy but also on the temperature dependence of the lithium activity of the Bi-Li reference electrode. Thus each sensor has to be individually calibrated to determine its sensitivity to the three variables. Although a_{Li} values can be obtained from the sensor using a knowledge of Bi-Li thermodynamics, [21] in normal operation the sensor emf is directly related to alloy composition by calibration.

The sensor design (Figure 6) [27, 30] comprises a solid sodium β-alumina electrolyte thimble (A) sealed in a cylindrical stainless steel housing (B). Both commercial ceramic cements and freeze seal joints have been used to seal the thimble in the concentric steel tube. The central electrode (C), composed of tantalum or stainless steel, dips into the reference electrode (D) and exits the thimble through a vacuum tight sealant situated at the top of the probe. The other electrical connection (E) is made to the external surface of the steel housing

(B) which makes contact with the Pb-17Li alloy (F), whether in a static pot or dynamic loop system.

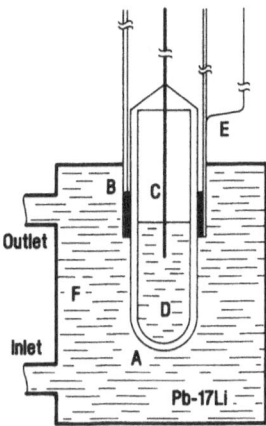

Fig. 6. Schematic representation of the electrochemical lithium sensor

Sensors have functioned satisfactorily in small scale experiments for periods up to 4000 hours. In general, the sensor signal stabilises within 10 minutes of immersion, and a reproducibility of ± 1 mV is achieved (Figure 7) [28]. They are, however, sensitive to thermal shock and, especially during the development stage, some failed immediately after addition to the liquid alloy [26]. A potential drawback to their use in loop systems is the susceptibility of the ceramic components and the seals to fracture at the relatively high pressures present at the bottom of the loops.

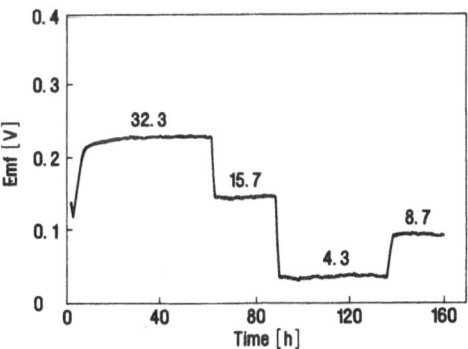

Fig. 7. Variation of the emf of the electrochemical lithium sensor as a function of time at 753 K as it is inserted in Pb-Li alloys of differing composition (quoted as atom % Li)

The influence of temperature on the sensor can be estimated from the temperature dependence of the lithium activities in Pb-17Li [26-28] and the biphasic {Li$_3$Bi(s) + Li-Bi(l)} system [21]. Using these data and the Nernst Law, a temperature coefficient of 0.26 mV K^{-1} is estimated for 723 K. A variation of 1 mV thus corresponds to a temperature change of 4 K. The influence of composition can be ascertained similarly from emf - composition data for Pb-Li alloys (Figure 8). Extrapolation to 723 K of Saboungi's data [21], which were determined at 770 K to 932 K, gives a composition coefficient of 4.6 mV.(at % Li)$^{-1}$. A variation of 1 mV thus corresponds to a composition change of 0.22 at % Li. Thus a temperature change of 1 K results in the same change in emf as a composition change of 0.057 at % Li.

The emf data obtained from the sensor can be verified by comparison with similar data obtained by Saboungi et al [21] using an electrochemical cell of the type:

$$Pb\text{-}17Li(l) \parallel \text{Molten LiCl-KCl} \parallel Li_3Bi(s) + Li\text{-}Bi(l)$$

The comparison is made in Figure 8. Emf data obtained from sensors at 653, 723 and 773 K are compared with Saboungi's data at 770 K. The two sets of results are comparable with a maximum discrepancy of 10 mV (equivalent to 2.2 at % Li). Although no obvious reason for this difference is given, the same deviation is found for the temperature dependence of the emf of the reference systems [28]. The discrepancy does not invalidate the sensor; since the emf behaviour of each separate sensor is consistent, they can easily be calibrated for use in Pb-17Li.

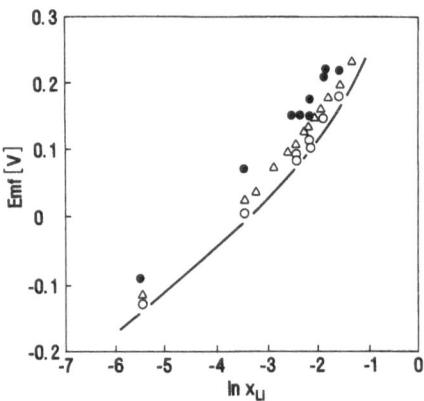

Fig. 8. Variation of the emf of the electrochemical lithium sensor as a function of Pb-Li composition at 773 K (o), 723 K (Δ) and 653 K (•): comparison with Saboungi's data at 770 K (—)

3.2 Electrochemical Oxygen Meter

The second electrochemical method of determining lithium content, developed by Adelhelm et al [16,31], is less widely applicable as it depends on the Pb-Li alloy being saturated with Li_2O; clearly, in non-isothermal systems (loops / rigs) the alloy will only be saturated at the coldest locations. The sensor, an electrochemical oxygen meter, is of the type:

In / In_2O_3 reference electrode $\parallel ThO_2/Y_2O_3 \parallel$ oxygen activity in Pb-17Li:
for which the cell reaction is:

$$2/3 \text{ In(s)} + 2 \text{ O}_{(Pb\text{-}17Li)} \rightarrow 2/3 \text{ In}_2O_3(s) \tag{6}$$

and the cell emf is given by:

$$E = \{2/3 \ \Delta G_f^o(In_2O_3) - 2RT \ln a_O(Pb\text{-}17Li)\}/4F \tag{7}$$

Since the equilibrium:

$$2Li_{(Pb\text{-}17Li)} + O_{(Pb\text{-}17Li)} \rightarrow Li_2O(s) \tag{8}$$

exists in the alloy, the oxygen and lithium activities in the alloy are related by the expression:

$$\Delta G_f^o(Li_2O) = 2 \ RT \ln a_{Li}(Pb\text{-}17Li) + RT \ln a_O(Pb\text{-}17Li) \tag{9}$$

$$\text{i.e., } RT \ln a_O(Pb\text{-}17Li) = \Delta G_f^o (Li_2O) - 2RT\ln a_{Li}(Pb\text{-}17Li) \tag{10}$$

Hence, the lithium activity in the liquid alloy saturated with Li_2O is related to the cell emf E (Volts) by the expression:

$$E = \{2/3 \, \Delta G_f^o(In_2O_3) - 2\Delta G_f^o(Li_2O) + 4RT \ln a_{Li}\}/4F \qquad (11)$$

Electrochemical oxygen meters based on CaO/ZrO_2 as well as ThO_2/Y_2O_3 were found to be suitable for measurement of a_{Li} in Pb-Li alloys; Y_2O_3/ZrO_2 electrolytes were not as effective. Temperature dependent a_{Li} values obtained using these meters in Pb-17Li saturated with Li_2O are compared with literature data in Figure 9. The agreement is highly satisfactory suggesting that oxygen meters might be used to monitor the depletion of lithium from Pb-17Li resulting from ingress of oxygen bearing gases. However, calibration data for alloys depleted of lithium have not, as yet, been reported. These meters, like the electrochemical lithium sensors, are susceptible both to thermal shock and to fracture at the high pressures pertaining near the bottom of Pb-17Li loops / rigs.

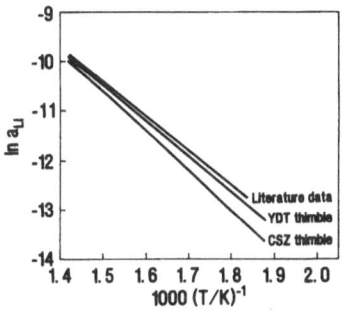

Fig. 9. Temperature dependence of lithium activities in oxygen-saturated Pb-17Li derived from electrochemical oxygen meter measurements

3.3. Electrical Resistivity Meter

The last monitor to be considered in this section is the resistivity meter developed by Hubberstey et al [32,33]. Its operation depends on the fact that the electrical resistivity of the Pb-Li alloy is dependent on composition. Typical resistivity - composition $(\rho$-x$)$ isotherms are shown in figure 10. The resistivity rises with lithium content with a slightly increasing gradient. For compositions close to Pb-17Li the gradient of the 725 K isotherm is $1.27.10^{-8} \; \Omega m \; (at\% \; Li)^{-1}$. Since the resistivity can be determined to an accuracy of $\pm 0.20.10^{-8} \; \Omega m$, changes in the composition of the alloy of $\pm 0.16 \; at\% \; Li$ can be detected. Consideration of resistivity-temperature data for Pb-17Li [32] gives a temperature coefficient of $0.054.10^{-8}\Omega m \; K^{-1}$ at 725 K. A variation of $0.1.10^{-8} \; \Omega m$ thus corresponds to a temperature change of 2 K. Thus a temperature change of 1 K results in the same change in resistivity as a composition change of 0.042 at % Li.

A potential disadvantage of the resistivity meter is the fact that it is non-specific, resistivity changes occurring on dissolution of most solutes [32]. Fortunately, none of the potential impurities, which include non-metals from adventitious gases (oxygen, carbon, nitrogen, hydrogen isotopes), containment metals (e.g., iron, chromium, nickel, manganese) and activation products (e.g., bismuth, polonium) are sufficiently soluble in the liquid alloy to change its resistivity by a measurable amount.

The electrical resistivity meter [32] is shown in figure 11; detailed dimensions have been reported [32]. Of all-welded construction in stainless steel, it is extremely robust. It consists of a capillary loop (C) within which the resistance measurements are made, and a miniature electromagnetic pump (B) which is used to sample metal continuously from the bulk source.

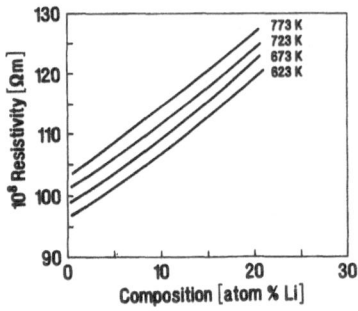

Fig. 10. Resistivity-composition isotherms for Pb-Li alloys

The capillary loop (C) is bridged by two plates, attached to each of which is a pair of silver electrical leads. The pump duct, fabricated from a flattened section of tube, is situated between the poles of a permanent magnet (0.55 T). Attached to the narrow faces of the duct are two silver leads for connection to a current source. The bulk source (A) is connected to the capillary directly and to the pump duct (B) via a short section of tube (o. d. 9 mm, i. d. 7 mm, length 20 mm) to provide an inlet and outlet to the meter. Passage of a low voltage (2 V) direct current (max. 40 A) through the duct at right angles to the magnetic field results in the continuous sampling of the liquid metal through the capillary section (flow rate ~10 mm s^{-1}).

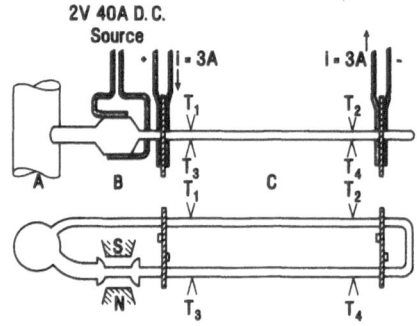

Fig. 11. Schematic representation of the electrical resistivity monitor

Resistance data are determined by a version of the four terminal method. A constant current (3A) is passed through a standard resistance (0.01 Ω) and the monitor arranged in series. The resistance of the monitor, R_t, is calculated from the potential differences measured across the standard, V_s and monitor V_t using the expression:

$$R_t = 0.01 \ V_s/V_t \tag{12}$$

The resistance of the Pb-Li alloy, R_m, is then calculated using the expression for parallel conductors:

$$R_m = R_c R_t/(R_c - R_t) \tag{13}$$

where R_c is the resistance of the empty capillary. The resistivity of the liquid metal, ρ_m, is given by the relationship:

$$\rho_m = R_m \cdot A/l \tag{14}$$

where A is the cross-sectional area and l the length of the capillary between the two plates. The constant A/l, which is specific to each monitor, can be derived by direct measurement or by calibration using a metal or alloy of known resistivity. Owing to the convoluted geometry of the capillary, the latter method is usually adopted. Prior to use, the monitor must be calibrated to determine both the resistance of the empty capillary and the constant A/l. This involves accurate measurement, as a function of temperature $(600 < T(K) < 800)$ of the resistance of the monitor, both empty and filled with metal of known resistivity.

Although the operation of the resistivity monitor has yet to be assessed in a loop or rig system, its response to ingress of oxygen [9,32,33], hydrogen [9,33], nitrogen [9,33], and water vapour [9,10,33] to Pb-Li alloys has been tested on a laboratory scale. Neither hydrogen nor nitrogen had any effect confirming their unreactivity towards Pb-Li alloys containing less than 50 at% Li. The meter did, however, respond to all the oxygen bearing gases as they abstracted lithium from the liquid alloy forming Li_2O. The results of a typical experiment are depicted in Figure 12; the variation in the resistivity (and hence composition) of an alloy of initial composition Pb-17.45Li is shown as a function of time for the addition of a nitrogen / oxygen (4/1) mixture. The response time was limited solely by the rate of sampling / mixing in the reaction vessel.

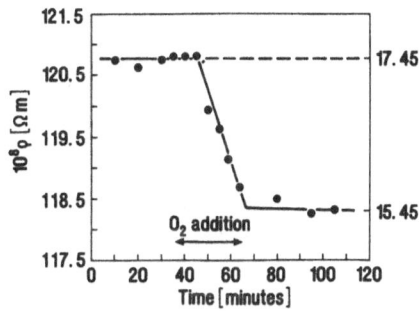

Fig. 12. Variation in the resistivity (composition) of Pb-Li alloys on addition of oxygen

5. CONCLUSIONS

The salient features of the various methods of monitoring the composition of Pb-17Li are collated in Table 1. Batch methods are attractive owing to their simplicity; their disadvantage is the fact that either samples have to be taken from the loop for remote analysis, or a section of the loop has to have a facility for thermal cycling. The resistance-temperature method is very simple to operate. It does not require calibration and is extremely robust. Thermal analytical methods are equally simple to operate; their main drawback is the fact unless the sample can be isolated from the loop, they are somewhat inaccurate.

Continuous methods are more attractive owing to their modus operandi; they are, however, considerably more complex than the batch methods. All three require extensive calibration prior to insertion in the loop system. That based on the electrochemical oxygen meter is of limited applicability, since it can only be operated in systems saturated with Li_2O, a situation which will not be realised in normal non-isothermal breeder blanket operation.

The major advantage of the electrochemical lithium sensor over the resistivity meter is the fact that it is specific to lithium. Its disadvantage is the fact that it is much more fragile, being sensitive to thermal shock and susceptible to fracture under high pressure; it can only be used near the top of loops. The resistivity meter is extremely robust, and can be used throughout the loop system. Although nominally sensitive to any solute, potential impurities have such small solubilities in Pb-17Li alloy that they will not influence its operation.

The electrochemical lithium sensor is less superior to the resistivity meter in the fact that it is more sensitive to temperature, the temperature changes required to bring about a change in measured parameter (emf, resistivity) equivalent to that caused by a composition change of ± 0.1 at % Li being ± 1.7 and ± 2.4 K, respectively (Table 1). The oxygen meter is particular poor in this respect, the corresponding temperature change being ± 0.5 K (Table 1). Furthermore, the resistivity meter is slightly more sensitive to composition change than the electrochemical sensor, the minimum detectable being ± 0.16 and ± 0.22 at % Li, respectively (Table 1).

The monitors with greatest potential are the electrochemical lithium sensor and the resistivity meter. The next stage in their assessment is a direct comparison in a loop system which is to take place in the CEA laboratories at Fontenay-aux-Roses, Paris.

Table 1. Lithium monitors: a comparison

Method	Calibn	Parameters	Sensitivity	Accuracy	Detection	Equivalent*
Chemical analysis	No	[Li]			± 0.3	
Thermal analysis (TA & DTA)	No	T vs t	0.17 at % Li per 1 K	± 2K	± 0.35	
Resistance-temperature analysis	No	R vs T	0.17 at % Li per 1 K	± 2K	± 0.35	
Electrochem-ical lithium sensor	Yes	E vs t	0.22 at % Li per 1 K	± 1 mV	± 0.22	± 1.7
Electrochem-ical oxygen meter	Yes	E vs t	0.21 at % Li per 1 K	± 1 mV	± 0.15	± 0.5
Resistivity monitor	Yes	ρ vs t	0.08 at % Li per 10^{-8} Ωm	$\pm 0.2.10^{-8}\Omega$m	± 0.16	± 2.4

* This parameter is the temperature change required to bring about a change in measured parameter equivalent to that caused by a composition change of ± 0.1 at % Li.

Acknowledgements

We would like to acknowledge the debt the Pb-17Li scientific community, in general, and we at Nottingham, in particular, owe to the late Vittorio Coen. His vision, drive and enthusiasm were critical to the development of the European collaborative effort on liquid lithium and liquid Pb-17Li which has resulted in much of the extensive data base now available on materials behaviour and physical chemistry in Pb-17Li.

References

1. Y. Severi, P. Labbe, L. Giancarli, L. Barraer, J. Quintric-Bossy, C. Bubief and J. Mercier, Fusion Technol. 1 (1990) 812.
2. G. Casini, P. Labbe, M. Reiger, L. Barraer, M. Biggio, F. Farfaletti-Casali, G. Gervaise, L. Giancarli, M. Roze, Y. Severi, J. Quintric-Bossy, S. Tominetti, J. Wu and M. Zucchetti, Fus. Eng. Des. 14 (1991) 353.
3 S. Malang, P. Leroy, G.P. Casini, R.F. Mattas and Yu. Strebkov, Fus.Eng. Des. 16 (1991) 95.
4. L. Giancarli, Y. Severi, L. Barraer, P. Leroy, J. Mercier, E. Proust, and J. Quintric-Bossy, Proc. 10th. Topical Meeting "Technology of Fusion Energy", Boston, Mass., June 1992.

5. G. Casini, L.. Petrizzi, V. Rado and M. Zucchetti, Proc. 17th. Symp. of "Fusion Technology", Rome, Italy, Sept. 1992.
6. S. Malang, H. Deckers, U. Fischer, H. John, R. Meyder, P. Norajitra, J. Reimann, H Reiser and K Rust, Fus. Eng. Des. 14 (1991) 373.
7. S. Malang, E. Bojarsky, L. Buhler, H. Deckers, U. Fischer, P. Norajitra and H. Reiser, Proc. 17th. Symp. of "Fusion Technology", Rome, Italy, Sept. 1992.
8. P. Hubberstey, T. Sample and M.G. Barker, J. Nucl. Mater. 191-194 (1992) 283.
9. P. Hubberstey and T. Sample, J. Nucl. Mater. 191-194 (1992) 277.
10. P. Hubberstey and T. Sample, J. Nucl. Mater. 199 (1993) 149.
11. J.D. Gordon, J.K. Garner and N.J. Hoffman, Proc. Int. Conf. "Liquid Metal Engineering and Technology", BNES, London, 1 (1984) 329.
12. H.U. Borgstedt, Kernforschungszentrum Karlsruhe Report KfK 4620 (1989).
13. H.J. Ache, Angew. Chem. Int. Ed. Engl. 28 (1989) 1.
14. V. Coen and T. Sample, Fusion Technol. 1 (1990) 248.
15. G. Casini and J. Sannier, J. Nucl. Mater. 179-181 (1991) 47.
16. C. Adelhelm, H.U. Borgstedt, D. Linder and G. Streib, Fus. Eng. Des. 14 (1991) 235.
17. M.G. Barker and T. Sample, Fus. Eng. Des. 14 (1991) 219.
18. J.B. Macchesney and P.E. Rosenberg, in: Phase Diagrams: Materials Science and Technology, vol. 1, ed., A. M. Alper (Academic Press New York, 1970) pp 114-165.
19. G.V. Raynor, in: Physical Metallurgy, ed. R W Cahn, (North Holland, Amsterdam, 1965) pp 291-363.
20. P. Hubberstey, T. Sample and M.G. Barker, Fusion Technol. 2 (1990) 949.
21. M.L. Saboungi, J. Marr and M. Blander, J. Chem. Phys. 68 (1978) 1375.
22. A.I. Demidov, A.G. Morachevskii and L.N. Gerasimenko, Elektrokhimia 9 (1973) 848.
23. R. Ihle, A. Neubert and C.H. Wu, Fusion Technol. 2 (1978) 639.
24. A. Neubert, J. Chem. Thermodyn. 11 (1979) 971).
25. D.L. Smith and G.D. Morgan, USDOE Report ANL/FPP-84-1, vol 2 (1984) p. 6-1.
26. F.De Schutter, J. Dekeyser, H. Tas and S.De Burbure, J. Nucl. Mater. 155-157 (1988) 744.
27. F.De Schutter, J. Dekeyser, and H. Tas, Proc 4th Int. Conf. "Liquid Metal Engineering and Technology", Avignon, 1988 pp 601-1 -12.
28. F. De Schutter, J. Dekeyser, J. Luyten and H. Tas, Fusion Eng. Des. 14 (1991) 241.
29. F. De Schutter, J. Dekeyser, J. Luyten and H Tas, Proc. 5th. Int. Conf. "Solid State Sensors and Actuators", Montreux, Switzerland, June 1989.
30. J.Dekeyser, personal communication at the European Workshop on Lithium and Lithium-Lead Corrosion, Karlsruhe, 1992.
31. N.P. Bhat, C. Adelhelm and H.U. Borgstedt, Proc 4th Int. Conf. "Liquid Metal Engineering and Technology", Avignon, Oct. 17-21, 1988 pp 623-1 - 6.
32. P. Hubberstey, M.G. Barker and T. Sample, Fus. Eng. Des. 14 (1991) 227.
33. P. Hubberstey, T. Sample and M.G. Barker, J. Nucl. Mater. 179-181 (1991) 886.
34. H. Feuerstein, L. Hörner, J. Oschinski and S. Horn, Proceedings of this conference. and M G Barker, J. Nucl. Mater. 179-181 (1991) 882.

DISSOLUTION OF SODIUM OXIDE IN LIQUID SODIUM :
A NEW KINETICS IN A LARGE TEMPERATURE RANGE

G. Rouviere, J. Letextier. L. Pignoly

DER / STML - CEA - Cadarache
F-13108 Saint-Paul-lez-Durance, France

1. GENERALITIES

The analysis of the plugging risk in narrow sections through which sodium may possibly flow, enables two plugging mechanisms to be distinguished :
- plugging by transport of solid, insoluble particles such as crytallized agglomerates, oxidized fims, etc. to the narrow section of interest ;
- plugging by crystallization in the cold narrow section of sodium oxide (or hydride) from the oxygen (or hydrogen) dissolved in the sodium and super-saturated by cooling.

In the past many situations have been encountered during reactor or test loop operation in which the interpretation of an incident required reference to one or the other of the two plugging mechanisms mentioned above. In each case, the time required for unplugging had to be evaluated ; to do so, the plugging mechanism had to be known so that a model could be defined.

We have identified four essential unplugging mechanisms :
- unplugging by mechanical means (increasing the pressure, entrainment of the particles by hydrodynamic shearing) ;
- unplugging by melting the obstructing impurity ;
- unplugging by thermal decomposition of the obstructing impurity ;
- unplugging by dissolution of the obstructing impurity.

In most cases, only the fourth mechanism can be envisaged, especially when the plugging is located within a large volume of sodium. As regards the dissolution of the sodium oxide crystals, studies of dissolution kinetics in cold traps have been performed and have allowed mass transfer coefficients to be established that include both the dissolution phenomenon at the oxide-Na interface and the diffusion phenomenon through the boundary layer of the sodium. According to Mac Pheeters [1], the transfer kinetics is limited by the dissolution kinetics, which amounts to saying that the kinetics (mass transfer coefficients) established for the 140 °C to 240 °C temperature range, are those of the dissolution phenomenon. Considering that the dissolution phenomenon is an interface phenomenon, it is unlikely that the established kinetics are still valid as regards the dissolution of the oxide present in an oxidized film floating on the surface of the sodium. When the Na_2O deposit is located within the sodium, it is advisable to estimate whether the sodium can or cannot circulate in contact with the Na_2O deposit in order to define the unplugging rate : in the first case, it is the Na_2O

dissolution kinetics that will allow the unplugging rate to be know ; in the second, it is advisable to take into account not only the dissolution kinetics, but also the diffusion rate between the surface of Na_2O and the sodium surface in contact with the dynamic sodium, the two surfaces being separated by a volume of stagnant sodium.

The coefficient of oxygen diffusion in the sodium was established by S. SIEGEL and L. F. EPSTEIN [2] :

$$D_{O(Na)} = 73.7 \cdot 10-5 \cdot \exp [- 2430/RT] \; [cm^2s^{-1}]$$

(with T in K) in which the activation energy is 2430 kcal.mol-1

This coefficient was also defined by VOLCHKOV and NALIMOV [3] :

$$D_{O(Na)} = 0.213 \cdot \exp [- 9450/T] \; [cm^2s^{-1}]$$

with T in K.

This latter value is lower than that of SIEGEL and EPSTEIN by an order of 2. According to our bibliographical investigation, there are no publications reposting Na_2O (crystal) dissolution kinetics at temperatures higher than 250 °C. This is why we shall describe the study that enabled us to define one.

2. EXPERIMENTAL PROCEDURE

The principle of these tests was to measure the evolution of the oxygen content in dynamic under-saturated sodium in contact with sodium oxide crystals which are at a temperature T (test temperature). This was done by means of the ECRIN test loop (see figure 1), described in [4].

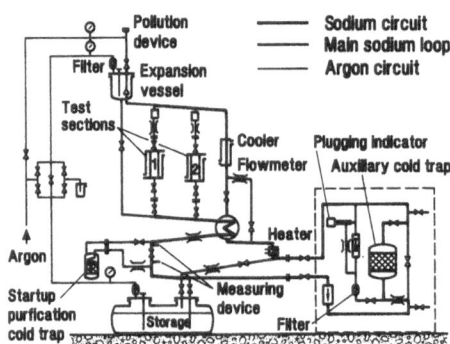

Fig. 1. The ECRIN test loop

Let us recall that this test loop contains about 0.06 m^3 of sodium and that it is equipped in particular with a testing device, which is in fact a cold trap containing steel wool packing loaded with oxide (figure 2), as well as a Westinghouse electrolytic oxygen-meter (reference electrode : air).

The experimental cold trap was loaded with sodium oxide by trapping oxide inserted manually into the loop in such a way that the crystals formed on the steel wool packing were evenly distributed. The sodium oxide mass inserted was 0.397 kg of Na_2O.

The electrochemical oxygen-meter operated at 354 °C then at 400 °C, after having been calibrated by means of the classic method : variation of temperature at the cold trap cold point, checking by manual measurement of the saturation temperature using a plugging indicator [4].

The test as a whole was interpreted using the solubility of oxygen in liquid sodium as modelled by NODEN [5]

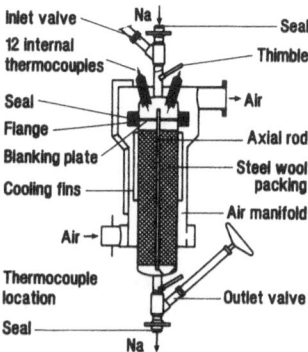

Fig. 2. The cold trap of the ECRIN test loop

3. ANALYSIS OF THE RESULTS

The unprocessed recording of the oxygen-meter signal S, as a function of time is reported in figure no.3. As the loop contains about 0.060 m^3 and the sodium flowrate was $\approx 1\ m^3h^{-1}$, the presence of a volume of saturated sodium in the experimental cold trap when the sodium was put into circulation created a discontinuity which was noted each time this particular volume of sodium went through the oxygen-meter. The experimental measurement points of interest to us are those corresponding to the low inflexions of the curve and therefore to the passage of a very under-saturated volume of sodium in the experimental cold trap.
Once the curve $S = f(t)$ is known, the curve $[0] = f(t)$ can be deduced from it. All the data obtained by means of these measurements will enable us to deduce a kinetics law.

4. METHODOLOGY FOR ESTABLISHING THE DISSOLUTION KINETICS OF SODIUM OXIDE

Let us first recall the general form of a law for the kinetics of the dissolution of sodium oxide which is:

$$dm/dt = K_O\ A\ \Delta C\ .\ \exp\ [-E_a/RT]$$

With dm/dt = oxygen dissolution rate in sodium that is under-saturated in oxygen (mass · time^{-1}) (g [O]·h^{-1}) ; K_O = coefficient of the reaction rate in : oxygen mass · surface $^{-1}$ · ppm $^{-1}$ · time $^{-1}$ (g O/m^{-2} · ppm^{-1} · h^{-1}) ;
A = surface of sodium oxide crystals available for dissolution (m^2) ;
ΔC = under-saturation of sodium : that is the difference between the oxygen saturation content at a temperature T of the sodium oxide crystals, and the oxygen content in the sodium circulating in contact with the crystals.
E_a = activation energy of the process (J) ;
R = constant of the perfect gases (R = 8.32 J/K)

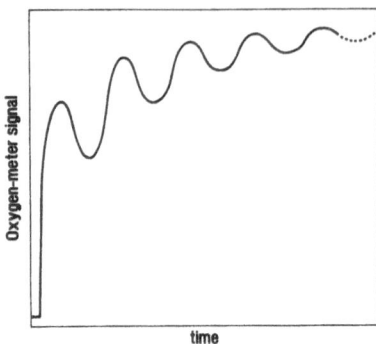

Fig. 3. Record of the signal of the Westinghouse oxygen meter vs. time

T = absolute temperature of the sodium oxide crystals (K).
(N.B. : The units in brackets are the ones commonly used. Our tests enabled us to measure the variation of the dissolved oxygen content (c) as a function of time and temperature.)
Now (c) is proportional to the dissolved mass of oxygen according to the relation :

$c = m / (vp) (g[O] / kg [Na]) = 10^3 m / (vp)$ p.p.m.
m : dissolved mass of oxygen (g [O])
v : volume in circulation in the loop (m^3)
p : density of sodium (kg / m3)

We therefore obtain :

$$dc/dt = \alpha K_O A \Delta C \exp [- E_a/RT]$$

We know that :
$\Delta C = C^* - C$
Where : C^* : content of dissolved oxygen at the test saturation temperature T
C : content of dissolved oxygen in the sodium circulating in contact with the sodium oxide crystals at the temperature T.
hence : $d \Delta C = - dc$
Therefore :

$$-d\Delta C/\Delta C = \alpha K_O A \exp [- E_a/RT] dt$$

which gives :
$Ln \Delta C = Ln \Delta C_0 - \alpha K_O A \exp (- E_a / RT) t$
This means that the slope of the straight line gives us values of the evolution rate of the content expressed in ppm / ppm.s, as a function of T, with :

$$K (T) = - \alpha K_O A \exp (- E_a / RT)$$

The determination of the dissolution kinetics amounts to the determination of the activation energy (E_a) and of K_O, which can be done by linearizing $Ln (-K (T))$ as a function of 1/T.

Estimation of the steel wool packing surface :

With this approach, the hypothesis is made that the crystalline surface available for dissolution is equal to the packing surface, assumed to be covered by the crystals, i.e. 2,1 m^2. The experimental results (Kexp) and the modelled results (K mod) are reported in Table 1.
with : T = absolute temperature
K_{CEA1} = rate calculated using the CEA.1 law
K_{exp} = rate determined using the experimental results

$$\varepsilon = [(K_{MP} - K_{exp})/K_{exp}] \cdot 100 \text{ (relative error in \%)}$$

Table 1. Comparison of the experimental and modelled results

T [K]	k_{exp}	k_{mod}	ε
*213	0.086	0.118	+ 37.20
*233	0.187	0.163	- 12.83
237	0.162	0.173	+ 6.179
284	0.347	0.331	- 4.61
309	0.383	0.449	+ 17.23
321	0.671	0.514	- 23.40
354	0.752	0.728	- 3.19
389	0.888	1.014	+ 15.32

* Results obtained by Mac Pheeters [1].

$$\sigma = 7.7 \cdot 10^{-2}$$

Figure 3 shows an example of a curve $[0] = f(t)$ $(T = 237 \,°C)$

The activation energy is determined using the slope of $Ln\ [K_{exp}] = f(1/T)$ (see Fig.4).

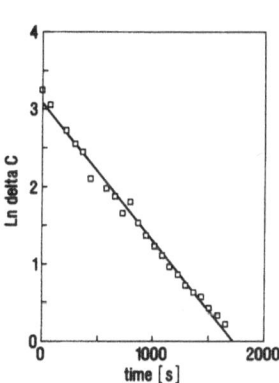

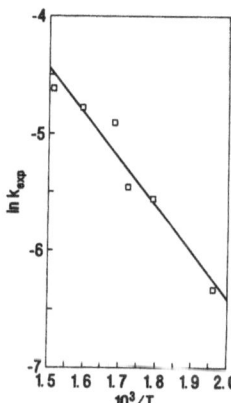

Fig. 4. Evolution of the oxygen content at 237 °C   Fig. 5. Determination of the activation energy

The law deduced according to the methodology described in paragraph 4 is as follows (expressed in Standard International Units) :

$$dm/dt = 1.067 \cdot 10^{-4} \cdot \exp[-32707.6/RT] \cdot A\ \Delta C \quad (kg\ [O]/s)$$

5. CONCLUSION

The study of sodium oxide dissolution kinetics at a temperature of 237 to 389 °C showed that :
- Considering the dissolution surface to be the steel wool packing surface, the law describes our experimental measurements as well as those of Mac Pheeters.
- The use of our experimental measurements led to a new estimate of the activation energy involved in the process.

It appeared that E_a = 32707.6 (J) instead of 79420 J. This allowed a new dissolution law to be defined that is valid from 200 to 400 °C.

The activation energy of the process, that is to say 32.7 (kJ · mol^{-1} · K^{-1}) is of the same order as that obtained for the sodium oxide growth phenomenon (45 kJ · mol^{-1} · K^{-1}) [4] ; this is logical in so far as it has been shown that this growth phenomenon, in the case of Na_2O, is basically diffusional.

References:

[1] C.C. McPheeters and J.M. Williams, in "Alkali Metal Coolants", IAEA, Vienna 1967, p.429.

[2] S. Siegel, L.F. Epstein, "The diffusion of Na_2O in sodium in the range 900-1000 °F" USAEC Rep. GEAP 3357 (1958)

[3] L.G. Volchkov, Yu.P. Nalimov., Technical Exchange CEA-GKAE, May 1983

[4] C.Latgé, in "Liquid Metal Engineering and Technology", British Nucl. Energy Soc., London, 1984, vol 2, 395.

[5] J.D. Noden, J. Br. Nucl. Energy Soc. <u>12</u> (1973) 57-62 and 329-331.

DEVELOPMENT OF A NEW PLUGGING INDICATOR AND EXPERIMENTAL TESTS IN A LITHIUM LEAD FACILITY

J. Desreumaux, C. Latgé, J. Le Texier, S. Poinsot

DER/STML - CEA-Cadarache
F- 13108 Saint-Paul-lez-Durance (FRANCE)

1.INTRODUCTION

In the framework of the European Programme on Fusion Reactor Technology, one of the options chosen for extracting the energy produced is to place a tritium breeding blanket of modules containing the eutectic Li17Pb83 around the plasma.

The LCDEV task (components development) relative to the water-cooled Li-Pb blanket is to study the components essential to a lead-lithium circuit. These components were mounted on the PABLITO circuit.

The objective of this study was to :

- become acquainted with filling, defreezing, draining and sampling techniques as well as with lead-lithium circuit control ;

- examine the characteristics of the components taken from sodium circuits : the electromagnetic pump, the electromagnetic flowmeter, the flux distortion flowmeter, pressure transducers, level gauges and valves ;

- study cold trap operation ;

- test a lead-lithium plugging indicator. Interpret plugging measurements ;

- carry out sampling and sample analyses.

2. DESCRIPTION OF THE CIRCUIT AND CHARACTERISTICS OF MAIN COMPONENTS

A diagram of the PABLITO circuit is shown in figure 1. Its capacity is 25 l, corresponding to a mass of about 250 kg. The different parts of the circuit are :

2.1   The melting tank which surmounts the whole facility is a 316 L stainless steel tank equipped with a flange. It has received the solid Li17 Pb83 ingots.

2.2   The filling tank (R2) and the draining tank (R1).
These are two identical and interchangeable tanks of about 50 l. in capacity. They are equipped with thimbles for level gauges.

2.3   A main vessel or expansion vessel in 316 L stainless steel.

Liquid Metal Systems, Edited by H.U. Borgstedt
and G. Frees, Plenum Press, New York, 1995

Its total capacity is about 5 l.

The vessel is equipped with a sampling system.

2.4 The Foucault current level gauges are used in the thimbles. They enable draining or filling of the different reservoirs to be easily monitored.

2.5 The electromagnetic pump.

This is an ET 196 type conduction pump.

A variator allows the intensity percentage to be set from 0 to 100%.

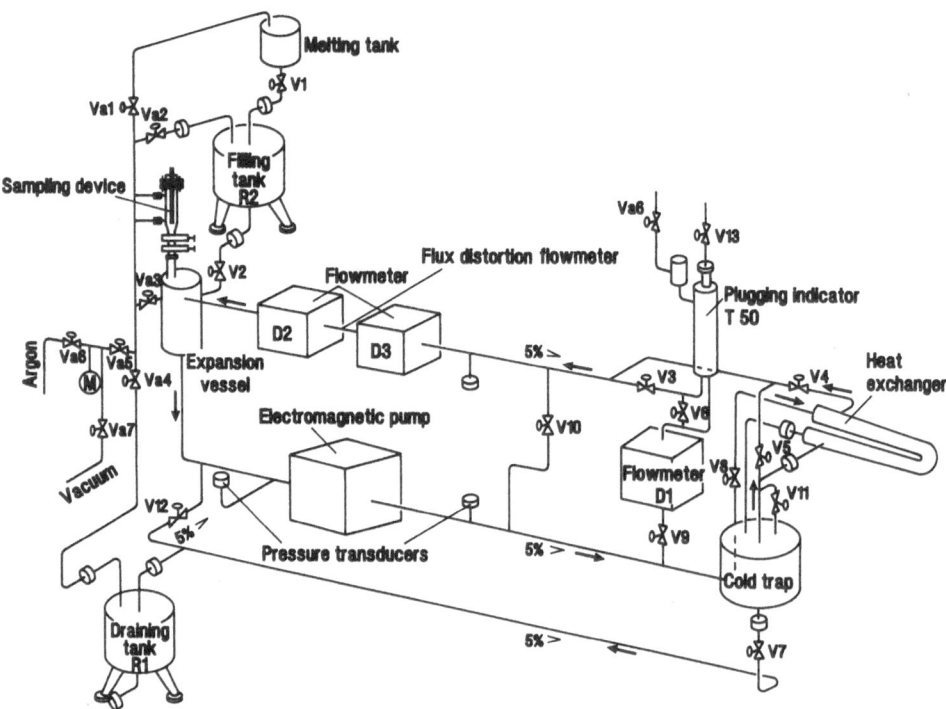

Fig. 1. Flow scheme of the PABLITO loop

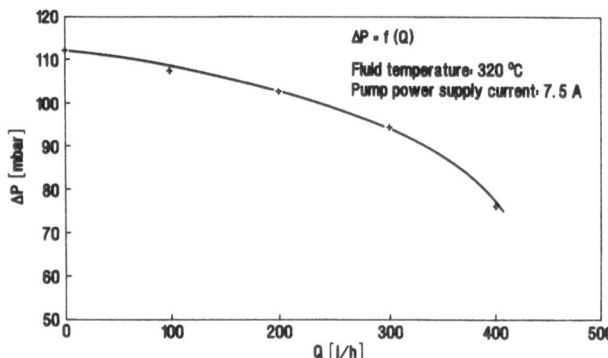

Fig. 2. Flow rate of the pump as a function of the differential pressure at entrance and exit of the pump

The figure 2 shows the results of measurements of the flowrate as a function of the differential pressure at the pump terminals.

2.6   Two pressure transducers are mounted on the pump, one upstream and one downstream. They enable the differential pressure to be measured at the pump terminals.

2.7   The electromagnetic flowmeters.

The circuit is equipped with 3 permanent magnet electromagnetic flowmeters (characteristics : 0 to 500 l/h) pre-heated by heater resistors.

D1 measures the flow in the plugging indicator, D2 and D3 are mounted in series on the return flow pipe.

D2 was chosen as the reference flowmeter.

D1 and D2 measurements were not only recorded in mV, but also given by digital voltmeter in $l \cdot h^{-1}$

The tests were satisfactory. The figure 3 shows measurement of the flowrate Q as a function of the pump power supply current Ip. The relation between flowrate and voltage supplied by the flowmeter is expressed as $Q = K \cdot e$; (with LiPb at 350 °C K = 0,146 $m^3/h \cdot mV$)

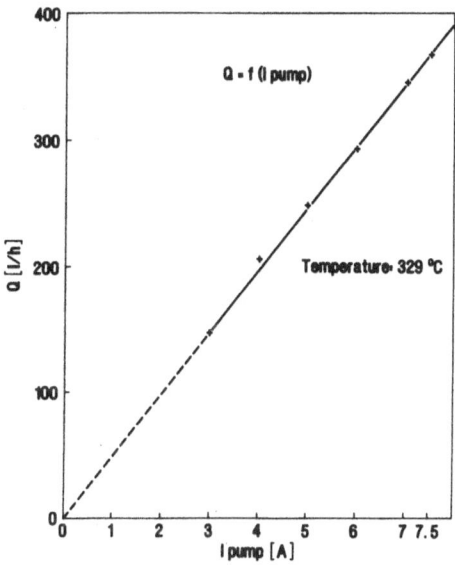

Fig. 3. Flow rate Q as a function of the pump current I

2.8   The flux distortion flowmeter is set between the D2 and D3 flowmeters. It is composed of three identical coils wound side by side on the flow pipe.

The central or primary coil is powered by alternating current and the two adjacent coils receive a voltage the difference of which is proportional to the liquid metal flow rate.

The tests were satisfactory. Figure 4 shows the measurements of voltage as a function of flowrate Q, at 318 °C.

2.9   The Cold Trap.

It is a 316 L stainless steel, air cooled tank of a capacity of about 13 l. ; inside there is the Pb-Li inlet distributor and steel knitted wire packing.

It consists of topped by a flange that enables it to be disassembled for deposit investigation.

The cold trap is equipped with 2 provisions for endoscopy.

2.10 The heat exchanger is made of 316 L stainless steel. It is composed of 2 U-shaped concentric tubes.

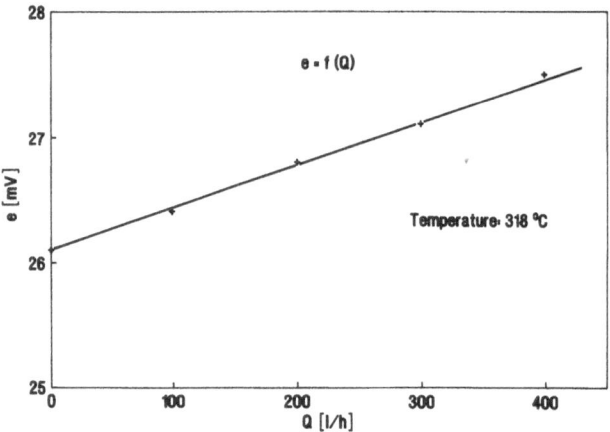

Fig. 4. Voltage of the flow meter as a function of the flow rate

2.11 The plugging indicator is a 316 L stainless steel cylinder of 1,2 l. It is based on the same principle as those perfected for sodium ; the alloy flows into a narrowed section (called pellet) of a by-pass tube situated on the main circuit. The temperature is slowly lowered until a drop in the flowrate occurs which indicates that the alloy is beginning to solidify; this cooling is carried out by means of a fan.

The plugging indicator is equipped with a pellet that can be withdrawn by means of a valve (V13) allowing unplugging. It is set up vertically so it can be drained and is topped by a container for gas bleeding (figure 5).

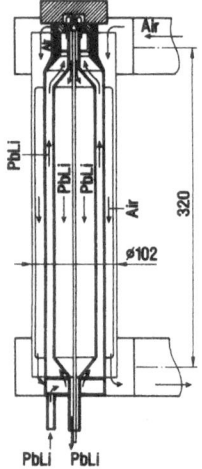

Fig. 5. Plugging indicator of the PABLITO loop

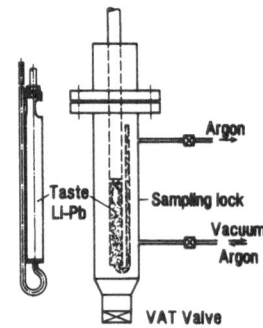

Fig. 6. TASTELI -Pb sampling system

A thermocouple in a thimble is placed very close to the pellet and the temperature is recorded and displayed on a digital voltmetre.

2.12 The Tasteli-Pb (Tasteli type) sampling system is placed over the main vessel flange and dips into the vessel. It goes through a sampling lock under argon that comprises two gate valves (VAT valves) (figure 6).

2.13 The argon circuit
N56 quality argon is used as cover gas in the expansion vessel, the draining and filling tanks, as well as the melting tank and is used for sweeping the sampling lock.
The pressure remains below 1,5 bar relative.

3. FILLING OF THE CIRCUIT.

The ingots were supplied by METAUX SPECIAUX SA.

Lithium analyses were carried out on the lower, middle and upper parts of several of these ingots by means of atomic absorption spectrophotometry. These analyses showed that there was important heterogeneity between the bottoms and tops of the ingots, the upper parts always proving to be enriched in lithium.

On average : 0,68 wt% Li for the lower and middle parts of the ingots.

0,81% Li for the upper parts.

This roughly means an average of 0,72 % for one ingot (The theoretical content of the eutectic is 0,68 wt%).

The ingots were placed in the melting tank under argon. The whole facility was put under vacuum. All the components were progressively heated up to 350 °C. Filling was monitored by means of the level gauge in the expansion vessel. Circulation of the Pb-Li continued for 48 hours. Then, the first sample was taken : Tasteli Pb no.1.

4. PLUGGING INDICATOR OPERATION - PLUGGING MEASUREMENTS.

The plugging indicator is shown in Figure 5. The plugging measurements were made under various temperature, flowrate and air circulation conditions in the plugging indicator. The Pb-Li was progressively cooled with a vane in the cooling circuit, that could be more or less widely opened.

Recorded values were :

the temperature of the thermocouple T50 in the thimble at the plugging indicator pellet level,

the flowrate D1 in the plugging indicator.

Thirteen tests were carried out :

- three tests with air and Pb-Li circulation in the same direction (concurrent),
- five tests with air and Pb-Li circulation in the opposite direction (counter-current),
- five tests with air and Pb-Li circulation in the opposite direction after addition of lead.

4.1 Tests with Air and Pb-Li Circulation in the Same Direction.

Air and Pb-Li circulation was in the same direction, upwards (concurrent). (Figure 7a). The pellet was not exactly in front of the inlet flux of cold air. It cooled slowly at the same time as the upper part of the plugging indicator.

The flowrate remained steady until the temperature reached 253 °C to 255 °C at the pellet. Then it fell drastically to 0, while the temperatures dropped to 234 - 235 °C. This is the solidification temperature of the Li17Pb83 eutectic.

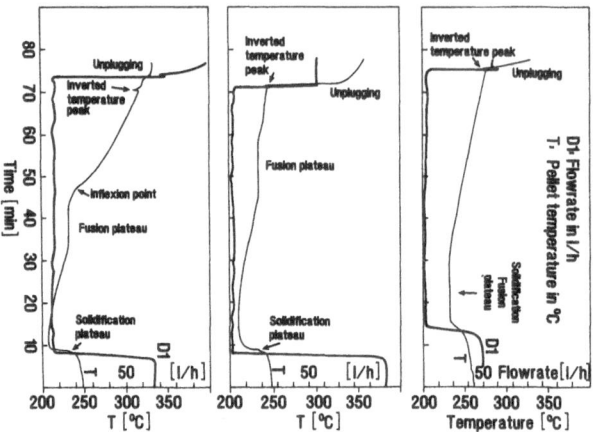

Fig. 7. Results of plugging meter tests in the PABLITO loop

No other anomaly on the flowrate curve was noted. The temperature remained at a plateau for 18 to 22 minutes with a nil flowrate for more than one hour. Unplugging took place when the pellet temperature reached 282 °C to 290 °C.

This shows that the Pb-Li had solidified in all the plugging indicator, most probably because of the concurrent air circulation that cooled the pellet last.

-       The small inverted temperature peak at the unplugging instant shows that there was Pb-Li melting in the vicinity of the pellet expressed by an uptake of energy.

Test conditions and results are listed in Table 1.

4.2 Tests with air and Pb-Li circulation in the opposite direction.

From these tests on, air cooling of the plugging indicator was done counter-current (air and Li-Pb circulating in opposite directions). The vicinity of the pellet was thus all cooled at the risk of freezing occurring in the double envelope before doing so at the pellet.

For the five tests, plugging took place at 243 °C - 247 °C (Figure 7-b). This temperature was lower than in the previous tests with concurrent cooling.

A very short solidification plateau of 20 to 30 seconds at 235 °C was observed and the temperature continued to drop until it reached 210 to 220 °C. It only rose slowly ; during the reheating phase, a fusion plateau at 235 °C was observed, that lasted from 10 to 13 minutes. Tests conditions and results are grouped in table 1.

In fact, the Pb-Li freezes completely or partially in the double envelope, before it freezes at pellet level. These differences are certainly due to the respective temperature profiles of the Pb-Li and
the air during cooling.

-       The solidification and fusion plateaus of 234-235 °C correspond exactly to the indicated melting point for the Li17Pb83 eutectic, but we were unable to determine a liquidus.

Table 1. Plugging measurements

Conditions	Concurrent cooling no Pb addition	counter current cooling no Pb addition	Counter current cooling after Pb addition
Flow rate D1 (l/h)	50	60 to 100	60
General flow rate (l/h)	218	150 to 238	220
Pellet temp. (Plugging indicator) °C	313	277 to 334	273 to 342
Expansion vessel temp. (°C)	332	290 to 344	281 to 359
Cooling rate (K/min)	2	0.3 to 0.5	0.6 to 1
Plugging temperature (°C)	253 to 255	243 to 247	242 to 246
Solidifiaction pateau temp. (°C)	234	234 to 235	234
Duration of this plateau (min.)	18 to 22	0.3 to 0.6	0.2 to 0.3
Minim. temp after solidification (°C)	234	208 to 220	207 to 218
Fusion plateau temp. (°C)	id.	235	234
Duration of fusion plateau (min)	solidification	10 to 13	8 to 10·
Inverted temperature peak (K)	1	1	2 to 10
Unplugging temperature (°C)	(at unplugging instant) 282 to 290	(at unplugging instant) 240 to 244	(3-5 min before plugging) 325 to 351

5. TASTELI Pb SAMPLING.

The purpose of the sampling is to measure the amount of lithium in the alloy and consequently to know its composition.

Two samples were taken from the initial volume of Pb-Li.

TASTELI Pb no1 : 48 hours after filling of the circuit.

TASTELI Pb no2 : 116 days after filling.

The diagram of the sampling device is given in figure 6.

It is introduced into the expansion vessel by means of a lock and touches the bottom.

The sampling device was filled properly. The Pb-Li was extracted in a glove box by removing the stainless steel casing. An ingot 70 mm high and 10 mm in diameter was thus obtained. The lithium measurements carried out by atomic absorption spectrophotometry using the standard additions method are given in Table 2.

A selection of samples was made in a glove box and the samples were taken from the lower and upper parts of the ingot.

Table 2. Determination of lithium in Pb83Li17 samples (wt %)

Sample	Position	Li content (wt %)
No. 1	Upper part	0.722
	Lower part	0.720
No 2	Upper part	0.728
	Lower part	0.721

The results show that the samples were homogeneous and that there was no lithium enrichment between the lower part and the middle of the expansion vessel.

The lithium content did not change during the 116 days of circuit operation. It was the same as that determined by measurement of the solid ingots (0,72% mass, that is to say 17,8 at.%).

6. MODIFICATION OF ALLOY COMPOSITION BY ADDITION OF LEAD

The plugging measurements showed that the expected eutectic melting point was found, but they did not allow us to determine the liquidus.

How do the plugging curves look like, if this composition is changed? In order to change the composition of the alloy, we could add either lithium or lead. We have chosen the pollution by lead and the system allowed us to add only a very small amount.

The polluting device (figure 8) was screwed onto the sampling system tube in place of the TASTELI Pb.

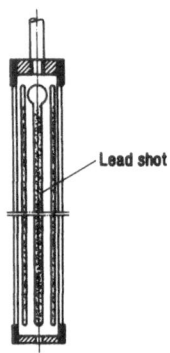
Lead shot

Fig. 8. The polluting device for inserting lead into the alloy

It was filled with lead shot, introduced into the expansion vessel at 408 °C and left at the bottom for 30 minutes so that the lead could melt and mix with the alloy. The addition of the lead was done in 2 stages : A total amount of 177,2 g.

This amount should have theoretically made the lithium content drop from 0,723% to 0,722%.

Will this very slight pollution have any effect on :
- the plugging measurements,
- the lithium content in the alloy dosed by atomic absorption spectro-photometry ?

7. PLUGGING MEASUREMENTS AFTER LEAD ADDITION

In all the following tests :
-   Cooling in the plugging indicator was carried out by counter-current circulation of air (air and Pb-Li in opposite directions).
    The conditions and results are given in Table 1.
-   These measurements confirmed the conclusions on previous tests Pb-Li freezing in all the plugging indicator and not only in the narrow section (Figure 7c).
    After modifying the composition of the alloy by adding lead, the following remarks can be made as regards the plugging measurements (compared to previous tests).
-   The average plugging temperature for the 5 tests is 244.2 °C (246 °C for the previous tests) ; there is therefore no significant difference ;
-   The eutectic melting and solidification temperatures were found to be the same for all the tests,

- The average duration of the solidification plateau is 14 s (26 s for all the previous tests). The solidification plateau is therefore shortened after lead addition.
- The average duration of the fusion plateau is 9 min. (12 min. for the previous tests). The fusion plateau is therefore slightly shortened,
- The inverted temperature peak that occurred at the unplugging instant (a few seconds before) occurred 3 to 5 minutes before unplugging in these tests,
- In the 5 tests, the unplugging temperature was equal to, or above, 327 °C (lead melting point temperature). It can be supposed that some lead precipitates during cooling at pellet level and that this temperature has then to be reached for unplugging to occur. Unplugging operations were more difficult than in previous tests.
- After the fusion plateau, the recordings show inflexion points that are much more accentuated than in the previous tests without lead addition. The first inflexion point is situated 3 to 5 K over the eutectic melting point. It was sometimes visible in previous tests. The second inflexion point was situated 10 K over the eutectic melting point, that is to say 244-245 °C. This second inflexion point did not exist in the previous tests.
- The very slight alteration of the alloy composition therefore had a detectable effect on the plugging measurements. Other more extensive pollutions can be envisaged to amplify the phenomena, especially in order to determine the liquidus.
  It seems that the phenomena at pellet level are hidden by the overlarge mass of Pb-Li in the plugging indicator that freezes or melts before and, or at the same time.

8. TASTELI Pb SAMPLING AFTER LEAD ADDITION

Six days after the second addition of lead, the Tasteli Pb No 3 sampling was carried out at 398 °C. The lithium measurement by means of atomic absorption spectrophotometry gave a content of 0,718 % in the upper part of the sample and 0,713% in the lower part. There is, therefore, a slight drop in the proportion of lithium compared to the Tasteli Pb 1 and 2 samples. These values are absolutely consistent with the mass of added lead.

9. CONCLUSION

The operation of all the components derived from sodium circuits : electromagnetic pump, electromagnetic flowmeter, flux distortion flowmeter, pressure transducers, level gauge, valves... was satisfactory.

The circuit filling, defreezing and draining operations did not pose a problem, the plugging measurements showed that the plugging indicator gives the eutectic Li17Pb83 melting point temperature (= 235 °C). The plugging temperature is about 245 °C with counter-current circulation.

The plugging curves are well reproducible if the tests were performed in the same range of conditions. The drop in the flowrate is extremely fast. No other anomaly on the flowrate curve is noted.

After changing the alloy composition by adding lead, the plugging temperature remains unchanged, but a few differences were noted concerning the duration of the solidification and fusion plateaus, the unplugging temperatures and the aspect of the temperature recording during reheating : there is an inflexion point at 243-245 °C in particular, that may indicate the liquidus. This shows that the plugging indicator is sensitive to very slight variations of composition without quantifying them. Further additions of lead should confirm these observations.

OXIDATION OF ZIRCONIUM-TITANIUM ALLOYS IN LIQUID SODIUM : VALIDATION OF A HOT TRAP, DETERMINATION OF THE KINETICS

C. Latgé, S. Sellier

DER/STML - CEA - Cadarache
F-13108 Saint-Paul-lez-Durance, France

1. INTRODUCTION

The principle of hot trap or "getter" operation is based on the capacity of the chosen material to oxidize when it is placed in the presence of sodium containing some amount of oxygen. A hot trap, therefore, is made of a highly reducing, metallic material likely to form, with the oxygen contained in the sodium, a more stable oxide than sodium oxide. This material can be soluble or insoluble as is the case in most industrial applications.

The physical form of the oxidized product can be :
- either an adherent, protective or porous oxidation film,
- on an adherent or non-adherent crust.

Hot trap oxygen getter capacity is high in the two following cases :
- if there is formation of an adherent but porous oxidation film,
- if there is formation of a non-adherent oxide crust.

In the latter case, the hot trap should be equipped with a filtering device able to retain solid oxide particles so as not to pollute the circuit on which it is installed.

Sodium purification kinetics is determined by the oxidation kinetics of the chosen material; the kinetics of the latter is determined by the limiting stage of the global, four-stage oxidation mechanism :
- oxygen diffusion in the sodium through the boundary layer,
- oxygen diffusion in the oxide layer formed,
- oxygen diffusion in the metal,
- oxide formation.

Obviously, in the case that the limiting stage is not the formation of an oxide, a certain number of parameters will influence the kinetics, such as :
- the hydraulics of the hot trap,
- the temperature,
- the oxygen content,
- the physical form of the oxidized product, etc ...

The choice of material for the hot trap must therefore be made knowing :
- its capacity,
- its purification rate,

and trap sizing must take into account the values of the parameters to be ensured so as to obtain the efficiency required (temperature, flowrate, ...).

Liquid Metal Systems, Edited by H.U. Borgstedt
and G. Frees, Plenum Press, New York, 1995

2. BIBLIOGRAPHIC STUDY

Reported here are the main studies that were carried out on materials likely to be used or that have actually been chosen for hot traps. It can be noted first that these studies, most of which were carried out in the sixties, and sometimes earlier, can be classified into two categories of contrasting, but mutually enlightening, objectives :

- Studies on materials that are not very oxidizable, for use in fuel cladding fabrication [1],[2];
- Studies on materials that are easily oxidizable, more reducing than the above, for use in the oxygen content in the sodium is low; when it is high, there should be formation of the hot traps so as to protect the latter from corrosion [3], [4], [5].

There are no recent studies on the subject since purification by cold trap has replaced that by hot trap because of its much easier implementation, its non-specificity for oxygen and the fact that it was superfluous to reach oxygen contents of less than 10^{-6} ppm in order to limit the corrosion of activation products.

The most reducing materials are zirconium and zirconium-based alloys : thus HINZE estimated the partial pressure of oxygen in equilibrium with zirconium oxide : 10^{-52} atm at 650°C compared with 10^{-33} for sodium oxide [6], [7]. A hot trap made up of zirconium was set up on the primary cooling system of the SRE reactor (U.S.A) [7]. The zirconium was in the form of grooved pellets and foil and was kept at a temperature of 650°C. The hot trap - with a surface area of 400 m^2 and a flowrate of 1.2 $m^3.h^{-1}$ - enabled the content to be reduced from about 3 ppm to 1 ppm in 200 hours (sodium mass treated : $22.7 \cdot 10^3$ kg). The oxide formed did not come away from the mass of zirconium.

In the USSR, it was shown that with high oxygen contents, a thick oxide layer formed on the zirconium surfaces, hindered oxygen diffusion and therefore slowed down the purification rate [6]. Moreover, it was shown that oxygen diffusion in the sodium played a significant role on transfer kinetics when the oxygen content was very low [5] ; this stage therefore appears to be limiting.

Tests on zirconium alloys with a view to using them as sub-assembly tube material showed that alloys containing aluminium, tin and molybdenum (zircaloys) could oxidize, but not enough to be used as hot traps [2].

J.M. WILLIAMS [4] studied several materials with a view to their possible use as hot traps : Zr, Ta, Ta-10W, Nb and 50 Ti - 50 Zr. This author thought that the stage limiting the kinetics of the Zr and 50 Ti - 50 Zr oxidation mechanisms was a combination of the formation kinetics of an adherent film and the diffusion kinetics of oxygen in the oxide and the metal. Lastly he classified the materials in decreasing order in regards to their trapping efficiency :

$$50 \text{ Zr - 50 Ti} > \text{Zr} > \quad > \text{Nb} > \text{Ta} > \text{Ta - 10 W}$$

L. CHAMPEIX [1] studied the corrosion kinetics of vanadium alloys (V- 20% Ti, V-10% Al, V- 6% Cr, V- 5% Nb, V- 1.5% Zr and V- 0.6% Si) and showed that only alloys with Zr did not bring about the formation of a protective oxidized layer able to limit oxygen diffusion.

R.L. KLUEH and J.H. DEVAN [8] also studied the corrosion of vanadium and certain of its alloys and drew the following conclusion from their investigation : Cr, Mo, Fe, Ta and Nb reduce the solubility of oxygen in the vanadium whereas Zr and Ti increase it and thus favour vanadium corrosion ; oxygen is present in 2 forms : combined in Zr, Ti on V oxides or dissolved in the alloyed phase.

For a description of the corrosion of Zr and Ti by oxidation, M.G. BARKER and D.J. WOOD [9] can also be referred to Zr and Ti which should oxidize in the form TiO and ZrO_2 when the oxygen content in Na is low; when it is high, there should be the formation of ternary oxides Na_2ZrO_3 and Na_4TiO_4, but in the case of Zr only after being exposed to

226

the sodium for a long period of time. The component Na_4TiO_4 would appear to be non-adherent whereas ZrO_2 would seem to be adherent.

One of the most interesting studies is the one reported by J.M. Mc. KEE [3] as it gives the oxygen trapping efficiencies of a great many highly reducing materials that could be used as hot traps: zirconium alloyed with Y, Al and Ti at different atomic concentrations. He verified that it was with Zr 0.87 - Ti 0.13 (% in atoms) that the highest oxygen purification rates were obtained. He was able to confirm that the rate was much higher for Zr 0.87 - Ti 0.13 than for Zr 0.5 - Ti 0.5 (ratio from 5 to 10 at 650 °C after 12 h. for [O] = $150 \cdot 10^{-6}$) and attributed this discrepancy to a difference of metallographic structure.

He ascribed the high performances of these two materials, however, to the fact that the corrosion kinetics is not apparently limited by oxygen diffusion through the oxidized layer as in the case of pure zirconium (which limits its capacity). He showed moreover that temperature seems to have little effect on the kinetics.

In the light of this brief bibliographical review, it can be noted that the alloy Zr 0.87 - Ti 0.13 has definite advantages: a high oxidation rate at a relatively low temperature (from 430 °C to 500 °C) ; the cracking of the oxide layer however, obliges the designer to plan for a particle filter. This is why we decided to test this alloy in order to confirm its performance (purification rate, getter capacity).

3. EXPERIMENTAL STUDY

The tests were performed on the ECRIN test loop, described in [10], figure 1, which has the following main characteristics : a volume of sodium in circulation of about 0.050 m^3, an Na_2O inlet device, an oxygen-meter for continuous monitoring of the evolution of the oxygen content, a cold trap specifically loaded with oxide. The hot trap that was tested contained 0.321 kg of Zr 0.87 - Ti 0.13 alloy and was equipped downstream with a filter able to retain the particles produced.

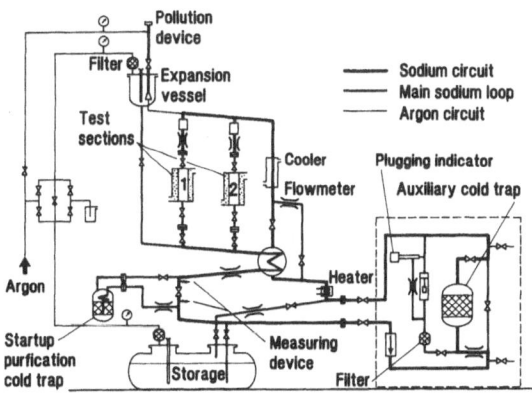

Fig. 1. Sketch of the ECRIN test loop

We investigated the influence of four essential parameters on the purification rate vs. the oxygen and vs. the final oxygen content at the end of the purification process. These four parameters are :

Tests 1 - Hot trap temperature (500°C, 460°C, 420°C, 380°C) ;
Tests 2 - Sodium flowrate (250 l.h^{-1}, 175 l.h^{-1}, 100 l.h^{-1}) ;
Tests 3 - Initial oxygen content (~ 10 ppm, ~ 250 ppm) ;
Tests 4 - Zr 0.87 Ti 0.13 oxidation rates.

The first two tests were considered to have been performed on a trap with a low oxidation state, which is not strictly true: the oxidation rate varied from 0 to about 8% (100% corresponding to a retention of 75.23 g. of oxygen, the final value obtained at the issue of the 4 tests).

4. MAIN FINDINGS

Thirteen tests were carried out ; their main characteristics are given in Table 1.
• Influence of the temperature (Table 1, Test 1):
In figure 2, the evolution of the oxygen content as a function of time is shown for the tests aimed to study the influence of the hot trap temperature.

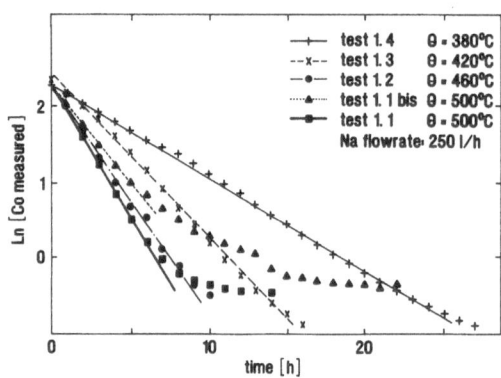

Fig. 2. Change of the oxygen content in sodium as a function of the reaction time at various temperatures

The following facts were clearly noted :
- the temperature has considerable effect on the oxidation kinetics: oxidation rate increases when the temperature increases ;
- the oxydation rate, which is very high for contents of over 3 ppm, decreases considerably below this value. The lower the temperature of the hot trap, the lower   the content at which this change of kinetics occurs (Fig. 2).
- the final content obtained was very low : < 0.3 ppm (under-saturated sodium at sodium melting point temperature).
• Influence of the sodium flowrate (Table 1, Test 2):
Let us consider the specific purification rate VS expressed in content variation per unit of time :
From tests 2.1, 2.2, and 2.3, it was deduced that :
at 250 l.h-1 VS = 360.10-3   ppm h^{-1}
at 175 l.h-1 VS = 355.10-3   "
at 100 l.h-1 VS = 189.10-3   "
In view of the small number of tests, it is difficult to draw a conclusion from this result. Nevertheless, if the mass transfer kinetics is limited by the diffusion kinetics of the ions 0= towards the Zr 0.87 - Ti 0.13 surface, logically the flowrate should significantly affect the global transfer kinetics, which does not seem to be the case here when 175 l/h is exceeded.
It can however be noted that a flowrate of 250 l · h^{-1} allows the highest purification rate (DNa . VS) to be obtained.
• Influence of the oxidation rate (Tests 3.2, 3.3, 3.4 and 3.5):
These tests were carried out at a temperature of 550°C and with a sodium flowrate of 250

$1 \cdot h^{-1}$. So as to study this influence, we compared the purification rates expressed in g O s^{-1} of the different tests in different content ranges (Fig. 3). We observed that the oxidation rate had no significant effect on the purification rate; this result is consistent with the fact that the oxide formed is not a common one and is likely to flake off.

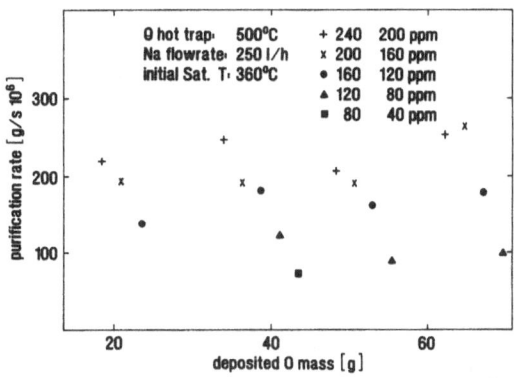

Fig. 3.: Comparision of purification rates at different ranges of contents of oxygen

• Influence of the Initial Oxygen Content:
No significant influence of this parameter was noted. This is consistent with the expression of the oxidation kinetics.

5. OXIDATION KINETICS

The kinetics was deduced from the integration of the mass balance as regards the oxygen in the ECRIN loop.

$$M (dc/dt) = -k_1 (T). C^n \tag{1}$$
$$\text{with } k_1 (T) = k_0 \exp (-E/RT)$$

where :

M designates the mass sodium in the circuit
C designates the oxygen content
n designates the order of the reaction
$k_1(T)$ designates the rate constant
k_0 designates a constant
E designates the "activation" energy of the phenomenon $(J.mol^{-1})$
R designates Boltzman's constant $(J.mol^{-1}.K^{-1})$
T designates the hot trap temperature (K).

In view of the shape of the curves ln C = f(t), we were able to verify that the kinetics order in regards to the concentration of oxygen was 1.
The integration of the equation (1) leads to the following equation :

$$\ln C = - (k_1(T)/M) \cdot t + \ln Co \tag{2}$$

The graphic determination of the slope allows k_1 (T) to be determined for each curve. We can also write that :

$$\ln k_1(T) = \ln K_0 - [(E/R) \cdot (1/T)] \tag{3}$$

The slope of the straight line representing ln $k^1(T)$ = f(1/T) allowed the "activation" energy of the phenomenon to be calculated (figure 4).
Considering that the oxidation phenomenon is a surface phenomenon, this kinetics is

better expressed in kg [O].h^{-1}.m^{-2}.

$$V = 41.26 \cdot 10^{-3} \cdot \exp(-40.3 \times 10^3/RT).C \quad \text{kg [O].h}^{-1}\text{m}^{-2} \tag{4}$$

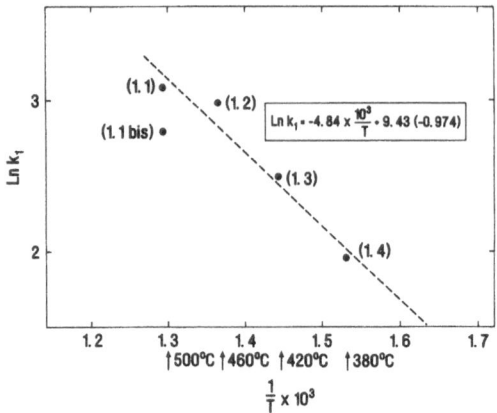

Fig. 4. Function of the rate constant vs. reciprocal temperature of the oxidation reaction

This first order kinetics was established in the content range 1 ppm - 10 ppm. We checked that it was still valid in the 10 ppm - 250 ppm range. A first order law is characteristic of a diffusion phenomenon, or of a phenomenon successively involving several stages, whose limiting stage is a diffusion stage. In our case, diffusion of the oxygen ions $O^=$, through the boundary layer of the sodium in contact with the alloy can be involved and/or the diffusion of $O^=$ in the oxide layer formed, and/or the diffusion of $O^=$ in the Zr 0.87-Ti 0.13 alloy.
Since :
   a)- the flowrate does not appear to have any significant effect on the oxidation rate beyond 175 l h^{-1} ; and
   b)- the oxide layer does not cover the whole surface and the oxidation kinetics does not depend on the oxidation rate,
it is highly probable that it is the stage of oxygen diffusion in the alloy itself that limits the overall oxidation kinetics beyond 175 l · h^{-1}.
   As is shown by the shape of the curves C = f(t), the kinetics does not seem to be of order 1 when the content is lower than 1 ppm ; as knowing the expression of the kinetics in this content range was of no particular interest, it was not determined.

6. OXYGEN RETENTION CAPACITY

Table 1 shows the amounts of oxygen retained in each test : the total mass of oxygen introduced was 75.23 g [O] that is to say 0.234 g [O]/g Zr 0.87 Ti0.13, a value very similar to that published by J.M. Mc KEE [3], i.e. 0.25 g [O]/g Zr 0.87 Ti0.13.
   Later evaluations showed that the larger part of the alloy had been turned into oxide, thus confirming that the capacity value obtained was in fact very near the maximal capacity. The filter set up downstream retained all the particles produced by the oxidation.

7. CONCLUSIONS

This study enabled the high efficiency of the Zr 0.87 - Ti O.13 alloy to be verified, from the point of view of the purification rate as well as of its retention capacity. Furthermore, it enabled the influence of several parameters on the purification rate to be defined; in particular it is to be noted that the temperature has a very considerable effect on the oxidation kinetics; on the other hand, the flow rate, the oxidation rate and the oxygen content have little effect on the oxidation kinetics.

Table 1. Qualification of the Zr0.87 Ti0.13 alloy: Characteristics of tests

Test	Flow rate of Na (l/h)	Temperature of hot trap (°C)	Initial content (µg/g)	Cumulated mass of oxygen (µg/g)
1.1	250	500	36.3	2.12
1.2	250	460	10.2	2.68
1.3	250	420	10.2	3.26
1.4	250	380	9.8	3.85
1.1 (2x)	250	500	10.2	4.41
2.1	250	420	9.3	4.94
2.2	175	420	10.2	5.52
2.3	100	420	9.36	6.06
3.1	250	420	243.9	17.76
3.2	250	500	264.8	33.06
3.3	250	500	254.4	47.36
3.4	250	500	244.3	61.43
3.5	250	500	244.3	76.23

Tests 1 : Study of hot trap temperature variations.
Tests 2 : Study of sodium flow rate variations.
Tests 3 : Study of the variation of the initial oxygen content and the variation of the oxidation rate of the alloy.

The oxidation kinetics being of first order, the limiting stage of the phenomenon is probably oxygen diffusion in the Zr 0.87 - Ti 0.13 alloy.

Finally, oxide flaking made it necessary to set up a filtering device downstream of the hot trap.

Acknowledgement

The authors would like to thank the metallographic analysis and consultancy laboratories of the LEPE, A. Deries and L. Pignoly for their help.

REFERENCES

[1] L. Champeix, R. Darras, and J. Sannier, "Alkali Metal Coolants", IAEA, Vienna 1967, pp. 45-62.
[2] T.L. Mackay, USAEC Report NAA-SR 6674, 1962.
[3] J.M. McKee, Technical Report AFAPL-TR 66-27, 1966.
[4] J.M. Williams Proc. Conf. Gatlinburg, Tenn., USA, 1965; USAEC Report CONF 650411.
[5] V.I. Subbotin and F.A. Kozlov, USAEC Report ORNL-TR-1611, 1967;Translated from: Alkali Metal Coolants; IAEA, Vienna,1967, pp. 535-551.
[6] O.J. Foust "Sodium-NaK Engineering Handbook." Vol. 5. Gordon and Breach, Inc.,New York 1972.
[7] R.B. Hinze, USAEC Report NAA-SR-3638, Atomics International, 1959.
[8] R.L.Klueh and J.H. DeVan, J. Less-Common Met. 30 (1973) 9 and 25.
[9] M.G. Barker and D.J. Wood, J. Chem. Soc., Dalton Trans. (1972) 2451.
[10] C. Latgé, "Liquid Metal Engineering and Technology", Brit. Nuclear Energy Soc., Lonson 1984, Vol. 2, p. 395.

HYDROGEN REMOVAL FROM NaK WITH MESH-PACKED AND MESHLESS COLD TRAPS

J. Reimann, R. Kirchner, M. Pfeff and D. Rackel

Kernforschungszentrum Karlsruhe GmbH, IATF
P.O.Box 3640, D-76021 Karlsruhe, Germany

1. INTRODUCTION

Oxygen and hydrogen removal by cold trapping is a standard technique for liquid metals such as Na and NaK, compare e.g. [1]. In order to achieve high efficiencies these cold traps are often filled with wire meshs. In the past these cold traps often plugged much earlier than expected. This was due to an improper design because of an insufficient knowledge of the precipitation kinetics. Although considerable progress has been achieved in the last years in the understanding of the impurity removal from sodium, there were still open questions concerning the impurity removal at very low concentration ranges.

For tritium removal from a liquid metal-cooled fusion reactor blanket, an intermediate liquid metal loop was foreseen where only very low tritium concentrations are acceptable for safety and other reasons [2,3]. Tritium removal by cold trapping from Na or NaK was proposed. The eutectic NaK appears more favourable because lower saturation concentrations c_{sat} exist at low temperatures as it can be seen from the following relationships (from [4]):

$$NaK: \ln c_{sat} \text{ (wppm)} = 15.06 - 6286/T \text{ (K)} \tag{1}$$

$$Na: \ \ln c_{sat} \text{ (wppm)} = 14.89 - 6961/T(K)$$

The minimum values are achieved close to the melting point: $c_{sat} = 1,2 \cdot 10^{-4}$ wppm for NaK (at $T_m = -12°C$) and $c_{sat} = 2,3 \cdot 10^{-2}$ wppm for Na (at $T_m = 100°C$).

However, this advantage of NaK compared to Na is only of practical importance if the precipitation kinetics are fast enough that these low values can be obtained in technical systems. One purpose of the article is to present the results of corresponding investigations.

For a fusion blanket cold trap, the aim is to recover the precipitated tritium twice a day. Therefore, the design goal is not a cold trap with a large capacity (large precipitated mass) but a cold trap with a high efficiency.

In the past, meshless cold traps were proposed besides mesh-packed cold traps, compare [1]. The meshless cold traps are supposed to be less sensitive to premature plugging and, therefore, could serve as high capacity cold trap. Such cold traps are also favourable if the operational conditions change (e.g. during initial loop purification periods or after an impurity increase due to leaks). Due to the small specific surface provided for precipitation, the efficiency of meshless cold traps was generally smaller than those of mesh-packed cold traps. It is the second goal of this article to describe a meshless cold trap which uses a technique to improve the mass transfer and to present the corresponding results.

In the following some features of the cold traps are briefly summarized. Cold trapping is based on the decrease of the impurity saturation concentration c_{sat} (in the following only hydrogen is considered) with decreasing temperature. If, therefore, a dilute solution is cooled down, c_{sat} is reached at the corresponding saturation temperature T_{sat} and for further decreasing of the temperature the actual concentration c becomes larger than c_{sat} and hydride precipitation might start.

For precipitation, at first the nuclei must be provided and then the precipitation rate is governed by the growth of the crystals. Nucleation can occur at solid-liquid interfaces (heterogenous nucleation) or within the liquid metal (homogenous nucleation). Heterogeneous nucleation requires a much lower supersaturation (concentration difference $(c-c_{sat})$) than homogeneous nucleation.

The importance of the heterogeneous nucleation process on the total precipitation process is presently not clear. Most of the researchers state that nucleation for hydrogen precipitation is negligible (see e.g. [5]). However, from the French work of CEA Cadarache (see e.g. [6]) it is concluded that the nucleation determines the location of crystals and therefore cannot be neglected.

The local mass transfer rate $\dot{m}$ due to crystal growth is expressed by

$$\dot{m} = k \, A \, (c-c_{sat})^n , \qquad (2)$$

where k is the mass transfer coefficient, A is the surface relevant for crystal growth and n is the order of reaction. Again, there are two groups of opinions: The growth rate is either diffusion controlled (n=1) (see e.g. [5]) or controlled by the integration of the hydride molecule into the crystal lattice (see e.g. [6]). Unfortunately, the publication [6] does not contain sufficient information for a quantitative use of the results.

In the following, some characteristic features of mesh-packed and meshless cold traps are outlined (Figs. 1a and 1b). In both cases, liquid metal flowing in an annular gap is countercurrently cooled from the outside and is heated up again in the inner tube (regenerative heat exchanger) before leaving the cold trap. For mesh-packed cold traps, numerous designs exist in respect to the geometry of the inner ducts and the position of the wire mesh. From the analyses of various designs, it was concluded [5] that the wire mesh should be located in the annular gap which should have a large cross section. Meshless cold traps were compared in another article [1]. Figure 1b shows a special cold trap design [7]. The liquid is fast cooled down in the narrow annulus to generate a high supersaturation. The bottom of the cold trap is designed as a settling zone where the liquid velocity is so small that the crystals growing within the liquid metal (homogeneous nucleation) are not transported in the subsequent heating-up zone. (This zone is often equipped with wire mesh serving as a filter).

Figure 1c shows a recently proposed cold trap [8] which is designed for the precipitation of hydrogen in a meshless part followed by a subsequent mesh-packed section for the precipitation of oxygen. Here, however, it is claimed that hydrogen precipitation only occurs due to the heterogeneous nucleation at the inner vessel wall caused by the high hydrogen diffusivity and low activation energy of heterogeneous nucleation. The assumption is questionable whether the hydrogen diffusivity is sufficiently large to transport the hydrogen through the large bulk to the wall; in our opinion, a recirculation flow occurs due to buoyancy effects which improves the hydrogen transport to the vessel wall but might have some undesired mixing effects, too.

An improvement of a meshless cold trap, based on the three zone concept (cooling, settling, filtration) was achieved by using a stator of an electromotor to induce a rotational flow field [9], see Fig. 1d. This induced flow should improve the mass transfer to the cold trap walls in the settling zone and thus reducing the amount of supersaturation due to homogeneous nucleation.

In the following, a recently proposed cold trap design is presented [10,11] which is designed to combine the advantages of both cold trap types, i.e. high efficiency due to heterogeneous nucleation and low sensitivity for plugging.

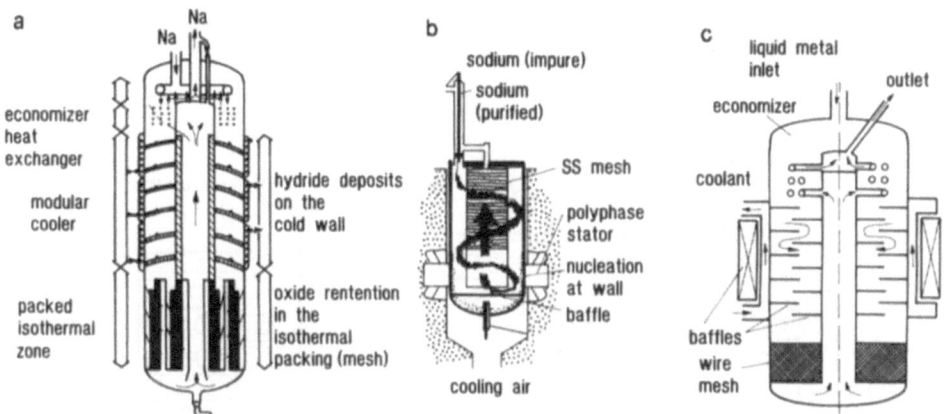

Fig. 1. Improved cold trap designs: a) Latgé et al (1988), b) Ray & Pohl (1972), c) new proposal

2. ROTATIONAL FLOW COLD TRAP (RCT)

Again, an electromagnetic stirrer is used to generate an angular velocity component within the fluid (the "rotational flow"). The annular duct is alternatively equipped with two types of concentric rings of metal sheet (or dense wire mesh), resulting in a meander-shaped flow in the axial direction, see Fig. 2a.

This pure meander-type flow is supposed to be not be very favourable for mass transfer because in many areas stagnant zones are expected to occur where the transport of hydrogen to the solid surfaces should be significantly deteriorated, see Fig. 2b.

If the bulk of the fluid is rotating due to electromotive forces, a secondary flow can occur which, for simplicity, is shown for the case that the flow rate through the cold trap is zero (Q=0), see Fig. 2c. This flow pattern is well known for rotating fluid systems (for example: the earth atmosphere). Boundary layers occur (generally called Ekman-layers) where the fluid has a velocity component v_r towards the rotational axis. (The reason for this is that within this boundary layer the centrifugal force is smaller than the pressure force.)

For characteristic cold trap flow rates, the corresponding mean velocity in the meander type ducts is about v_a=0.01 m/s. If an angular velocity ω of the fluid bulk exists, different mechanisms occur with increasing ω (compare e.g. [12,13]) which always improve mass transfer:

a)   At very low ω, the stagnant zones are swept away.

b)   At higher ω, the Ekman layers connected with the secondary flow system occur (laminar flow).

c)   The Ekman layers become unstable and Helmholtz-Taylor-vortices occur.

d)   Finally, the flow becomes turbulent which results in much thinner Ekman layers.

Presently, no theoretical analysis exists which can predict the superposition of the rotational and meander-type flow in this cold trap. Even the assessment of the rotational velocity field for the case of no throughput (Q>0) is connected with considerable uncertainties and requires the measurement of the existing magnetic field [14]. Therefore, specific experiments are required. If it is assumed that the liquid bulk rotates like a solid body, one can assess the radial velocity v_r [12]. For a frequency of 6 Hz, a mean radial velocity in the Ekman layer of $v_r = 0.2$ m/s is obtained at a characterstic radial position. This value is connected with the regimes b) or c). In this case v_r is much larger than v_a and one could suppose that the fluid rotation should govern the mass transfer.

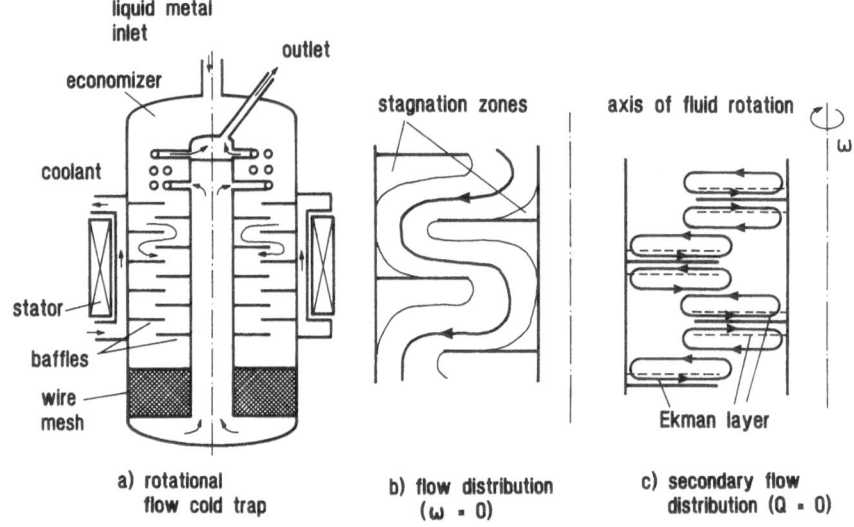

Fig. 2. Proposed meshless cold trap with rotational flow

3. EXPERIMENTS

The experiments were performed in the WAWIK facility [3] shown in Fig. 3. Hydrogen is permeated into the NaK through a coiled nickel tube. Two experimental cold traps (ECTs) were installed in the loop. At the cold trap inlets and outlets, the hydrogen concentrations were measured with a hydrogen meter based on the permeation through a nickel membrane and an ionic pump.

In most of the experiments the characteristic time periods were about 5 hours during which the cold traps were loaded with hydride. Then, the cold trap was drained from NaK, heated up to about 380°C, and the hydrogen was pumped off.

Figure 4 shows the different ECTs used in the present investigations: There is no internal back flow in order to have simpler temperature and velocity distributions. The cold trap cylinder has an inner diameter of 102 mm and a height of 500 mm. The mesh-packed cold trap (in the following designated with MCT) has an annular flow duct (the displacement body reduces radial temperature gradients and allows higher axial velocities). The specific surface of the wire mesh packing (wire diameter 0.3 mm) is $A_{spez}=1000$ m^2/m^3. This value is higher by a factor of about 3 compared to values currently used for high capacity MCTs.

Two types of cooling jackets were used: The MCT shown on the left side of Fig. 4 is characterized by a cooling zone which has the same axial length as the wire mesh packing whereas the cooling jacket of the other MCT is much shorter thus providing an extended isothermal zone downstream of the cooling zone. The first type of cooling jacket is better suited for the investigation of the precipitation kinetics because higher concentration differences ($c-c_{sat}$) should occur at the cold trap outlet due to the existing temperature gradient. Downstream of the cooling zone an isothermal zone reduces the concentration difference and, therefore, increases the efficiency. Therefore, the cold trap with an isothermal zone is recommended for technical applications.

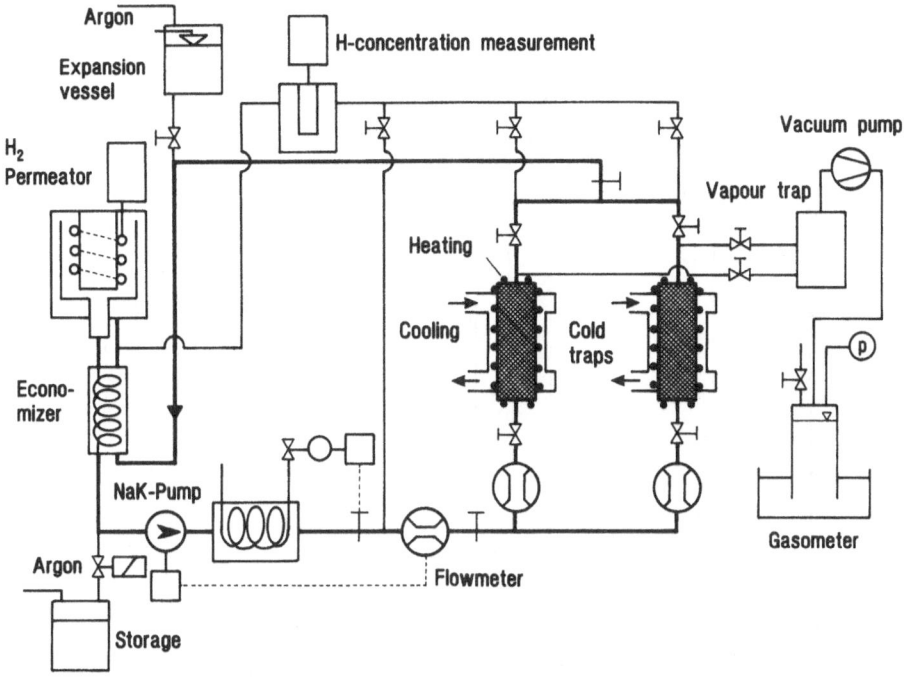

Fig. 3. WAWIK experimental facility

The rotational flow cold trap (RCT) uses a stator of an electromotor with two pairs of pole shoes. The frequency is 50 Hz, the power can be varied by means of a transformator. The specific surface of the RCT is about 17% of that of the MCT.

Each cold trap is instrumented with about 30 thermocouples for the measurement of the temperature distribution within the cold trap.

4. RESULTS

4.1 Mesh-packed Cold Trap (MCT)

The present experiments concentrated on the investigation of cold trap efficiencies at low concentrations; results on the behaviour of different cold traps at higher concentration ranges were published previously [15]. At low concentrations, the nucleation might play a significant role and cold trap efficiencies could be quite low. A model which does not take into account the nucleation could then predict too high values. Nevertheless, the results using such a model (Eq. (2)) are presented in Table 1. It was assumed that the mass transfer is dominated by diffusion (n = 1) and the mass transfer coefficient k was determined from the Sherwood number correlation given in [16]. A one-dimensional model was used (dependence of variables only in axial direction); the calculations were performed for a linear decrease of the wall temperature over the total axial length of the cold trap and an inlet concentration equal to the saturation concentration $c_i = c_{isat}$. The results are given for different inlet temperatures T_i keeping the temperature difference (T_i-T_o) constant. The corresponding saturation concentrations calculated from [4] are also listed. An additional parameter was the specific wire mesh surface A_{spec}.

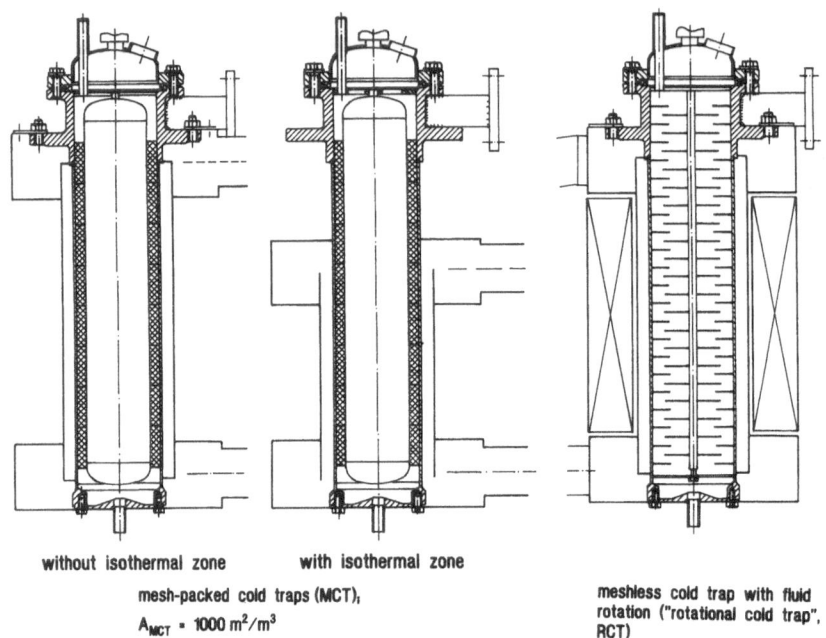

without isothermal zone   with isothermal zone

mesh-packed cold traps (MCT),

$A_{MCT} = 1000 \, m^2/m^3$

meshless cold trap with fluid rotation ("rotational cold trap", RCT)

Fig. 4. Experimental cold traps

The model calculates the outlet concentration c_o and with this the cold trap efficiency η, defined by

$$\eta = (c_i - c_o) / (c_i - c_{osat}) = 1 + (c_o - c_{osat}) / (c_i - c_o)^{-1} \qquad (3)$$

Table I shows that there is a significant decrease of η with decreasing specific surface of the wire mesh. With increasing throughput (characterized by the velocity v_{NaK}), the efficiency also decreases slightly. For a constant temperatures difference, η does not differ very much for different inlet temperature and concentration levels, respectively.

Figure 5 shows the experimental results for the MCT without isothermal zone obtained after a loading time period of 5 hours and the comparison with calculations which take into account the measured temperature distribution. It is seen that the experimental results are below the calculated values, however, the values are still surprisingly high. More experiments will be performed in the near future to increase the data base and to develop a more appropriate Sherwood number correlation.

Figure 6 shows the time dependency of the cold trap efficiency for an experiment where the cold trap was loaded continuously over a time period of 120 hrs. The efficiency increases fast within the first minutes. Here, the nucleation phase is of importance. After that, a slight increase occurs due to the increasing available surface for crystal growth.

Figure 7 contains results for the MCT with an isothermal zone for short loading periods. Compared to Fig. 5, the efficiencies a higher which is expected, as outlined before.

To summarize the results on MCTs one can state:

i)   Cold trapping of hydrogen from NaK can be very efficient at low concentration ranges. The nucleation process seems to be so fast that it does not play an important role for practical applications.

ii)   From these results it cannot decided yet if the crystal growth process is diffusion rate dominated. Future experiments where the hydride distribution within the cold trap will be determined will be very benificial.

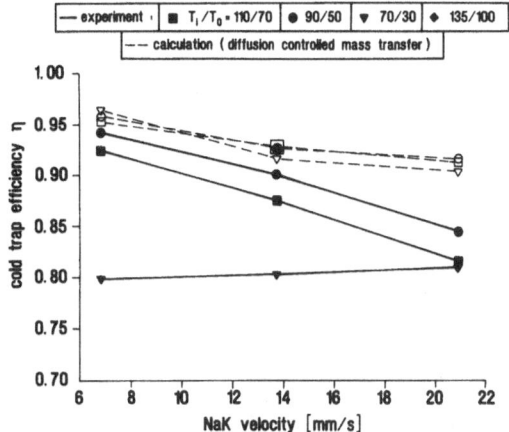

Fig. 5. Cold trap efficiencies after loading time periods of 5 hours (MCT without isothermal zone)

4.2 Rotational Flow Cold Trap (RCT)

Up to now only very few experiments were performed. Figure 8 shows a characteristic experiment: Before starting the hydrogen permeation (at about t=0 in Fig. 8), the loop was operated for about 10 hours with constant inlet and outlet temperatures and with an electrical stator power of about 600 W in order to obtain a concentration close to the saturation concentration $c_{0\,sat}$ at the cold trap outlet. About 9 minutes after the onset of permeation, the hydrogen front has travelled through the cold trap and a small bypass flow has reached the hydrogen meter. The outlet concentration c_0 increases; however, reaches a fairly constant value after some minutes. Then, the stator power is stepwise reduced down to zero which results in always higher levels of the outlet concentration.

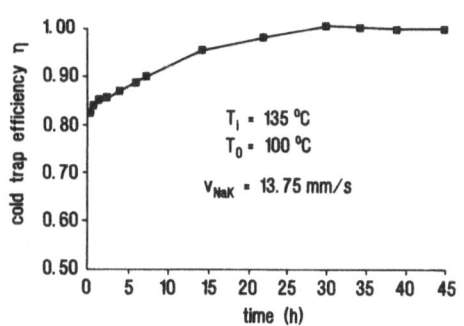

Fig. 6. Cold trap efficiency vs. time; MCT without isothermal zone

For the case of no flow rotation, the outlet concentration became so large that the hydride started very quickly to plug propably at the cold trap outlet. Therefore, the stator power was again increased to 600 W and nearly instantaneously the outlet concentration dropped closely to that value which existed previously.

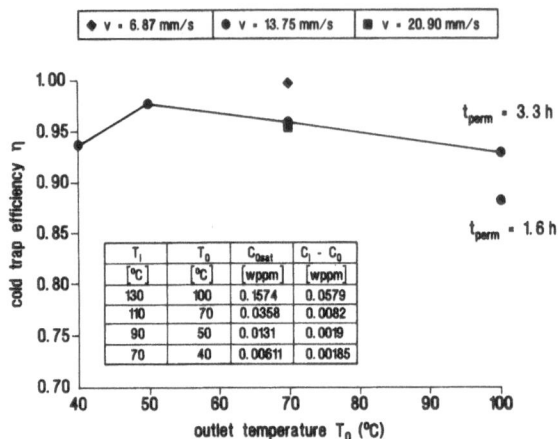

Fig. 7. Cold trap efficiencies after loading periods of 5 h or smaller (MCT with isothermal zone)

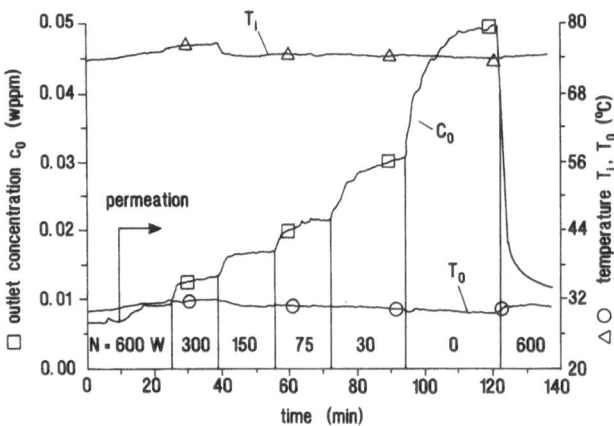

Fig. 8. Outlet concentration of RCT (starting with regenerated cold trap)

Figure 9 shows the cold trap efficiency for these different stator powers. This curve demonstrates clearly the improvement of mass transfer due to rotation.

Some comments should be given on the stator power measured as the product of current and voltage on the primary side of the transformator. This electric power consists of about 95% of "apparent power". Therefore, only about 5% of this value are transfered to the liquid metal. If the power is increased from zero to 600 W, a temperature rise and a reduction of the flow rate due to the increased pressure drop can be merely detected. (There is a small influence on the temperature distribution but this is attributed to the changed heat transfer characteristics).

The comparison between the MCTs and the RCT is difficult due to several reasons:

i)   The RCT experiment was performed with a regenerated cold trap and loading periods of less than one hour which results in lower efficiencies compared to the results given for the MCTs.

ii)   The results are compared with MCTs using a specific surface of A_{spec}=1000 m^2/m^3 which is much higher than the value currently used in MCTs for high capacity. For a more characteristic value of A_{spec}=300 m^2/m^3, the efficiencies are expected to be considerably smaller, compare Table 1.

iii) The fluid rotation is most pronounced in the area covered by the stator, compare Fig. 4. In the zone close to the cold trap outlet, the fluid rotation and with this the mass transfer improvement is supposingly not very significant. In this zone, the specific surface area of the investigated RCT is small which is unfavourable for achieving high efficiencies. For practical applications it is, therefore, recommended to use a wire mesh packing at the cold trap outlet, as indicated in Fig. 2. Then, the largest portion of the hydride would precipitate in the zone which is characterized by large ducts which are insensitive to premature plugging and only a very small mass flow rate would precipitate at the cold trap end which again would not result in premature plugging.

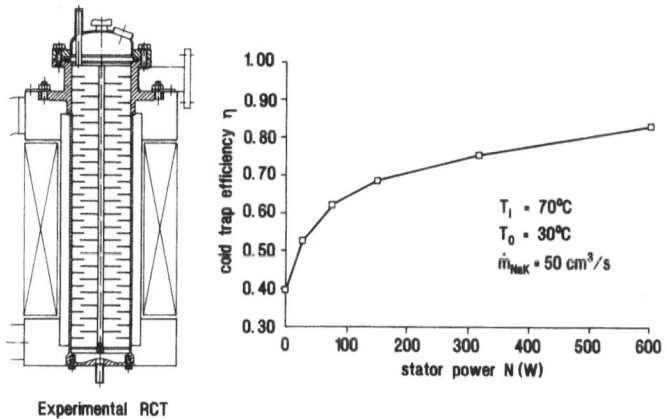

Experimental RCT

Fig. 9. Efficiency of RCT (regenerated cold trap)

One can conclude that this novel cold trap design might be of interest for several technical applications.

Acknowledgements

This work has been performed in the framework of the Nuclear Fusions Project of the Kernforschungszentrum and is supported by the European Communities within the European Fusion Technology Program.

Table I: Calculated efficiencies of mesh-packed cold traps (diffusion-limited mass transfer, mass transfer coeff. from [16])

T_i [°C]	T_o [°C]	c_{isat} [wppm]	c_{osat} [wppm]	A_{spec} [m2/m3]	v [mm/s]	$c_o - c_{osat}$ [wppm]	η [%]
140	100	0.857	0.168	1000	5	0.0230	96.7
		0.857	0.168	1000	10	0.0341	95.05
		0.857	0.168	1000	15	0.0434	93.7
140	100	0.857	0.168	300	5	0.0990	85.6
		0.857	0.168	300	10	0.150	78.25
70	30	0.0384	0.00343	1000	5	0.00130	96.3
		0.0384	0.00343	1000	10	0.00209	94.0
		0.0384	0.00343	300	5	0.00705	79.9
		0.0384	0.00343	300	10	0.0105	70.0

References

[1] R.B.Hinze, "Sodium-NaK Engineering Handbook", Vol. V, Ed. O.J. Foust, Gordon and Breach (1979).

[2] S. Malang, J. Reimann, and H. Sebening (Eds.), "Status Report KfK Contribution to the Development of DEMO-relevant Test Blankets for NET/ITER. Part 1: Self-cooled Liquid Metal Breeder Blanket. Vol. 1: Summary." KfK-4907 (December 1991).

[3] J. Reimann and S. Malang, Kernforschungszentrum Karlsruhe, Report KfK-4105 (October 1986).

[4] P. Hubberstey, in: "Handbook of Thermodynamic and Transport Properties of Alkali Metals", ed. R.W. Ohse, Blackwell Oxford, 1984 p. 885.

[5] C. McPheeters and D. Raue, "Liquid Metal Engineering and Technology", BNES, London (1984) vol. 2, p.371-378.

[6] C. Saint-Martin, C. Latgé, P. Michaille, and C. Laguerie, Fourth Int. Conf. "Liquid Metal Engineering and Technology", Paris 1988, Vol. 3 paper 617/1-10.

[7] R.L. Eichelberger, Sec. Int. Conf. Liquid Metal Techn. in Energy Production, Richland, USA, CONF-800401-p2 (1980) paper 20/33-38

[8] M. Latgé, Mme. Lagrande, M. Suraniti, and M. Ricard, Fourth Int. Conf. "Liquid Metal Engineering and Technology", Paris 1988, Vol. 3 paper 610/1-12.

[9] P. Roy and L.E. Pohl, Nuclear Technology 13 (1972) 284-288.

[10] J. Reimann, H. John, and S. Malang, Fusion Technology ?? (1988) 1306-1311.

[11] J. Reimann: patent pended

[12] F. Schulz-Grunow, Z. Angew. Math. Mech. 15 (1921) 191.

[13] M. Wimmer, Prog. Aerospace Sci. 25,(1988) 43-103.

[14] Davidson, P.A., J.Fluid Mech. 245 (1992) 669-699.

[15] J. Reimann, R. Kirchner, M. Pfeff and D. Rackel, Fourth Topical Meeting on "Tritium Technology in Fission, Fusion and Isotopic Applications", Albuquerque, New Mexico, USA, (1991) .

[16] H. Brauer, "Stoffaustausch einschließlich chemischer Reaktionen", Verlag Sauerländer, Hanau und Frankfurt (Main) (1971), p. 343.

STANDARD GIBBS ENERGY OF FORMATION OF K_3NbO_4 AND K_3TaO_4 AND THRESHOLD OXYGEN LEVELS FOR THEIR FORMATION IN POTASSIUM, NIOBIUM AND TANTALUM

N.P. Bhat* and H.U. Borgstedt**

* Indira Gandhi Centre for Atomic Research,
Kalpakkam 603 102, Tamil Nadu, India
** Kernforschungszentrum Karlsruhe GmbH
P.O.Box 3640, 76021 Karlsruhe, Germany

1. INTRODUCTION

Liquid potassium metal is being considered as a working fluid in heat pipes, heat transfer media in solar energy converters and as a thermodynamic working fluid in advanced Rankine-type power stations [1]. The high temperatures encountered in these systems necessitate the use of special containment materials for the liquid metals. Niobium, tantalum and their alloys have been investigated for this purpose. It has been shown that corrosion of these metals and alloys in liquid potassium is greatly influenced by oxygen present in the alloy or in the liquid metal [2, 3]. Corrosion is believed to be through the formation of the ternary oxides K_3NbO_4 and K_3TaO_4 which have been identified as corrosion products [4 - 6] of niobium and tantalum specimens, respectively, exposed to liquid potassium. A knowledge of thermodynamic stabilities of these ternary oxides would be essential to understand the corrosion mechanism of niobium and tantalum in liquid potassium. This investigation was undertaken in order to generate accurate data on Gibbs energy of formation of K_3NbO_4 and K_3TaO_4. This data could be utilised to predict threshold oxygen levels in the liquid metal (K) as well as in the solid metals (Nb, Ta) for the formation of these ternary oxides on the solid metals exposed to high temperature liquid potassium.

2. EXPERIMENTAL

Pure potassium metal sticks (Fluka AG) were cut free of oxide layer and grease and the metal was purified by vacuum distillation. Purified metal was kept molten at 100 °C in a stainless steel tray in an inert atmosphere glove box. Samples of potassium for the experiments were taken from this tray. The binary oxides Nb_2O_5 and Ta_2O_5 were of high purity (>99.99%; Alfa Ventron). Reaction studies of Ta_2O_5 with potassium were carried out in nickel crucibles enclosed in "Conoseal" capsules. About 0.25 g of Ta_2O_5 along with 5 g of potassium metal was taken in the nickel crucible enclosed in the "Conoseal" capsule and heated to the required temperature (673 K and 873 K) for 24 hours. The reaction product was isolated by distilling potassium under vacuum and characterised by x-ray diffraction.

Emf measurements were carried out with Harwell electrochemical oxygen meters (In, In_2O_3 reference) fitted to "Conoseal" capsules made of nickel [7]. About 0.5 g of binary oxide was taken in the "Conoseal" capsule along with 10 g of potassium. The capsule was assembled leak tight with the oxygen meter and slowly heated to the required temperature. The emf was measured with a high impedance digital voltmeter. Emf reading was taken when the value remained constant for atleast four hours. Temperature was raised in steps and emf measured at each temperature.

3. RESULTS

3.1 Reaction of Nb_2O_5 and Ta_2O_5 with Liquid Potassium

Addison et al. [8] have studied the reactions of oxides of niobium with liquid potassium at 673 and 873 K. At both the above temperatures, Nb_2O_5 reacts with liquid potassium according to the reaction

$$4\ Nb_2O_5 + 15K \rightarrow 3Nb + 5\ K_3NbO_4 \tag{1}$$

Tantalum pentoxide is expected to behave in a similar manner.

$$4\ Ta_2O_5 + 15K \rightarrow 3Ta + 5\ K_3TaO_4 \tag{2}$$

Since literature data was not available for this reaction, Ta_2O_5 was equilibrated with liquid potassium at 673 and 873 K. As shown in Table 1 the reaction products at both the temperatures, identified by XRD were found to be $Ta + K_3TaO_4$ confirming reaction (2).

Table 1. Results of equilibration experiments

Reactants	Temperature (K)	Duration (h)	Reaction products
$Ta_2O_5 + K$	673	24	$K + K_3TaO_4$
$Ta_2O_5 + K$	873	24	$K + K_3TaO_4$

3.2 Emf Measurements and Oxygen Potentials

On the basis of the above reactions, sample compartments of the cells I and II for emf measurements contain the three phase mixtures $K(l)$-$Nb(s)$-K_3NbO_4 and $K(l)$-$Ta(s)$ $K_3TaO_4(s)$, respectively. The two cells can be represented as

$$In,\ In_2O_3\ /\ YDT\ /\ K,\ Nb,\ K_3NbO_4 \quad Cell\ I$$

$$In,\ In_2O_3\ /\ YDT\ /\ K,\ Ta,\ K_3TaO_4\quad Cell\ II$$

The emf for these cells is related to oxygen potentials of the reference and sample electrodes by the Nernst equation

$$E\ (V) = 1/4F\ [\Delta\ \overline{G}_{O2}\ (ref) - \Delta\ \overline{G}_{O2}\ (sample)] \tag{3}$$

The results of emf measurements on the two cells are given in Table 2 and shown graphically in Fig. 1. Least square analysis of the data yielded following linear equations:

$$E \pm 1.2\ (mV)\ = 797.14 + 0.0038\ T(K) \tag{4}$$

$$E \pm 3.0\ (mV)\ = 584.23 + 0.2691\ T(K) \tag{5}$$

Substituting these values and literature data [9] for oxygen potential of the reference electrode into equation (3), oxygen potentials of the sample electrodes were calculated.

$$\Delta \overline{G}_{O2} + 1200 \text{ (J/mol)} = -931545 + 222.88 \text{ T(K)} \tag{6}$$
$$\Delta \overline{G}_{O2} + 1200 \text{ (J/mol)} = -682736 + 15783\text{T(K)} \tag{7}$$

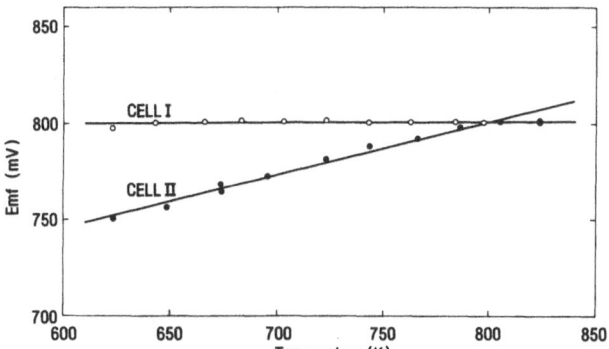

Fig. 1. Emf as a function of temperature for the cells (Cell I) In, In_2O_3 / YDT / K, Nb, K_3NbO_4 (Cell II) In, In_2O_3 / YDT / K, Ta, K_3TaO_4

Table 2. Emf data for the two cells

Cell I		Cell II	
In, In_2O_3 /YDT/ K, Nb,K_3NbO_4		In, In_2O_3 / YDT/ K, Ta, K_3TaO_4	
Temp. (K)	Emf (mV)	Temp. (K)	Emf (mV)
763	800.1	673	768,3
784	800.2	623	750.1
797	800.5	648	755.5
823	798.6	673	763.4
743	800.1	695	772.0
723	800.8	723	780.7
703	800.2	744	788.0
683	801.2	766	792.7
666	800.5	786	797.6
643	799.6	805	799.5
623	797.0	823	800.5

Table 3. Standard Gibbs energy of formation data for the ternary oxides

$\Delta G_{of,T}$ (J/mol) = (A+BT) + C				$\Delta G_{of,T}$ (kJ/mol)	
Ternary oxide	-A	B	C	673 K	873 K
K_3NbO_4	1863.090	445.76.	2400	1565.1	1475.9 this work
	1949.000	411.70	-	1671.9	1599.6 ref. [11]
K_3TaO_4	1698.744	240.98	3800	1536.6	1488.4 this work

These oxygen potentials corresponding to the three- phase fields K(l)-Nb(s)-K_3NbO_4(s) and K(l)-Ta(s)-K_3TaO_4(s), respectively, are plotted in Fig. 2 along with oxygen potentials for oxygen solution in liquid potassium [10]. Also plotted in the figure is estimated data of Lindemer et al. [11] for the phase field $K(l)$-$Nb(s)$-$K_3NbO_4(s)$. Oxygen potentials of the Nb-NbO and Ta-Ta_2O_5 binary phase fields [12] are compared with oxygen potentials of the ternary phase fields.

3.3 Standard Gibbs Energy of Formation Data for the Ternary Oxides

Standard Gibbs energy of formation data for the ternary oxides were calculated from the above oxygen potentials for the reactions

$$3K(l) + Nb(s) + 2 O_2(g) = K_3NbO_4(s) \tag{8}$$
$$3K(l) + Ta(s) + 2 O_2(g) = K_3TaO_4(s) \tag{9}$$

for which

$$\Delta \bar{G}_{O2} [K(l) - Nb(s) - K_3NbO_4(s)] = 1/2[\Delta G_{of,T} \, K_3NbO_4,(s)] \tag{10}$$
$$\Delta \bar{G}_{O2} [K(l) - Ta(s) - K_3TaO_4(s)] = 1/2[\Delta G_{of,T} \, K_3TaO_4,(s)] \tag{11}$$

substituting respective oxygen potential values from equations (6 and 7)

$$\Delta G_{ofT} \, K_3NbO_4 (s) + 2400 \, (J/mol) = 1863090 + 445.76 \, T(K) \tag{12}$$
$$\Delta G_{ofT} \, K_3TaO_4 (s) + 3800 \, (J/mol) = 1698744 + 240.98 \, T(K) \tag{13}$$

These data are represented in Table 3 along with estimated data of Lindemer et al. [11].

3.4 Threshold Oxygen Levels in Liquid Potassium for the formation of the Ternary Oxides

The following relation was derived previously on the basis of assessed data for Gibbs energy of formation of K_2O and recently measured oxygen solubility data in liquid potassium [10].

$$\Delta \bar{G}_{O2}(O_{soln.}, Na)(J/mol) = -704688 + 122,8 \, T + 38.286 \, \log C \, (O/K) \tag{14}$$

Equating this to respective oxygen potentials of the three phase systems from equations (6 and 7) and solving for $\log C(O/K)$ threshold oxygen levels in liquid potassium for the formation of K_3NbO_4 on niobium and K_3TaO_4 on tantalum metal respectively were obtained.

$$\log C \, (O/K) \, (ppm) = 2.6140 - 5925.3/T \, (K) \tag{15}$$
$$\log C \, (O/K) \, (ppm) = -0.062 \cdot - 3779/T \, (K) \tag{16}$$

These data are presented in Table 4 and are also evident in Fig. 2.

Table 4. Threshold oxygen concentrations C in liquid potassium and the solid metals for the formation of respective ternary oxides

Ternary oxide	Metal	$\log C$ (ppm) = A-B/T(K)		C (ppm)	
		A	B	673 K	873
K_3NbO_4	Potassium	2.6140	5925.3	$6.4 \cdot 10^{-7}$	$6.7 \cdot 10^{-5}$
	Niobium	5.7820	3821.5	1.23	25.4
K_3TaO_4	Potassium	-0.0627	3779	$2.1 \cdot 10^{-6}$	$4.1 \cdot 10^{-5}$
	Tantalum	2.8394	2603	0.09	0.72

3.5 Threshold Oxygen Levels in Niobium and Tantalum for the Formation of the Ternary Oxides

Interstitial oxygen in niobium and tantalum has been found to influence corrosion of these metals in liquid potassium [5, 6]. The ternary oxides K_3NbO_4 and K_3TaO_4 were identified respectively as corrosion products of the two metals. Threshold oxygen levels in these metals for the formation of the respective ternary oxide would be helpful to interpret the corrosion mechanism. A similar procedure as in sect. 3.4 above was adopted to calculate the

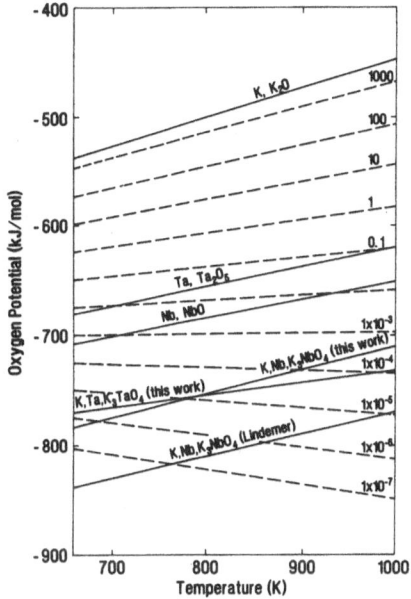

Fig. 2. Oxygen potentials and threshold oxygen levels in potassium for the formation of NbO, Ta_2O_5, K_3NbO_4 and K_3TaO_4

threshold oxygen levels. Literature data [13] or Gibbs energy of formation of NbO and Ta_2O_5 and solubility data [13] for oxygen solution in niobium and tantalum were utilised to calculate the oxygen potentials for oxygen solution in the two metals [12].

$$\Delta \overline{G}_{O_2}(O \text{ soln., Nb})(J/mol) = -785233 + 1.51\,T + 38.286\,T \log C(O/Nb) \qquad (17)$$

$$\Delta \overline{G}_{O_2}(O \text{ soln., Ta})(J/mol) = -749715 + 1.78\,T + 38.286\,T \log C(O/Ta) \qquad (18)$$

Equating these to respective oxygen potentials of the three phase systems from equations (6 and 7) results in:

$$\log C(O/Nb) \text{ (ppm)} = 5.7820 - 3821.5 / T(K) \qquad (19)$$

$$\log C(O/Ta) \text{ (ppm)} = 2.8394 - 2603 / T\ (K) \qquad (20)$$

These data are presented in Table 4 and is also shown in Figs. 3 and 4 along with oxygen potentials and threshold oxygen levels for the formation of respective ternary oxides of sodium with niobium and tantalum.

247

4. DISCUSSION

Tantalum pentoxide reacts with liquid potassium and liquid sodium in an identical manner giving tantalum metal and the respective ternary oxide [8, 14]. However, the reaction of Nb_2O_5 with liquid potassium differs from reaction of the binary oxide with liquid sodium. In the former case niobium metal and the ternary oxide K_3NbO_4 are formed whereas in the latter case NbO and Na_3NbO_4 are products of the reaction [15]. Oxygen potentials for the three phase fields K(l)-Nb(s)-K_3NbO_4 (s) and K(l)-Ta(s)-K_3TaO_4(s) as shown in Fig. 2 are much lower than those of the binary phase fields Nb-NbO and Ta-Ta_2O_5. Threshold oxygen levels for the formatioin of the two ternary oxides are also extremely low (1 to 5 parts per trillion). Ternary oxides will be formed of niobium and tantalum exposed to liquid potassium containing even extremely low levels of oxygen and the respective phase fields would buffer the oxygen potential in the liquid metal. Oxygen potentials for the formation of the binary oxides may not be reached at all. Hence there is no possibility of the binary oxide co-existing with liquid potassium. In this respect the K-Nb-O system differs from that of the Na-Nb-O system where the three phase field Na(l)-NbO(s) Na_3NbO_4(s) has been identified [15]. In the latter case the oxygen potential of Na(l)-NbO(s)-Na_3NbO_4(s) phase field is higher than that of the Na(l)-Nb(s)-NbO(s) phase field.

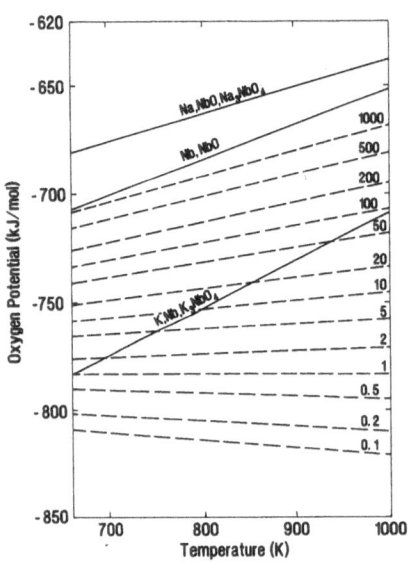

Fig. 3. Oxygen potentials and threshold oxygen levels in niobium for the formation of NbO, K_3NbO4 and Na_3NbO_4

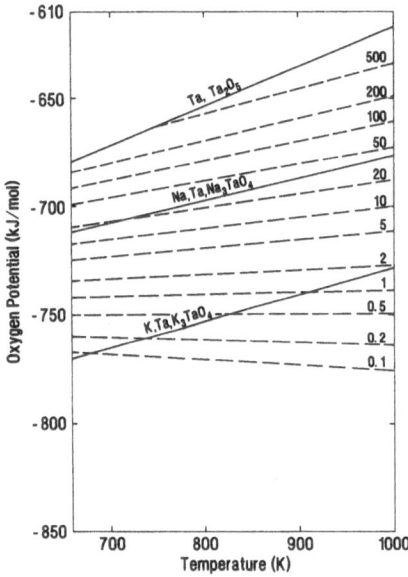

Fig. 4. Oxygen potentials and threshold oxygen levels in niobium for the formation of TaO, K_3TaO_4 and Na_3TaO_4

Extremely low oxygen levels in liquid potassium for the formation of the ternary oxides indicate that in normal operating liquid potassium systems where it is almost impossible to maintain the liquid metal to such high purity by cold trapping, corrosion of both the metals would take place. This has been observed in corrosion studies of niobium and tantalum in liquid sodium containing 15 to 20 ppm oxygen [5, 6]. Incidentally, niobium and tantalum metals appear to be excellent getter materials for oxygen in liquid potassium. The present experimental set up has shown the possibility of maintaining liquid potassium in a highly pure state with oxygen concentrations in the range of 10 to 50 parts per trillion. As shown in Figs. 3 and 4 comparatively low oxygen levels in the solid metals would result in corrosion of the metals in liquid potassium. These levels are much lower than those reported so far based on experimental observations where oxygen doped metals were exposed to liquid potassium. Stecura [5] identified the respective ternary oxides on niobium doped with 2800 to 4200 ppm oxygen and tantalum doped with 800 to 1930 ppm oxygen, when they were exposed to liquid potassium (15 ppm oxygen) at 1268 K. Hickam Jr. [6] reported the formation of K3TaO4 as corrosion product on tantalum (doped with 1600 to 3800 ppm oxygen) exposed to liquid potassium (20 ppm oxygen) at 1255 K. However, in their blank test, where tantalum metal containing 200 ppm oxygen was exposed to the same liquid potassium at 1255 K, no corrosion product was observed on the specimens. This may be due to limitations of the experimental technique used to identify the corrosion products.

Gibbs energy of formation data for K_3NbO_4 and K_3TaO_4 have been determined for the first time in this study. Only estimated data is available in literature [11] for Gibbs energy of formation of K_3NbO_4 and as shown in Fig. 2 this differs from the present data, as is observed with the Gibbs energy of formatioin data for Na_3NbO_4.

REFERENCES

1. H.U. Borgstedt and C.K. Mathews, "Applied Chemistry of the Alkali Metals", Plenum Press, New York, 1987.
2. L. Rosenblum, C.K. Scheuermann and T.A. Moss, "Alkali Metal Coolants", Int. Atom. Ener. Agency, Vienna 1967, p. 699.
3. J.R. Distefano and A.P. Litman, Corrosion 20 (1964) 382.
4. Coulson, C.K. Scheuermann and C.A. Barret, Report NASA-TN-D-3429 (1966).
5. S. Stecura, J. Less-Common Metals 25 (1971) 1.
6. C.W. Hickam Jr., J. Less Common Metals 14 (1968) 315.
7. N.P. Bhat and H.U. Borgstedt, Nucl. Technol 52 (1981) 153.
8. C.C. Addison, M.G. Barker and R.M. Lintonbon, J. Chem. Soc. (A) (1970) 1465.
9. T.J. Anderson and L.F.Donaghey, J. Chem. Thermodyn. 9 (1977) 617.
10. D. Krishnamurthy, A. Thiruvengadasamy, N.P. Bhat and C.K. Mathews, J. Less-Common Metals 135 (1987) 285.
11. J.B. Lindemer, T.M. Besmann and C.E. Johnson, J. Nucl.Mater. 100 (1981) 153.
12. N.P. Bhat and H.U. Borgstedt, Werkstoffe und Korrosion 39 (1988) 115.
13. O.M. Sreedharan and J.B. Gnanamurthy, J. Nucl. Mater. 89 (1980) 113.
14. M.G. Barker, A.J. Hooper and D.J. Wood, J. Chem. Soc. Dalton Trans. (1974) 55.
15. D. Krishnamurthy, N.P. Bhat and C.K. Mathews, Int. Symp. Thermochem. Chem. Process. The Indian Institute of Metals, Kalpakkam, 1991, p.301.

CAESIUM AND ITS MIXTURES: THEIR CHEMICAL REACTIONS WITH ALLOYS OF TRANSITION METALS USED TO CLAD REACTOR FUELS

R.J. Pulham, M.W. Richards, J.W. Hobbs

Chemistry Department,
University of Nottingham, Nottingham NG7 2RD
England, United Kingdom

1. INTRODUCTION

The elements Cs, Te and O_2 are considered to be the primary corrodants during burn-up of mixed oxide UO_2/PuO_2 fuel in Fast Breeder Nuclear Reactors, and the alloys PE16 and 12R72HV comprise advanced forms of fuel cladding to replace the originally employed M316 steel. The steels FV448 and DT2203Y05 are members of a ferritic class of alloys that might also play a role in the cladding under conditions of high burn-up. Most of the work [1-5] was devoted to determining the extent and nature of corrosion of the alloys PE16, 12R72HV, M316, FV448 and DT2203Y05 by $Cs/Te/O_2$ mixtures in sealed capsules for an arbitrary time of 168 h at 948 K, a temperature probably typical of the clad temperature in an operating fuel pin.

The oxygen potential, $\Delta \bar{G}_{O_2}$ increases with the fuel burn-up. For example, at 1300 K, the measured value for the fuel increases from about -500 to -400 kJmol^{-1} O_2 for burn-up from 3.8 to 11.2 at% (see [1]). To simulate this condition, the O_2 potential inside the capsule is controlled by metal/oxide couples. For example, the buffers Cr/Cr_2O_3, Cr (in M316 steel)/Cr_2O_3, Mo/MoO_2 and Ni/NiO give O_2 pressures corresponding to $\Delta \bar{G}_{O_2}$ values of -583, -575, -417 and -307 kJmol^{-1} O_2, respectively, at 948 K. The O_2 potential is controlled by the $\Delta_f G^o$ of the metal oxide; the less stable the oxide the higher is the O_2 potential.

The Cs:Te ratio, derived from the fission product yields, is relatively constant (ca. 7:1) throughout radiation, except during the first few days when Te is in large excess over Cs. Broadly, conditions range from Te rich initially to Cs rich with an overall increase in O_2 potential, and this background shaped the method and conditions of the laboratory work.

2. EXPERIMENTAL

The metal specimens (foils, 24-70 x 8 x 0.295-0.455 mm) were partly immersed in corrodant (Cs, Te or Cs-Te compounds, 1.5 g) contained in an Al_2O_3 crucible (20 mm high,

Liquid Metal Systems, Edited by H.U. Borgstedt
and G. Frees, Plenum Press, New York, 1995

11 mm diameter) sitting inside a Ni capsule (50 mm high, 20 mm diameter) which contained the O_2 buffer (ca. 3.0 g of an equimolar mixture of metal with its oxide). The loading of the corrodant, buffer and foil into the vertical Ni capsule was done in a glove-box filled with argon, and the capsule was sealed by welding on a lid. The assembly was mounted in a vertical furnace, and the internal gas pressure was balanced by an equal pressure of argon in the furnace. The temperature was raised quickly to 948 K and held constant for 168 h. At the end of the annealing the capsule was opened in a glove-box, and in some cases the surface of the foil was examined by x-ray diffraction. In other cases the foil was mounted in thermosetting Bakelite, metallurgically polished, and the cross section examined by optical and scanning electron/EDAX microscopy.

The compositions and origins of the alloys have been reported previously; PE16, M316 and FV448 [1], DT2203Y05 [4] and 12R72HV [5]. Briefly, 12R72HV (15.3 Cr, 14.8 Ni, 0.32 Ti and 65.9 Fe) resembles austenitic M316 (17 Cr, 13.5 Ni and 64.7 Fe), whereas Nimonic PE16 contains 16.5 Cr, 43.5 Ni and 33.7 Fe. Both FV448 (10.5 Cr and 86.1 Fe) and DT2203Y05 (13.2 Cr, 2.2 Ti and 82.3 Fe) are ferritic steels but the latter is an oxide (TiO_2/Y_2O_3) dispersion strengthened (ODS) form, manufactured by a special process.

To understand the significance of the Cs:Te ratios used it is necessary to show the nature of the corrodants when they are combined. This can be illustrated by a three dimensional phase diagram. Part of a probably diagram is shown in Fig. 1.

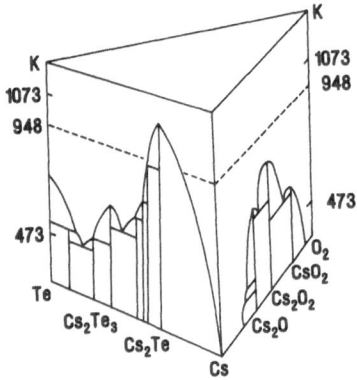

Fig. 1. Partial Cs - Te - O_2 phase diagram

There are two relevant basic corrodant systems, (Cs + O_2) and (Cs + Te) linked through Cs, so that these two phase diagrams comprise two faces of a trigonal prism; the third face is the (Te + O_2) phase diagram. The congruently melting compounds on the Cs + O_2 face are Cs_2O, Cs_2O_2 and CsO_2 but at 948 K, all of these would be in the molten state. Adding Te to Cs forms the congruently melting (1094 K) compound Cs_2Te, so that at 948 K the 4:1 composition is a mixture of solid Cs_2Te in equilibrium with its solution in Cs. Increasing the Te concentration forms the congruently melting (668 K) compound Cs_2Te_3 whereas the other phases decompose before they melt. At 948 K, the 2:1 ratio is solid Cs_2Te whereas both the 1:1 composition and elemental Te are liquid. The addition of O_2 to the Cs/Te mixtures from a buffer moves the composition into the interior of the prism, the nature of which is unknown.

3. RESULTS AND DISCUSSION

3.1 Cs + O_2

Generally, liquid Cs is inert to austenitic steels from 723 to 1273 K at O_2 potentials corresponding to 10 ppm O_2 in the metal for up to 10^3 h. Corrosion commences at ca. 500 ppm O_2 and increases with increasing O_2 potential and increasing temperature giving intergranular and transgranular corrosion. The molten salts and their mixtures Cs_2O-Cs_2O_2-CsO_2 which correspond to the highest O_2 potentials are extremely corrosive, and the dominant theme throughout is the formation of caesium chromium oxides (see refs. in [2]),

Table 1. Corrosion Data.

Alloy	Corrodant	ΔG_{O_2} kJ mol^{-1} O_2	Depth/µm	Type of Corrosion
PE16	Cs	-307	85	IG + Layer
M316	Cs	-307	>200	IG
FV448	Cs	-307	90	IG + TG
PE16	Cs:Te 4:1	-417	15-20	Layer
PE16	Cs:Te 4:1	-307	90	IG + Layer
M316	Cs:Te 4:1	-417	48	Layer
M316	Cs:Te 4:1	-307	200	IG
FV448	Cs:Te 4:1	-417	32	IG + TG
FV448	Cs:Te 4:1	-307	100	IG + TG
PE16	Cs:Te 2:1	-583	0	None
PE16	Cs:Te 2:1	-417	12	Layer
PE16	Cs:Te 2:1	-307	40	Layer
M316	Cs:Te 2:1	-417	30	Layer
M316	Cs:Te 2:1	-307	120	Layer
12R72HV	Cs:Te 2:1	-417	60	Layer
12R72HV	Cs:Te 2:1	-307	≥140	Layer
FV448	Cs:Te 2:1	-417	70-90(140)	IG + TG
FV448	Cs:Te 2:1	-307	120-130(180)	Layer
ODS	Cs:Te 2:1	-417	<10	TG
PE16	Cs:Te 1:1	-583	≥150	SIG
PE16	Cs:Te 1:1	-417	≥150	SIG
PE16	Cs:Te 1:1	-307	≥150	SIG
M316	Cs:Te 1:1	-417	≥200	SIG
FV448	Cs:Te 1:1	-417	150	SIG
FV448	Cs:Te 1:1	-307	150	SIG
ODS	Cs:Te 1:1	-417	50-70	S
PE16	Te	no buffer	20	S
PE16	Te	-417	20	S
M316	Te	no buffer	35	S
M316	Te	-417	33	S
M316	Te	<-575	43	S
FV448	Te	no buffer	25	S
FV448	Te	-417	no value	S

IG = Intergranular; TG = Transgranular; S = Solution.
SIG = Solution + IG; ≥ means that the foil was corroded through its entire thickness;
( ) = sporadic deeper corrosion.

The results [2] in Table 1 for M316 and Cs/Cs_2O confirm these earlier findings. In these later experiments the corrodant was metallic Cs exposed to an O_2 potential of -307 kJmol^{-1}

O_2 from the Ni/NiO buffer and it is estimated that the Cs is converted to a liquid containing up to ca. 33 mol% O, i.e. molten Cs_2O.

In particular, the steel is severely corroded giving intergranular penetration by Cs and O. PE16 is more corrosion resistant than is M316; the intergranular attack is less severe, and is accompanied by a layered type of corrosion. The chemistry involves the oxidation of Cr and Fe to Cs_2CrO_4/Cr_2O_3 and $CsFeO_2/Fe_3O_4$, respectively.

3.2 4 Cs + Te (+ O_2)

Disolute solutions of Te in liquid Cs are probably inert to the clad alloys in the absence of O_2. Caesium is inert and so also is solid Cs_2Te (Table 1, Cs:Te = 2:1, ΔG_{O_2} = - 583 kJmol^{-1} O_2) so that solutions of this salt in the metal should also be non-corrosive. The 4:1 mixtures, made up from Cs_2Te + 2 Cs components, become corrosive, however, when the O_2 potential increases. Thus at both of the experimental O_2 potentials, corrosion occurs and the depth of corrosion increases in the order PE16 < FV448 < M316 but is greater at -307 than at -417 kJmol^{-1} O_2.

3.3 2 Cs + Te (+ O_2)

In the absence of O_2, Cs_2Te is inert and this is consistent with thermodynamics (see [2]). The $\Delta_f G^o$ at 948 K (kJmol^{-1}) is -328 compared with the less negative value of -312 for Cr_2Te_3, which is the most stable known telluride of Cr. The driving force for Cs_2Te to react with the other transition metals is even less because the thermodynamic stability of their tellurides decreases in the order Cr > Ni > Fe.

Corrosion occurs, however, when the O_2 potential is increased to -417 kJmol^{-1} O_2. The depth increases in the order ODS < PE16 < M316 < 12R72HV < FV448. The extent of corrosion is more extensive at -307 kJmol^{-1} O_2 and the order is slightly changed to ODS < PE16 < M316 < FV448 < 12R72HV.

The corrosion of PE16, M316 and 12R72HV is characterized by the formation of alternate layers ostensibly of $Cs_3Cr(V)O_4/Cr_2Te_3/Cr_2O_3$ (dark) sandwiched between Ni/Fe (light). The Ni/Fe, or Ni in the case of PE16, is the residue of the steel after the Cr has been extracted by the corrodant. The increase in corrosion with increase in O_2 potential is associated with the formation now of $Cs_2Cr(VI)O_4$ and more extensive oxidation of Fe to $CsFeO_2$ and $FeTe_{0.9}$

$$3 \, Cs_2Te + 4 \, Cr + 4 \, O_2 \rightarrow 2 \, Cs_3CrO_4 + Cr_2Te_3$$
$$2 \, Cr + 3/2 \, O_2 \rightarrow Cr_2O_3$$
$$1/2 \, Cs_2Te + 14/9 \, Fe + O_2 \rightarrow CsFeO_2 + 5/9 \, FeTe_{0.9}$$

3.4 Cs + Te (+ O_2)

Compositions richer in Te than in Cs_2Te are very corrosive. Although Cs_2Te is inert to PE16 and M316 alloys, changing the composition to 1:1 (Cs_2Te + Te) causes solution and severe intergranular corrosion of PE16, M316 and FV448 alloys. The depth of corrosion increases in the order ODS < FV448 < PE16 < M316 at the O_2 potential of -417 kJmol^{-1} O_2, and there are signs that the extent is relatively insensitive to O_2 potential. Both PE16 and M316 alloys are penetrated through their entire thickness via grain boundaries, and some grains are rounded as if dissolving. Intergranular cohesion is so reduced that entire grains are rubbed out of the alloy surface during polishing. With PE16, Cs and Te are found in the distended grain boundaries which are enriched in Ni. With M316 the Ni is replaced by Cr. The corroded alloys carry a scale rich in Cs, Te and alloy components which appears to be the result of the solution process. With ODS steel, there is no intergranular corrosion but a scale grows at the expense of evenly consumed steel. With FV448 the scale is accompanied by sporadic intergranular corrosion.

3.5 Overview of the Depth of Corrosion.

A schematic trigonal prismatic diagram (Figure 2) can be drawn showing the extent of corrosion caused by Cs + O_2 along one face, and by Cs + Te along another; the third face is made up of Te + O_2.

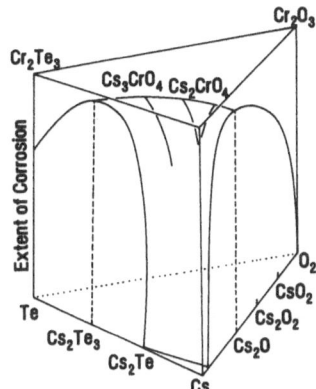

Fig. 2. Schematic representation of the variation in the extent of corrosion with corrodant composition

Caesium is inert but adding O_2 to Cs causes a steep increase in corrosion rising to a plateau of severe intergranular attack (M316 > PE16) probably at Cs_2O. Corrosion falls again to a very low value for O_2 due to the formation essentially of protective Cr_2O_3. Whereas adding O_2 to Cs causes corrosion almost immediately, adding Te to Cs (up to Cs_2Te) causes no corrosion. This is because Cs_2Te is more stable than Cs_2O and hence is less chemically active. Increasing the Te concentration, however, causes a large increase in corrosion (severe intergranular / solution; M316 > PE16) which centres on Cs_2Te_3 with Cr_2Te_3 being the dominant corrosion product. Corrosion then falls on progressing to Te but Cr_2Te_3 is the persistent product, and is analogous to Cr_2O_3 on the O_2 axis.

Although Cs_2Te is inert, adding O_2 to Cs_2Te causes immediate corrosion (alternate layering; M316 > PE16) and a ternary corrosion face can be constructed in the Cs corner of the Te-Cs-O_2 prism rising steeply to form a ridge of high corrosion linking Cs_2Te_3 with Cs_2O. Corrosion products along this ridge consist of Cr_2Te_3 + Cs_2Te at the starting point (Cs_2Te_3), but these are replaced by Cr_2O_3 + Cs chromates on moving towards Cs_2O. This system is not symmetrical because the high stability of Cs_2Te prevents the formation of the analogous caesium chromium tellurides. The scheme shows that the corrosion danger zones centre on Cs_2Te_3 (see later) and Cs_2O.

3.6 Implications

In a new fuel pin that has been operating briefly and where the Te:Cs fission yield is high, the corrosion is expected to be due to Te followed by the more serious solution intergranular attack by liquid Cs_2Te_3. Our results indicate that PE16 alloy is probably as susceptible as M316 steel to attack by the 1:1 mixtures and by liquid Cs_2Te_3 (see later). It is significant that Fission Product Induced Liquid Metal Embrittlement and the non-oxidative Clad Component Chemical Transport of M316 occur at compositions between Te and Cs_2Te and at Cs:Te rations between 0.7:1 and 2:1, respectively, both of which encompass the phase Cs_2Te_3 (see [3] for details).

In a pin that has undergone significant radiation and where the Cs:Te ratio has stabilised near 7.6:1 then severe corrosion is not expected but there might be a mild layering attack that is characteristic of Cs_2Te and $Cs_2Te + Cs$ at moderate O_2 potentials. These are the probable causes of the widely observed oxidative mild matrix attack that has been found by post radiation examinations (see [3] for details). Our results predict that PE16 alloy is more resistant than M316 to this type of corrosion.

In a high burn-up fuel pin where the O_2 potential has moved into the Cs_2O regime, then severe intergranular corrosion of M316 steel is expected but PE16 should be more resistant.

3.7 Nature of Corroding Species

There are two areas where the corrosion is severe sand where the species responsible are not obvious. These are (a) in the Cs:Te 1:1 region and (b) when Cs_2Te is exposed to O_2.

3.7.1 Region Cs : Te 1:1

It is suggested [3] that the very deleterious intergranular type of corrosion seen with Cs:Te 1:1 is caused by the congruently melting compound Cs_2Te_3 which persists in the liquid at 948 K (Figure 1). The composition of Cs:Te = 1:1 would not be molten Cs_2Te_3 solely but would consist of equimolar quantities of corrosive Cs_2Te_3 and inert Cs_2Te.

The next step has been to synthesize Cs_2Te_3 and to examine its corrosive effects on selected alloys. Capsule experiments were carried out as before at 948 K for 168 h using PE16, M316 and 12R72HV alloy foils, with and without Mo/MoO_2 buffers. The results are shown in Table 2.

Table 2. Corrosion by Cs_2Te_3

Alloy	Corrodant	ΔG_{O2} kJ mol^{-1} O_2	Depth/μm	Type of Corrosion
PE16	Cs:Te 2:3	no buffer	>200	Solution+Intergranular
PE16	Cs:Te 2:3	-417	>200	Solution+Intergranular
M316	Cs:Te 2:3	-417	>200	Solution
12R72HV	Cs:Te 2:3	no buffer	>200	Solution+Intergranular

The submerged ends (400 μm thickness) of all the foils had completely dissolved in the liquid Cs_2Te_3, and the higher regions of PE16 and 12R72HV were penetrated severely by solution (S) and intergranular attack (IG). The solution process with M316 appeared to be faster than the intergrain penetration; steel had been removed leaving an irregular edge.

With PE16 and 12R72HV, the corrodant had attacked preferentially between the grains but in addition the scanning electron microscope (SEM) showed some parallel transgrain lines of attack at the outer edge of the 12R72HV foil where the corrosion was most severe (Fig. 3). With both alloys the intergrain products contained Cs, Te and alloy components.

These results show that Cs_2Te_3 is indeed more corrosive than the Cs:Te 1:1 mixture, and support the premise that it is the Cs_2Te_3 component in the mixture that is responsible for corrosion. The increased corrosion is not due to elemental Te because this is less corrosive than Cs_2Te_3 and the nature is not intergranular.

Solid Cs_2Te_3 is a black, lustrous, ionic polytelluride with the K_2S_3-type of structure in which the V-shaped Te_3^{2-} ions are arranged in chains throughout the lattice (see refs. in [3]). Although the Gibbs free energy of formation is not known, it is expected to be less negative than that of Cs_2Te because of the disparity in their melting points. The driving force for the corrosion can be attributed to the formation of the very stable tellurides Cs_2Te (Δ_fG^0 = -328 kJmol^{-1}) and Cr_2Te_3 (Δ_fG^0 = -312 kJmol^{-1}). Thus the basic corrosion reaction is:

$$3 \; Cs_2Te_3(l) + 4 \; Cr(s) = 2 \; Cr_2Te_3(s) + 3 \; Cs_2Te(s)$$

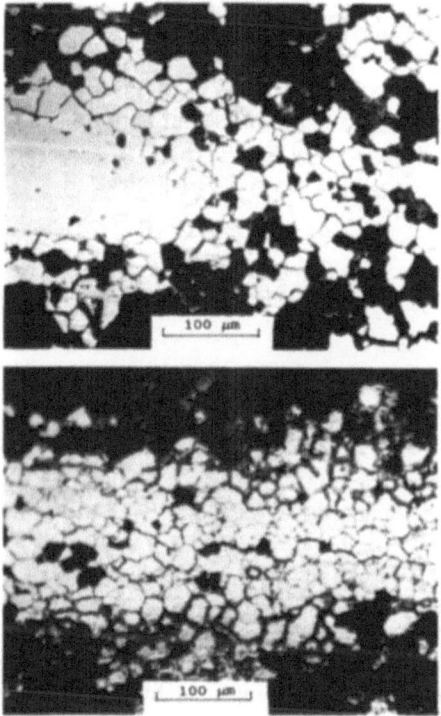

Fig. 3. optical photomicrographs of cross sections of PE16 (upper) and 12R72HV (lower) alloy showing intergranular corrosion by Cs_2Te_3

The nature of the corrosion was precisely the same as seen with the 1:1 Cs:Te mixtures. The extent of corrosion was much greater, however, there was no alloy remaining below the liquid, whereas with the 1:1 mixture there remained some coherence between the grains. The buffer made no detectable difference.

Assuming that $\Delta_fG^0(Cs_2Te_3)$ cannot be more negative than that of Cs_2Te, and knowing that the lattice energy has been partially overcome by melting, then $\Delta G^0(948 \; K)$ for the corrosion reaction must be at least as negative as -624 kJmol^{-1}.

The mechanism of this type of corrosion appears to involve the preferential diffusion of the large ions Cs^+ and Te_3^{2-} of the molten polytelluride through the grain boundaries, where the metal-metal bonding is weaker, forming transition metal tellurides that are soluble in the molten salt.

3.7.2 Region Cs_2Te/O_2

When the corrodant Cs_2Te is exposed to O_2 from the buffers Mo/MoO_2 and Ni/NiO (but not from Cr/Cr_2O_3), the yellow/green compound turns purple and partially liquefies to form an unknown species that is responsible for the corrosion. At very high O_2 potentials the mixture is converted to caesium tellurate, Cs_2TeO_4. This led us to believe [3] that Cs_2Te is oxidized at least in part to a caesium tellurite e.g. Cs_2TeO_3, at an O_2 potential slightly more negative than -417 kJmol^{-1} O_2. An alternative would be a mixture of the liquid compounds Cs_2Te_3 and Cs_2O:

$$Cs_2Te(s) + 3/2\ O_2 = Cs_2TeO_3(s)$$

$$3\ Cs_2Te(s) + O_2 = Cs_2Te_3(l) + 2\ Cs_2O(l)$$

The O_2 potential for the formation of Cs_2TeO_3 from Cs_2Te at unit activity is calculated to be -256 kJ·mol^{-1} O_2 [3] but this could be reduced for Cs_2TeO_3 towards -417 kJ·mol^{-1} O_2 by liquid solution in Cs_2Te.

The corrosion process then involves carrying O_2 to the alloy surface via the intermediate Cs_2TeO_3 and possibly also through the gas phase but both Cr and Cr_2O_3 are consumed by Cs_2TeO_3 at the surface to give the overall result:

$$Cs_2Te + 3/2\ O_2 = 3\ Cs_2TeO_3$$

$$2\ Cr + 3/2\ O_2 = Cr_2O_3$$

$$3\ Cs_2TeO_3 + Cr_2O_3 + 5\ Cr = 3\ Cs_2CrO_4 + Cr_2Te_3$$

It was necessary, clearly, to investigate the chemical and corrosive properties of Cs_2TeO_3.

3.7.3 Corrosion by Cs_2TeO_3.

Capsule experiments were carried out as before at 9948 K for 168 h using PE16, M316 and 12R72HV alloy foils partly immersed in Cs_2TeO_3, with Mo/MoO_2 buffers. Whereas this buffer served to add O_2 to Cs_2Te, it was expected to remove O_2 from Cs_2TeO_3 and come to the same equilibrium. The nature of the corrosion is shown in Fig. 4.

The main feature with all three alloys is the layered nature of the corrosion, and this is identical with that caused by $Cs_2Te + O_2$ [3]. This is good confirmation that the buffers absorb O_2 as well as release it. The clearest layering is seen with M316; the innermost damage is grey coloured and then this is replaced by alternate light and dark layers on progressing outwards from the substrate steel. The parallel zigzag morphology is apparent again.

The layering nature becomes progressively less obvious with 12R72HV and PE16 alloys, and PE16 showed some intergranular corrosion ahead of layering.

The layers were analysed by Electron Probe Micro Analysis (EPMA) at UKAEA Windscale and the overall finding was that the dark layers were rich in Cs, Te, O_2 and Cr, whereas the light layers were alloy denuded of Cr/Fe (PE16) or denuded of mainly Cr (12R72HV).

The transition metal analyses are the most revealing. Thus for PE16, whereas the part composition (%) of the original alloy is 16.5 Cr, 34 Fe, 43.5 Ni, a specific dark layer contained 25.5, 10.0, 27.8 showing enrichment in Cr. Conversely, a specific light layer had composition 7.42, 13.86, 74.8 showing that Cr and Fe are leached out of the alloy.

Similarly for 12R72HV; the original composition of 15 Cr, 65 Fe, 15 Ni is converted to 39.9, 14.3, 0.0 showing Cr enrichment in a dark layer, but a specific light layer contains 1.4, 50.3, 48.1 showing Cr depletion of the original alloy. Gradations are found on moving from the alloy outwards through the corrosion zone, but these do not affect the general picture.

The crystal structure of Cs_2TeO_3 shows that it is made up of Cs^+ and TeO_3^{2-} ions (see [3]) so that the mechanism might include diffusion of these species into the grain boundaries of the alloys. Reaction with Cr/Cr_2O_3 might convert these to Cs_2CrO_4 and Cr_2Te_3 (dark layers), and intergranular penetration is much favoured over transgranular attack. The Cr is replaced by diffusion from grains which become largely Fe/Ni (light layers). The parallel nature of the edges of each zigzag layer may be due to the geometry of the original grain boundary in which product grows at the expense of adjoining grains.

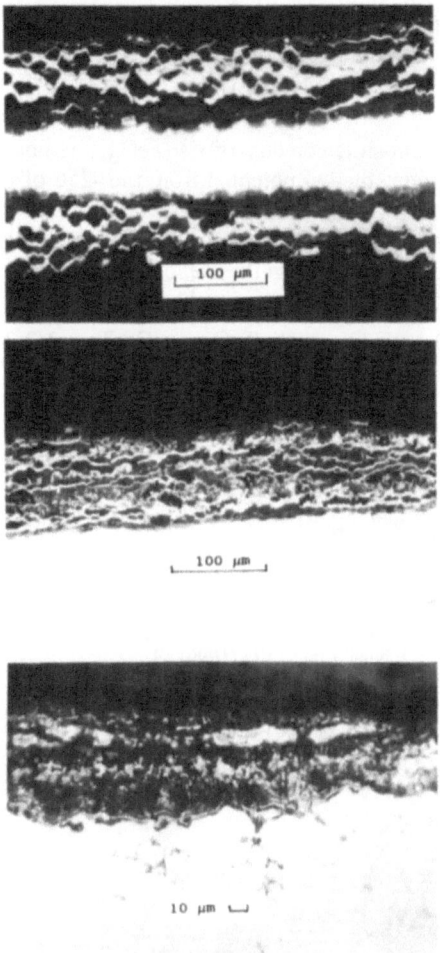

Fig. 4. Optical photomicrographs of cross sections of M316 (top), 12R72HV (middle) and PE16 (bottom) alloys showing layered corrosion by Cs_2TeO_3. For M316 the remaining metal is in the centre of the frame; for 12R71HV and PE16 the substrate metal is at the bottom of each frame.

The nature of the Cs_2TeO_3, however, had changed during the corrosion process from a white powder to dark purple mass that had been partially liquid. This indicated that the corroding species was not solely Cs_2TeO_3. Many experiments were carried out subsequently with Cs_2TeO_3 enclosed in capsules with a variety of buffers but in the absence of alloys. In all cases the Cs_2TeO_3 turned purple and glassy and the buffers gained O_2. Analysis by Atomic Emission for Cs, and Atomic Absorption for Te showed that the ratio of Cs:Te was still 2:1 in the residue. X-ray powder diffraction showed that the purple material was still basically Cs_2TeO_3. Precise measurements of the hexagonal unit cell, however, showed that there was a small but real change in cell dimensions.

$$a = b = 6.775, c = 7.931 \text{ Å}; V = 315.353 \text{ Å}^3$$

$$a = b = 6.732, c = 7.981 \text{ Å}; V = 313.33 \text{ Å}^3$$

The unit cell had decreased in volume, and there was an increase in the c dimension but a decrease in the a and b dimensions. It seems that the loss of O_2 leaves vacant sites, reduces

TeIV to TeII, decreases the cell volume, and the mixed oxidation state of Te is probably responsible for the purple colour.

$$Cs_2TeO_{3-x} = Cs_2Te(IV)_{1-x}Te(II)_xO_{3-x} + x/2\,O_2$$

The precise amount of non-stoichiometry of Cs_2TeO_{3-x} is not known at this stage, but appears large enough to reduce the O_2 potential from the -256 of Cs_2TeO_3 to at least -417 kJmol^{-1} O_2 (Mo/MoO$_2$).

The inference at this stage in the investigation is that the non-stoichiometric Cs_2TeO_{3-x} is the species responsible for the corrosion in these experiments and that this is formed whether Cs_2TeO_3 is reduced or whether Cs_2Te is oxidized by the buffers.

$$Cs_2TeO_3 - x/2\,O_2 \rightarrow Cs_2TeO_{3-x} \leftarrow Cs_2Te + 1/2(3-x)\,O_2$$

Acknowledgement

This work was sponsored by UKAEA Windscale and by the SERC.

References

[1] R.J. Pulham, M.W. Richards, J. Nucl. Mater. 171 (1990) 319.
[2] R.J. Pulham, M.W. Richards, J. Nucl. Mater. 172 (1990) 47.
[3] R.J. Pulham, M.W. Richards, J. Nucl. Mater. 172 (1990) 206.
[4] R.J. Pulham, M.W. Richards, J. Nucl. Mater. 172 (1990) 304.
[5] R.J. Pulham, M.W. Richards, J. Nucl. Mater. 187 (1992) 39.

REACTIONS OF LIQUID CAESIUM WITH
THE OXIDES OF URANIUM

S. E. Smith* and M. G. Barker

Department of Chemistry, University of Nottingham
Notttingham, NG7 2RD, U.K.
* Current Address:
School of Chemistry, University of Birmingham
Birmingham, B15 2TT, U.K.

1. INTRODUCTION

Interest in the Cs-U-O system stems from the fact that during burn-up of the fuel in fast breeder reactors caesium is produced as one of the major fission products, which may then react with the $(U,Pu)O_2$ fuel.

Post -irradiation studies of fuel pins have shown that caesium migrates both axially and radially along temperature and oxygen potential gradients which exist within the fuel pins. This migration results in the largest concentration of caesium being found at the fuel-breeder interface and at the fuel-cladding gap [1,2].

The reaction of the caesium with the mixed-oxide fuel or the UO_2 breeder pellets results in the production of low density caesium plutouranates and uranates, which can cause deformation and cracking of the fuel pin cladding. In addition the caesium, either by itself or combined with other fission products, can attack the steel cladding of the fuel pin, which could also lead to cladding failure and coolant-fuel interactions.

An extensive solid state study of the Cs-U-O system undertaken by Cordfunke and Van Egmond [3-7] revealed the presence of some ten compounds, several of which showed a variety of structural forms.

Of the caesium-rich compounds the monouranate Cs_2UO_4 was found to be hygroscopic and decomposed on heating in air at 650°C (923 K) to $Cs_2U_2O_7$ and $Cs_4U_5O_{17}$, which slowly decomposed at 1000°C (1273 K) yielding $Cs_2U_4O_{13}$.

Three forms of the diuranate $Cs_2U_2O_7$ were found to be stable towards air at room temperature. The β-phase was formed at temperatures in excess of 600°C (873 K), which on slow cooling to 300°C (573 K) reverted to the α-phase. The β-phase structure could be frozen in by rapid cooling from 600°C (873 K). The hexagonal γ-phase was metastable, readily converting to the β-form and its synthesis was considered fortuitous.

All these compounds were prepared by heating various ratios of caesium carbonate and uranium trioxide in gold boats at 600°C (873 K) for periods of several weeks. The most caesium-rich phase Cs_4UO_5, which had previously been proposed by Efremova [8] was, however, not observed in the study by Van Egmond.

Liquid Metal Systems, Edited by H.U. Borgstedt
and G. Frees, Plenum Press, New York, 1995

Several studies into the reactions of the uranium oxides and caesium uranates with liquid caesium have been carried out, but there exists far more uncertainty as to the stoichiometry of the phases produced, than in the case of the solid state reactions.

Various authors have reported a phase which was only stable in the presence of liquid caesium and that distillation of the caesium or exposure to oxygen or moisture (even the small concentrations in a glove box) resulted in the formation of Cs_2UO_4.

Cordfunke and Westrum [9] obtained a previously unreported phase, X, from the reactions of several caesium uranates with an excess of liquid caesium in sealed nickel capsules in the temperature range 550-800°C (823-1073 K). Reactions of liquid caesium with the caesium uranates (VI), $Cs_2U_2O_7$ or $Cs_2U_4O_{12}$, yielded X together with uranium dioxide, however the reaction of liquid caesium with Cs_2UO_4 produced only the X phase. Cordfunke and Westrum proposed that X may be a "reduced" phase (i.e. with a uranium valency of less than six) with the formula Cs_2UO_{4-y}.

Fee and Johnson [10] studied the reactions of liquid caesium with the uranium oxides, U_3O_8 and UO_3 and observed a similar phase, which the authors believed existed in two forms. The X-ray diffraction pattern obtained for the alpha phase consisted of the five strongest lines observed by Cordfunke and Westrum [9]. The beta phase was characterised by two additional lines at d = 4.37 and 2.18 Å. The authors proposed the formula $Cs_2UO_{3.56}$ for the phase.

A series of reactions of liquid caesium with the binary oxides of uranium and ternary caesium uranates performed by Lorenzelli et al [11] also yielded a similar phase. The X-ray powder diffraction pattern of this phase was temperature dependent. At high temperature, 1100°C (1373 K), the pattern resembled a face-centred cubic (f.c.c.) structure, with a unit cell dimension a_0 = 5.70 Å, however at lower temperatures, 680-900°C (953-1173 K), the pattern also consisted of a f.c.c. structure with a_0 = 5.63 Å. The authors suggested that this phase was Cs_3UO_4; analogous to the sodium uranate which exists in equilibrium with liquid sodium and uranium dioxide.

The highly reactive nature of this phase has so far prevented positive characterisation.

2. PRESENT STUDY

This study extends the previous work on the reactions of liquid caesium with the oxides of uranium and attempts to elucidate the identity of the unknown phase. We have also investigated the system by means of an electrochemical cell.

As part of this investigation we have also repeated some of the solid state reactions by Van Egmond and can confirm that the phase Cs_4UO_5 may not be prepared by the reaction of uranium trioxide with caesium carbonate. Reactions of the appropriate molar ratio yielded a mixture of Cs_2UO_4 and excess caesium carbonate. This does not however, rule out the existence of such a phase, which may possibly be prepared by the use of the more reactive Cs_2O.

In reactions leading to the formation of caesium diuranate, $Cs_2U_2O_7$, the metastable γ-phase was always obtained. This observation is somewhat surprising in the light the repeated attempts by Van Egmond to obtain the hexagonal γ-phase after its initial, fortuitous preparation [7].

3. EXPERIMENTAL

Our experimental approach was to react the binary uranium oxides with an excess of liquid caesium; and any caesium metal remaining at the end of the reaction was removed by vacuum distillation.

The reactivity of caesium towards the atmosphere necessitated the use of inert atmosphere handling techniques. The high vapour pressure of caesium meant it was necessary to carry out the reactions in sealed crucibles to avoid the premature distillation of the caesium during the reactions. Thus the filling of the reaction crucibles with liquid caesium was carried out in an evacuable, argon-filled glove box fitted with electrical connections, enabling the stainless steel crucibles to be welded under inert conditions.

Caesium metal was melted and slowly pipetted into the stainless steel crucibles (diameter 10 mm, length 100 mm) containing the oxide reactants. The lids were argon-arc welded on to the crucibles and removed from the glove box. The crucibles were heated at 600°C (873K) for 2-5 days. Individual reaction details are shown in Table 1.

Table 1. Reactions of Excess Liquid Caesium with the Binary Uranium Oxides

Reaction Number	Reactants	Temperature [°C]	[K]	Time [h]	Products
1	$UO_2 + Cs$	600	873	120	UO_2
2	$U_3O_8 + Cs$	600	873	120	UO_2
3	$UO_3 + Cs$	600	873	120	Cs_2UO_4
4	$U_3O_8 + Cs$	600	873	48	$UO_2 + X$
5	$UO_3 + Cs$	600	873	48	$UO_2 + X$

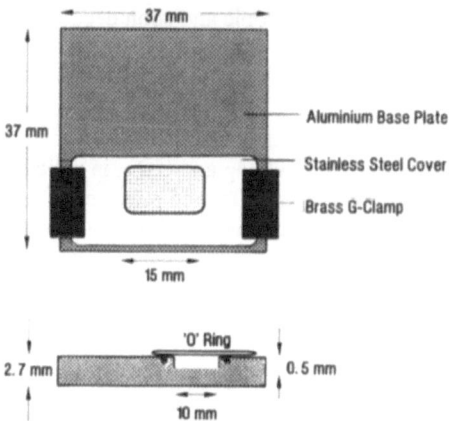

Fig. 1. Air-sensitive sample holder for X-ray powder diffraction

After cooling the capsules were returned to the glove box and cut open using a cutting wheel. The opened crucibles from Reactions 1-3 were placed in a distillation vessel and the excess caesium was removed by vacuum distillation at 250°C. The products retrieved from the crucibles were analysed by X-ray powder diffraction.

The products from Reactions 4 and 5 were not separated from the liquid caesium, however X-ray powder diffraction patterns were recorded of the product slurry, utilising the specially designed air-tight sample holder, described below and shown in Figure 1.

The holder comprised of an aluminium base (38 x 38 x 2.75 mm) which had an oblong, sample depression (0.5 x 15 x 10 mm) outside which a groove held an 'O'- ring. Mylar film and a steel sheet (0.5 mm thick), with a hole (15 x 10 mm) to reveal the sample, were held on to the 'O'- ring to provide an air-tight seal, by two small brass G-clamps. This holder was loaded into the diffractometer sample chamber in a flow of dry nitrogen. This new design was successful in obtaining the X-ray powder diffraction patterns of very air-sensitive compounds.

4. RESULTS

From Table 1 it can been seen that Reaction 1 showed the stability of uranium dioxide in liquid caesium and Reaction 2, led to the reduction of U_3O_8 to the dioxide by liquid caesium. The Reaction 3 of uranium trioxide led to the formation of caesium monouranate, Cs_2UO_4 as follows;

$$2\,Cs + UO_3 + 1/2\,O_2 \text{ (dissolved)} \rightarrow Cs_2UO_4$$

The X-ray powder diffraction patterns obtained for the products from Reactions 4 and 5 were of poor quality. In spite of this, however, it was possible to identify peaks which corresponded to the most intense lines for the X phase reported by Cordfunke and Westrum [9] as well as those for UO_2. These most intense peaks also corresponded with the two cubic structures proposed by Lorenzelli et al [11], with unit cell dimensions $a_o = 5.62, 5.70$ Å. The two additional lines of the β-phase reported by Fee and Johnson [10] were not observed. It may well be that the reduced phase has two cubic structures observed by ourselves and Lorenzelli et al [11], with the weak lines seen by Cordfunke and Westrum resulting from decomposition products.

The X-ray powder diffraction pattern of the product from Reaction 4 is compared with that of the reduced phase obtained by Cordfunke and Westrum in Table 2. The four strongest peaks were also observed in the X-ray powder diffraction pattern from Reaction 5.

The product slurries were very hygroscopic and soon decomposed.

5. DISCUSSION

The reactions of liquid caesium with U_3O_8 and UO_3, in which the liquid caesium was not removed by distillation, are further evidence for the presence of a reduced caesium uranate in equilibrium with liquid caesium. Unfortunately the instability of the phase prevented further analysis and characterisation.

Considering the initial Reactions 1-3 it is likely that the reduced phase was produced, however the distillation process lead to its decomposition and the formation of the monouranate, Cs_2UO_4.

Comparing with the analogous Na-U-O system, the proposal of Lorenzelli et al [11] that the reduced phase is Cs_3UO_4 seems the most favourable interpretation. The decomposition of this phase under vacuum, and its stability in liquid caesium could be expressed by the simple equilibrium;

$$Cs_3UO_4 \leftrightarrow Cs_2UO_4 + Cs$$

There are many similarities between this equilibrium and that observed for the compound Na_3WO_4, previously investigated at Nottingham. At 700°C, Na_3WO_4 was found to lose sodium, under vacuum, to give Na_2WO_4; however at temperatures below this distillation temperature Na_3WO_4 could be isolated from the reverse reaction of Na_2WO_4 with liquid sodium [12].

$$\overset{700°C}{Na_3WO_4 \leftrightarrow Na_2WO_4 + Na}$$

Na_3WO_4 ($a_o = 4.574$ Å) is isostructural with the cubic phase Na_3UO_4, and therefore it is reasonable to suggest that Cs_3UO_4 could also have a cubic structure, as observed by ourselves and Lorenzelli et al [11].

Table 2. Comparison of X-ray powder diffraction patterns

Reaction 4 Cs + U_3O_8		Cs_2UO_{4-y} [9]		Literature data UO_2 [JCPDS]		Reaction 4 Cs + U_3O_8		Cs_2UO_{4-y} [9]		Literature data UO_2 [JCPDS]	
d(Å)	I/I_0	d(Å)	I/I_0	d(Å)	I/I_0	d(Å)	I/I_0	d(Å)	I/I_0	d(Å)	I/I_0
								2.982	5		
3.28	90	8.032	40					2.860	80		
3.25	75	7.889	10			2.85	40	2.815	100		
		.546	<5			2.81	100			2.735	48
3.16	35	6.074	30			2.73	15	2.693	10		
		5.614	5					2.650	15		
		4.651	10					2.614	5		
		4.632	10	3.157	100			2.426	5		
		4.111	10					2.407	5		
		3.904	20					2.387	5		
		3.626	5					2.334	10		
		3.573	5					2.293	5		
		3.447	5					2.243	20		
		3.373	<5					2.226	10		
		3.339	<5					2.214	5		
		3.286	100			2.00	65	2.203	5		
		3.252	100			1.97	10				
		3.188	30			1.93	20			1.934	49
		3.046	10			1.70	30				
		3.033	5			1.64	20			1.649	47
		3.013	5			1.56	15				
		2.992	5								

6. ELECTROCHEMICAL CELL MEASUREMENTS

In recent years electrochemical cells have become recognised as a valuable means of monitoring non -metals, especially oxygen, dissolved in liquid alkali metals. Previous work has predominantly concentrated on oxygen solutions in liquid sodium, howecver studies have also been extended to potassium, rubidium and caesium [13-21]. In the present study a liquid caesium cell apparatus was developed and used to investigate the reaction of liquid caesium with U_3O_8.

The cell may be represented in the following way:

Cs, O (dissolved in Cs) |Oxygen-ion conducting solid electrolyte |Oxygen reference electrode.

The reference electrode used in this study was the In/In_2O_3 couple. The solid electrolyte widely used in the measurement of oxygen potentials in liquid alkali metals has been yttria doped thoria (7 wt% Y_2O_3), YDT. However, its high cost and limited availability make the use of YDT prohibitive, and for this reason the established alternative material, calcia stabilised zirconia, CSZ was used. The difference in the chemical potential of oxygen across the cell gives rise to an e.m.f. (E), which is related to the difference in the oxygen potentials of the reference electrode and B the sample such that;

$$n\,F\,E = \Delta\,\overline{G}^{ref}_{O_2} - \Delta\,G^{sample}_{O_2}$$

where F is the Faraday constant and n is the number of electrons involved in the cell reaction, which will equal 4 in the case of the following cell reaction;

$$4/3\ In + 2\ Cs_2O \leftrightarrow 2/3\ In_2O_3 + 4\ Cs$$

265

Using the literature expression for the free energy of formation of In_2O_3 [22] the above expression may be written in the following manner:

$$\Delta \overline{G}_{O_2} = -601240 + 202.589\, T - 386076\, E$$

where $\Delta \overline{G}_{O_2}$ is in J (mol $O_2)^{-1}$, T is the temperature in Kelvin and E is in volts. Figure 2 shows the cell designed specifically for the caesium work. This was an "all-in-one" design intended to reduce caesium distillation onto the walls of the outer jacket. Assembly of the apparatus was carried out in an evacuable, argon-filled glove box.

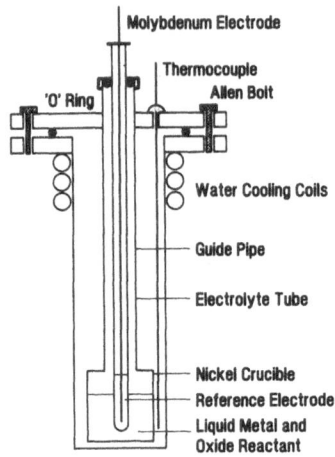

Fig. 2. Experimental set-up for measurements of emf of liquid metal and oxide reactant

The U_3O_8 reactant was placed in the crucible, and liquid caesium pipetted down the guide pipe into the crucible. The electrolyte tube, containing the reference electrode mixture and fitted with the molybdenum wire, was lowered down the guide pipe, and the couplings tightened. The outer steel jacket was then sealed and the apparatus removed from the glove box. The O-ring seal was cooled by a continuous flow of water passing through the copper cooling coils on the outer vessel. The cell apparatus was heated in a vertical electrical furnace controlled by a linear temperature programmer.

The cell was slowly heated to 377°C (650 K) at which point an electrometer was attached and an e.m.f. of 0.550 V recorded. The temperature was increased and the cell allowed to equilibrate over a period of 48 hours. After which time, the temperature was once again altered and a new e.m.f. reading taken. This procedure was carried out for the next 10 days, with readings being recorded in the temperature range 405-575°C (678-848 K).

On increasing the temperature towards 600°C (873 K) the cell voltage dropped almost to zero, indicating cell failure. Upon glove box examination of the cell it was found that the electrolyte tube had cracked leading to a short-circuiting of the cell with the majority of the liquid caesium distilling out of the crucible. The products were extracted from the cell for X-ray powder diffraction. The cell e.m.f. and temperature measurements, along with the equilibrium oxygen potentials are shown in Table 3.

The temperature dependency of the oxygen potential is given by the following expression:

$$\Delta \overline{G}_{O_2} = -854.063 + 0.239\, T\ [\text{kJ mol}^{-1}]$$

It was assumed that the U_3O_8 would react with the liquid caesium to form UO_2 and a ternary oxide in equilibrium with oxygen in the caesium. The X-ray diffraction of the orange-brown product showed only the presence of UO_2, however the colour may indicate the presence of a phase containing uranium in an oxidation state greater than 4+.

The oxygen potentials of a short-lived cell, monitoring the Cs + UO_2 + Cs_2UO_4 equilibrium, were measured by Adamson et al [17]. The authors reported that the oxygen potentials were more negative than the theoretical values calculated for the above equilibrium and attributed this discrepancy to the conversion of Cs_2UO_4 to the reduced caesium uranate which was more stable in liquid caesium at low oxygen potentials. However, the theoretical values used by Adamson et al [17] were calculated on the basis of the free energy of formation of Cs_2UO_4 obtained by Osbourne et al [23] of -1774.8 $kJmol^{-1}$, which is less negative than the more recent free energy of formation values reported by Cordfunke and O'Hare [24], - 1803.7 $kJmol^{-1}$ and Wagman et al [25], - 1806.2 $kJ\ mol^{-1}$.

Table 3. E.m.f. and oxygen potentials for the liquid caesium-U_3O_8 cell

T (°C)	T (K)	E.m.f. (V)	ΔG_{O_2} [kJ mol $^{-1}$]
447	720	0.594	- 684.705
500	773	0.584	- 670.107
554	827	0.576	- 656.079
575	848	0.573	- 650.667

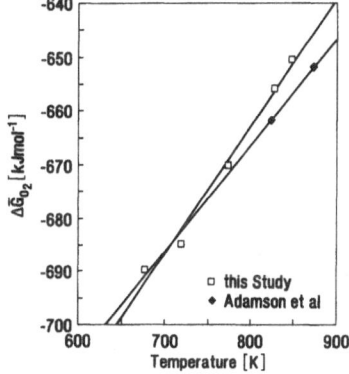

Fig. 3. Temperature dependence of the Gibbs energy of formation of Cs_2UO4 from emf measuremente

Figure 3 shows the data of Adamson et al [17] are in agreement with those obtained in the present investigation and would suggest that both cells were measuring the oxygen potential of the same equilibrium.

If the reaction did lead to the formation of the reduced phase, the following cell equilibrium could be written;

$$3\ Cs + UO_2 + O_2 \leftrightarrow Cs_3UO_4$$

It would then be possible to derive the free energy of formation for Cs_3UO_4, knowing that of uranium dioxide [26];

$$\Delta G_f^o\ (Cs_3UO_4) = \Delta G_f^o\ (UO_2) + 3\ \Delta G\ (Cs) = -1936.3 + 0.410\ T\quad (kJ\ mol^{-1})$$

Extrapolating this expression gives a value of - 1814.2 $kJmol^{-1}$ at 298 K. Although this value is more negative than the recent values for Cs_2UO_4, suggesting the formation of the reduced phase, the sheer closeness of the data makes it impossible to state which equilibrium the cells have been monitoring.

In fact, rather than clarifying the situation, the electrochemical cells have set new questions and challenges about the Cs-U-O system and showing the need for further high temperature thermodynamic studies of the system to be conducted.

REFERENCES

1. J.R. Phillips, G.R. Waterbury and N. E. Vanderborgh, J. Inorg. Nucl. Chem. 36 (1974) 17-23.
2. D.C. Fee and C.E. Johnson, J. Nucl. Mater. 96 (1981) 71-9.
3. E.H.P. Cordfunke, A.B. van Egmond and G. Van Voorst, J. Inorg. Nucl. Chem. 37 (1975) 1433-6.
4. A.B. Van Egmond J. Inorg. Nucl. Chem. 37 (1975) 1929-31.
5. A.B. Van Egmond, J. Inorg. Nucl. Chem. 38 (1976) 1645-7.
6. A.B. Van Egmond, J. Inorg. Nucl. Chem. 38 (1976) 1649-51.
7. A.B. Van Egmond, J. Inorg. Nucl. Chem. 38 (1976) 2105-7.
8. K.M. Efremova and Yu P. Simanov, Vestnik Mosk. Univ. Khim. Ser., 24 (1969) 57.
9. E.H.P. Cordfunke and E. F. Westrum Jnr, Proc. Symp. on Thermodynamics of Nuclear Materials, Julich 1979, IAEA SM-236/34, 125-41.
10. D.C. Fee and C.E. Johnson, J. Nucl. Mater., 99 (1981) 107-16.
11. R. Lorenzelli, R. Le Dudal and R. Atabeck, Proc. Symp. on Thermodynamics of Nuclear Materials. Julich 1979, IAEA SM-236/87 539-64.
12. A.J. Hooper, Ph.D. Thesis, University of Nottingham, 1971.
13. R.C. Asher, D.C. Harper, T.B.A. Kirstein, F. Leach and R.G. Taylor, Proc. 4th Inter. Conf. Liquid Metal Engineering and Technology, 17-21 October 1988, Avignon, Vol 3, Paper 602.
14. J. Jung and H. Runge, Ibid., Vol 3, Paper 603.
15. H.H. Stamm and K.Ch. Stade, Ibid., Vol 3, Paper 707.
16. R.G. Taylor and R. Thompson, U.K.A.E.R.E. Report AERE-R 12619 (1987).
17. M.G. Adamson, E.A. Aitken and D. W. Jeter, Proc. Inter. Conf. Liquid Metal Technology in Energy Production, Champion, PA, CONF-760503 (1976) 866-74.
18. P.G. Gadd and H.U. Borgstedt, J. Nucl. Mater., 119 (1983) 154-61.
19. P.G. Gadd and H.U. Borgstedt, Liquid Metal Engineering and Technology, London (1984) 107-11.
20. V. Ganesan and H.U. Borgstedt, J. Less-Common Met., 114 (1985) 343-54.
21. D.K. Chamberlain, Ph.D. Thesis, University of Nottingham, 1990.
22. D. Chatterji and R.W. Vest, J. Am. Ceramic Soc., 55 (11) (1972) 575-80.
23. D.W. Osborne, P.A. Brletic, H.R. Hoekstra and H.E. Flotow, J. Chem. Thermodyn., 8 (1976) 361-5.
24. E.H.P. Cordfunke and P.A.G. O'Hare, The Chemical Thermodynamics of Actinide Elements and Compounds, Part 3; Miscellaneous Actinide Compounds, IAEA, Vienna 1978.
25. D.D. Wagman, W.H. Evans, V.B. Parker, R.H. Schumm, I. Halow, S.M. Bailey, K.L. Churney and R.L. Nuttall, N.B.S. Tables of Chemical Thermodynamic Properties, J. Phys. Chem. Ref. Data. 11 Suppl. 2 (1982).
26. M.H. Rand, R.J. Ackermann, F. Gronvold, F.L. Oetting and A.Pattoret, Rev. Int. Hautes Temper. Refract. Fr. 15 (1978) 355-65.

THERMOCHEMISTRY OF Na-Fe-O SYSTEM AND ITS RELEVANCE TO CORROSION OF STEELS IN SODIUM

R. Sridharan, T. Gnanasekaran, G. Periaswami and C.K. Mathews

Materials Chemistry Division, Chemical Group,
Indira Gandhi Centre for Atomic Research,
Kalpakkam 603 102, Tamilnadu, India.

1. INTRODUCTION

Iron is the major constituent of austenitic stainless steels which is used as structural material in fast reactors. Oxygen present in sodium has been found to influence the corrosion of stainless steels [1-3]. This is due to the chemical interaction between oxygen, sodium and alloying elements in steels. In Na-Fe-O system Na_4FeO_3 is known to be the reaction product between sodium containing excess oxygen and iron [4-6]. Another compound Na_2FeO_2 was reported by Wu et al [7]. But this was not observed by other workers [5, 8, 9]. From results of DTA and EMF measurements, Dai et al derived data on the ternary oxides of iron [10,11]. Based on calculated thermodynamic data on various compounds of sodium, iron, and oxygen, Lindemer and Besmann proposed a tentative phase diagram for the Na-Fe-O system [12]. Since some of these calculated results were not consistent with experimental observations, they expressed the need for more experiments which permitted true equilibria. Such experiments were designed and carried out in our laboratory, the results of which are presented in this paper.

2. EXPERIMENTAL

Approximately 0.2 g of Fe_2O_3 taken in an alumina crucible was equilibrated with 10 ml of liquid sodium in a completely sealed reaction vessel at the desired temperature for extended lengths of time. The reaction products were isolated from excess sodium by vacuum distillation and characterised by XRD.

For oxygen potential measurements, excess liquid sodium was equilibrated with iron oxide in a nickel container at the desired temperature. A sample of sodium was taken by filtering it through a fine nickel filter and the excess sodium was distilled off under vacuum. The residue was analysed for sodium by AAS and correlated to the dissolved oxygen concentration. Further details of the experiment are described elsewhere [13].

In pseudo-isopiestic equilibrations, FeO, Fe_2O_3, and $NaFeO_2$ were equilibrated with sodium vapour of known partial pressures at 923 K and 773 K. The details of the

Liquid Metal Systems, Edited by H.U. Borgstedt
and G. Frees, Plenum Press, New York, 1995

experimental set-up and the general procedure are described in an earlier publication [14].

DTA assembly used in this work is described in reference 15. Approximately 1g of FeO or Fe_2O_3 and 2 cm^3 of liquid sodium was taken in the sample capsule made of nickel. The capsule was closed by welding. The reference capsule contained only liquid sodium. DTA runs were carried out after heating the samples separately at a temperature of 873 K for a few hours in a furnace. These runs employed a heating rate of 4 K/min. The cooling rate varied from 3 K/min at high temperatures to 2 K/min at lower temperatures.

A Setaram model DSC-111 Calorimeter was used to investigate the reaction between sodium and Fe_2O_3. In the sample capsule made of stainless steel, approximately 30 mg of sodium and 13 mg of Fe_2O_3 were loaded and the capsule was closed leak tight by placing a nickel gasket and crimping a special lid placed over it. The reference capsule was an identical empty capsule. Prior to the DSC run, the sample capsule was tested by heating it at 923 K in a closed vessel to ensure its leak tightness. The DSC run was then carried out at a constant heating and cooling rate of 5 K/min, from 573 K up to 873 K. Argon flow was maintained over the capsules to prevent oxidation.

3. RESULTS

The results of the in-sodium equilibrations are given in Table 1. The reaction products isolated from excess sodium by vacuum distillation, with no control of temperatures were iron, $Na_2O(s)$ and $Na_4FeO_3(s)$. When excess sodium was distilled out at a controlled temperature of 623 K, the products were found to be only $Na_2O(s)$ and $Fe(s)$. Since only two condensed phases are expected to exist in equilibrium with liquid sodium at a given temperature, $Na_4FeO_3(s)$ could have formed from the reaction between $Na_2O(s)$ and $Fe(s)$ towards the end of the distillation, in those experiments in which temperatures were not controlled.

Table 1: Results of in-sodium equilibrations

Reactants	Temperature (K)	Time (h)	Distillation condition	Products *
$Fe_2O_3(s)$ + ex. Na(l)	923	50	u	$Na_4FeO_3(s)$ + Fe(s) + $Na_2O(s)$
$Fe_2O_3(s)$ + ex. Na(l)	873	65	u	$Na_4FeO_3(s)$ + Fe(s) + $Na_2O(s)$
$Fe_2O_3(s)$ + $Na_2O_2(s)$ + ex. Na(l)	823	187	u	$Na_4FeO_3(s)$ + Fe(s) + $Na_2O(s)$
$Fe_2O_3(s)$ + ex. Na(l)	723	36	u	$Na_4FeO_3(s)$ + Fe(s) + $Na_2O(s)$
$Fe_2O_3(s)$ + ex. Na(l)	723	170	c	$Na_2O(s)$ + Fe(s)
$Fe_2O_3(s)$ + $Na_2O_2(s)$ + ex. Na(l)	673	580	c	$Na_2O(s)$ + Fe(s)
$Fe_2O_3(s)$ + ex. Na(l)	623	720	c	$Na_2O(s)$ + Fe(s)

* - analysed by XRD;   c - controlled distillation at 623 K;
ex - excess   u - distillation temperature uncontrolled

Results of the oxygen potential measurements as a function of temperature are shown in Fig. 1. At temperatures below 626 K, the measured oxygen potentials correspond to the saturation of $Na_2O(s)$ in $Na(l)$. Above this temperature, the slope is different indicating the appearance of a new equilibrium phase at this temperature.

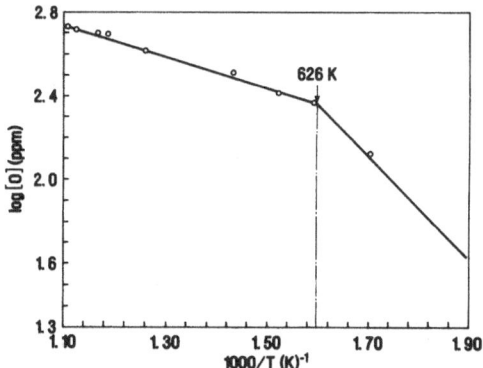

Fig. 1. Oxygen concentration in sodium as a function of temperature after equilibration with iron oxide

The results of the pseudo-isopiestic equilibrations are shown in Table 2. The compound $Na_4FeO_3(s)$ and $Fe(s)$ were the only products obtained when different iron oxides were equilibrated at 923 K with sodium vapour at pressures of 1408 and 880 Pa and at 773 K with a sodium vapour pressure of 4.7 Pa.

Table 2: Results of pseudo-isopiestic equilibrations*

Sodium	Sodium Temperature (K)	Sodium Pressure (Pa)	Equilibration Temperature (K)	Products Analysed by XRD
$Fe_2O_3(s)$	798	880	923	$Na_4FeO_3(s) + Fe(s)$
$FeO(s)$	798	880	923	$Na_4FeO_3(s) + Fe(s)$
$NaFeO_2(s)$	798	880	923	$Na_4FeO_3(s) + Fe(s)$
$Fe_2O_3(s)$	823	1408	923	$Na_4FeO_3(s) + Fe(s)$
$Fe_2O_3(s)$	598	4.7	773	$Na_4FeO_3(s) + Fe(s)$
$FeO(s)$	598	4.7	773	$Na_4FeO_3(s) + Fe(s)$

* - Equilibrations generally carried out for a minimum period of 720 h

DTA runs showed a reversible transition, the temperatures of which during heating and cooling runs are shown in Table 3. Identical transition temperatures were obtained irrespective of the iron oxide used. They were reproducible in either heating or cooling runs, differing from each other by about 15 K. The transition temperature, obtained as the average of 6 runs, was 760 $\pm$ 6 K.

Table 3: DTA results of Na-Fe-O system

| Sample | Run | Transition Temperature (K) | |
		Heating	Cooling
	1	765	749
	2	766	757
Fe_2O_3(s) + ex.Na(l)	3	766	748
	4	-	752
	5	764	760
FeO(s) + ex. Na(l)	1	766	757

The DSC curves obtained in this work are shown in Fig. 2. The results confirmed the reversible transition observed in DTA runs at 760 K. In addition, the DSC curve also shows shallow endothermic peaks followed by drastic change of slope in the base line in the temperature range of 621 K to 723 K. These changes are reproducible during cooling.

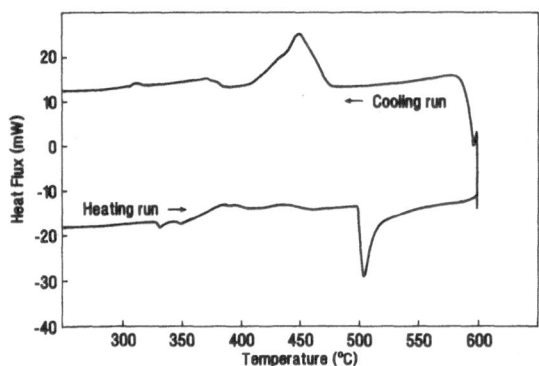

Fig. 2. DSC trace of a sample containing liquid sodium with excess oxygen and iron metal

The results of solid-state reactions between oxides of iron and sodium are given Table 4. The XRD pattern obtained for the products of reaction between Na_2O (s) and FeO (s) taken in 1:1 ratio and Na_2O_2 (s) and Fe (s) in 1:1 ratio showed the products to be a mixture of Na_3FeO_3 (s) and Fe (s) and not the expected compound, Na_2FeO_2. The results of other solid state equilibrations established the existence of the binary lines involving the pairs, Na_4FeO_3(s)-Na_3FeO_3(s), Na_4FeO_3(s)-Fe(s), Na_3FeO_3(s)-Fe(s), Na_3FeO_3(s)-$NaFeO_2$(s), $NaFeO_2$(s)-Fe(s) and the two phase fields bound by these lines, namely Na_4FeO_3(s) - Na_3FeO_3(s) - Fe(s) and Na_3FeO_3(s) - $NaFeO_2$(s) - Fe(s). The results of the equilibrations carried out in hermetically sealed capsules confirmed these two phase fields.

4. DISCUSSION

From the results of in-sodium equilibrations it is clear that below 623 K only iron and sodium oxide coexist with liquid sodium. This is also in agreement with the equilibrium oxygen potential measurements, which showed a change of slope at 626K. The oxygen concentration measured below 626 K corresponded to the saturated solubility of oxygen in sodium indicating that the appropriate phase field is Na(l)-Na$_2$O(s)-Fe(s) upto 626 K.

Table 4. Results of solid-state reactions

ratio of reactants	reactants	* equilibrium products observed (analysed by XRD)
1:1	Na$_2$O$_2$: FeO	Na$_3$FeO$_3$ + NaFeO$_2$ + Fe
3:2	Na$_2$O$_2$: Fe	Na$_4$FeO$_3$ + Fe
2:1	Na$_2$O$_2$: Fe	Na$_4$FeO$_3$ + Na$_3$FeO$_3$ + Fe
1:2	Na$_2$O$_2$: Fe	NaFeO$_2$ + Fe
1:1	Na$_2$O$_2$: Fe	Na$_3$FeO$_3$ + NaFeO$_2$ + Fe
3:2	Na$_2$O : Fe	Na$_4$FeO$_3$ + Na$_3$FeO$_3$ + Fe
1:1	Na$_2$O : FeO	Na$_3$FeO$_3$ + NaFeO$_2$ + Fe
# 3:2	Na$_2$O : FeO	Na$_4$FeO$_3$ + Na$_3$FeO$_3$ + Fe
# 2:3	Na$_2$O : FeO	Na$_3$FeO$_3$ + NaFeO$_2$ + Fe

* Equilibration temperature: 923 K   # Equilibration in hermetically sealed capsules.

The results of pseudo-isopiestic equilibrations of iron oxides with sodium vapour show that Na$_4$FeO$_3$ (s) is the ternary compound that coexists with Fe(s) and Na(l) at temperatures above at least 773K.

The results of DSC and DTA experiments have confirmed the presence of a reversible process at 760 $\pm$ 6 K which is corroborated by the recent in-sodium oxygen potential measurements using an EMF technique reported by Borgstedt et al [16]. These results in conjunction with the in-sodium and the pseudo- isopiestic equilibration results reveal that the large reversible peak corresponds to a phase transition of first order type namely the appearance of the compound Na$_4$FeO$_3$ at 760 K. The small endo peaks and consequent shift in base line suggest a phase change of second order type getting initiated at around 620K. These changes can be attributed to the simultaneous dissolution of sodium oxide phase (which is endothermic) and the formation of associates or clusters involving iron, sodium and oxygen ions. This line of argument gains support from the work of Thompson [17]. In applying the lattice solvation model to Na-Fe-O system, Thompson has suggested a switch in solvation pattern, from the dilute solution of oxide ions solvated by surrounding sodium ions to a more concentrated solution where iron ions are solvated by surrounding oxygen ions in liquid sodium. This results in an increase in the solvation energy of oxygen ions and a lowering of its chemical activity. This process thus involves a definite change in the oxygen potential of the system with entropy change playing a major role rather than enthalpy change. The result of oxygen potential measurements does indicate a change in oxygen potential at 626K. DSC curve shows the initiation of such a change at 621 $\pm$ 3 K which probably continues upto 723K and at 760K a complete reconstruction takes place to form Na$_4$FeO$_3$.

The products observed in solid-state equilibrations in closed conditions confirmed the existence of the following two phase fields at 923 K:

$$Na_4FeO_3(s) - Na_3FeO_3(s) - Fe(s)$$

$$Na_3FeO_3(s) - NaFeO_2(s) - Fe(s)$$

From the above discussions the partial phase diagram of Na-Fe-O system at 773 K <T< 923 K was deduced and is given in Fig.3.

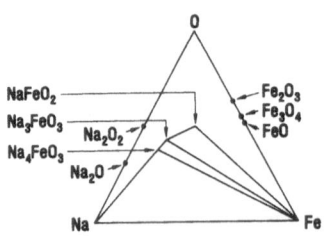

(773 K< T< 923 K)

Fig. 3. Partial phase diagram of Na-Fe-O system at 773<T<923 K

Gibbs energies of formation of relevant ternary oxygen compounds that can coexist with liquid sodium -austenitic stainless steel system are given in Table 5. The present work has shown that no ternary oxygen compound of iron would be stable in sodium below 626 K and $Na_4FeO_3(s)$ appears as an equilibrium phase only above 760 K. $Na_2FeO_2(s)$ is proposed to be the coexisting phase at temperatures above 1073 K by Lindemer et al [12]. In our experiments, however, this phase could not be synthesised by all reported methods. Nickel does not form a ternary oxygen compound that would be stable in sodium. To consider the stabilities of these ternary oxygen compounds in an operating sodium system, oxygen potentials of following equilibria were calculated:

$$Na(l) + x \, [M]_{ss} + y/2 \, O_2(g) \leftrightarrow NaM_xO_y(s) \tag{1}$$

$$[M = Fe,Cr,Mn \text{ and } Mo]$$

$$\Delta G_{O2} = 2/y \, [\, \Delta G°_{f,T} - x \, RT \ln a_M] \tag{2}$$

In addition, the oxygen potentials for the following equilibria were also computed:

$$[Mn]ss + 1/2 \, O_2(g) \leftrightarrow MnO(s) \tag{3}$$

$$3 \, Na(l) + Nb(s) + 2O_2(g) \leftrightarrow Na_3NbO_4(s) \tag{4}$$

$$2 \, Na(l) + 1/2O_2(g) \leftrightarrow [Na_2O]_{Na} \tag{5}$$

For these calculations, chemical activities of the alloying elements in SS 316 measured by Azad et al. [20] were used. The compound $Na_4Mn_2O_5(s)$ has not been considered in this regard due to non-availability of its thermodynamic data.

Table 5. Gibbs free energy of formation of ternary oxygen compounds in sodium-stainless steel system

Ternary oxygen compound	$\Delta G°_f = A + B *T$ J/mol		Reference
	A	B	
$Na_4FeO_3(s)$	- 1211000	337.644	12
$Na_2FeO_2(s)$	- 780000	199.216	12
$NaCrO_2(s)$	- 870773	193.171	18
$NaMnO_2(s)$	- 775098	155.590	19
$Na_4MoO_5(s)$	- 1907223	436.130	14
$Na_3NbO_4(s)$	- 1504667	181.127	19

274

The results of the calculations are shown in Fig. 4. It is seen that all ternary oxygen compounds involving Fe, Mn and Mo would form only at oxygen concentrations above 1000 ppm at temperatures above 900K. At oxygen concentrations below 50 ppm, $NaCrO_2(s)$ would be more stable than all other ternary oxygen compounds. Threshold oxygen levels for its formation (which can be calculated from data in Fig. 4) would be valid only during the start up of the system. At these low levels of dissolved oxygen concentration, selective leaching of the alloying elements, (Ni,Cr and Mn) and a simultaneous enrichment of Fe and Mo at the corroding surface in the hot leg regions would take place. This increase in chemical activity of Fe and Mo cannot however bring about the formation of ternary oxygen compounds involving them because much higher oxygen potentials would be required, as seen in Fig. 4. The reduction in the chemical activity of chromium would raise the threshold oxygen concentrations for the appearance of $NaCrO_2(s)$.

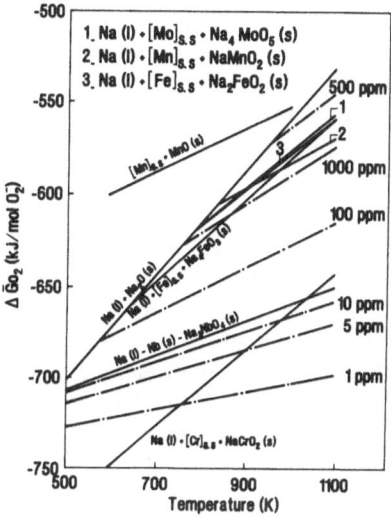

Fig. 4. Equilibrium oxygen potentials of various species in sodium-oxygen-stainless steel system

At the cold regions of the circuit, chemical activities of metallic constituents (that are leached out from the high temperature sections) would be high because of the temperature dependence of their solubility in sodium. For the same reason, carbon activity also would be higher at these low temperature sections. Under these circumstances formation of metallic carbides would become possible. This has been borne out by experimental observations in sodium loops. The precipitates observed in deposition regions of sodium loops were found to contain both carbide and $NaCrO_2(s)$ [18]. Interference due to carbon does not occur at the corroding hot legs, when it is present at decarburising levels.

Enhancement of corrosion by dissolved oxygen in sodium has to be considered by taking into account of the existence and formation of ternary oxygen compounds. Steady-state corrosion of steels, either austenitic or ferritic, is essentially the corrosion of iron in sodium because it constitutes the bulk of the steel. The other alloying elements may influence this process by changing the mechanism.

Based on experimental observations in sodium loops, rate expressions for the corrosion of austenitic stainless steels in sodium have been reported in literature [21-24]. The dependence of corrosion rate on the oxygen content in sodium in these correlations has been expressed in terms of oxygen concentration instead of oxygen activity. Since the activity of the species involved is the appropriate term to be incorporated to derive the activation energy values, the expressions were suitably modified. Data on solubility of oxygen in sodium from reference 25 was utilised for this purpose. The modified expressions are shown in Table 6. It can be seen that the activation energies of the corrosion lie in the range of 25-50 kcal mol^{-1}.

Table 6. Modified corrosion rate expressions

Reference	Corrosion rate expression
Thorley and Tyzack	S (mil/y) = $1.5 \cdot 10^4 \cdot a_{(Na_2O)}^{1.5} \cdot \exp(-34776/RT)$
Weeks and Isaacs	R (mil/y) = $1.089 \cdot 10^{12} \cdot a_{(Na_2O)}^{1.456} \cdot \exp(-50300/RT)$
Kraev	$R(mg/cm^2/h) = 1.99 \cdot 10^3 [C^o_{sur}/C^o_{vol}] 1.2 \cdot a_{(Na_2O)}^{1.2} \cdot \exp(-39021/RT)$
Kolster	$I_{st}(mg/cm^2/y) = 1.2 \cdot 10^{11} \cdot \exp(-48500/RT) + 54.75 \cdot 10^2 \cdot a_{Na_2O} \cdot \exp(-25184/RT)$

It is to be noted that the formation of ternary oxygen compounds involve the reaction of dissolved oxygen with the metallic constituents in steel as indicated below:

$$x\,[Na_2O]_{Na} + y\,[M]_{ss} \rightarrow NaM_yO_x(s) + (2x-1)\,Na(l) \qquad (6)$$

$$[\,M = Fe,\ Cr,\ Mn\ and\ Mo\,]$$

The standard enthalpy of the reactions, calculated with the data in Table 5, are listed in Table 7. The derived activation energies for corrosion are much higher than these enthalpy changes indicating other kinetically controlled processes. Corrosion involves: a) diffusion of any interacting impurities from the bulk of sodium on to the surface of the structural materials through the laminar boundary layer (between the surface and bulk of sodium), b) corroding reaction at the interface and c) the diffusion of corrosion products from the structural materials' surface into the bulk of sodium. Hence the activation energies would be higher than these enthalpy values.

Table 7. Enthalpy of reactions for ternary compound formation

Reaction	$\Delta H^0{}_{Rx,\ 298}$ per mole of Na_2O (kJ mol^{-1})
$3\,Na_2O(s) + Fe(s) \longrightarrow Na_4FeO_3(s) + 2\,Na(l)$	-13.433
$2\,Na_2O(s) + Cr(s) \longrightarrow NaCrO_2(s) + 3\,Na(l)$	-18.287
$5\,Na_2O(s) + Mo(s) \longrightarrow Na_4MoO_5(s) + 6\,Na(l)$	+35.655
$2\,Na_2O(s) + Mn(s) \longrightarrow NaMnO_2(s) + 3\,Na(l)$	+29.551

Formation of a stoichiometric ternary oxygen compound in the rate determining step is not expected since the exponent on oxygen concentration in the reported corrosion rate expressions would then be observed as > 2. The observed values are between 0.8 and 1.5. In the case of corrosion of pure iron, one expects the exponent to be 3 if $Na_4FeO_{3(s)}$ formation is involved in the rate determining steps. The observed value in this case is 2. These indicate that in the rate determining step, some complex species are involved instead of the ternary oxygen compounds. Weeks and Isaacs proposed a model based on a solvated Na-Fe-O complex to explain the oxygen concentration dependence of corrosion of steels in sodium [26]. Kolster has proposed another similar model for corrosion of austenitic stainless steels in sodium from corrosion experiments carried out in a molybdenum loop [27]. The results of DSC experiments of the present work are to be viewed in this context.

REFERENCES

1. Borgstedt, H.U., and Mathews, C.K., Applied Chemistry of Alkali Metals, Plenum Press, New York, 1987.
2. Weeks, J.R., and Isaacs, H.S., Advances in Corrosion Science and Technology, Plenum Press, New York, Vol. 3, 1973, p.1.
3. Borgstedt, H.U.,Reviews on Coating and Corrosion, 2 (1977) 121
4. Horsley, G.W., J.Iron and Steel Inst. 182 (1956) 43.
5. Gross, P., and Wilson, G.L., J. Chem. Soc. (A), (1970) 1913.
6. Addison, C.C., Barker, M.G., and Hooper, A.J., J. Chem. Soc., Dalton Trans., (1972) 1017.
7. Wu, P.C.S., Chiotti, P., and Mason, J.T., Proc. Int. Conf. on Liquid Metal Technology in Energy Production, Champion, Pennsylvania, USA, CONF-760503-P2, 1976, p.638.
8. Tschudy, A., Kessler, H. and Hatterer, A., in: Proc. Intern. Conf. on Liq. Alkali Met., British Nuclear Energy Society, London, 1973, p. 209.
9. Knights, C.F., and Philips, B.A., in High Temp. chem. Inorg. and Ceramic Mater., Glasser, F.P., and Potter, P.E., (Eds.), The Chemical Society, London, 1977, p.135.
10. Dai, W., Seetharaman, S., and Staffansson, L.-I., Met. Trans., 15B (1984) 319.
11. Dai, W., Seetharaman, S., and Staffansson, L.-I., Scan. J. Met., 13 (1984) 32.
12. Lindemer, T.B., Besmann, T.M., and Johnson, E., J. Nucl. Mat., 100 (1981) 178.
13. Sridharan, R., Krishnamurthy, D., and Mathews, C.K., J. Nucl. Mater. 167 (1989) 265.
14. Gnanasekaran, T., Mahendran, K.H., Periaswami, G., Mathews, C.K., and Borgstedt,H.U., J. Nucl. Mater., 150 (1987) 113.
15. Gnanasekaran, T., Mahendran, K.H., Sridharan, R., Periaswami, G., and Mathews, C.K., Proc. of Fourth Int. Conf on Liquid Metal Engineering and Technology, LIMET-88, 1988, Avignon, France, Vol. 2, paper 521.
16. Bhat, N.P. and Borgstedt, H.U., J. Nucl. Mater. 158 (1988) 7.
17. Thompson, R., AERE Report - 9172 (1979).
18. Gnanasekaran, T. and Mathews, C.K., J. Nucl. Mater., 140 (1986) 202.
19. Frankham, S. A., Ph. D., Thesis, University of Nottingham, U.K., 1982.
20. Azad, A. M., Sreedharan, O. M., and Gnanamoorthy, J.B., J. Nucl. Mater., 144 (1987)94.
21. Thorley, A.W., and Tyzack, C. in Ref.8, pp. 257.
22. Weeks, J. R. and Isaacs, H.S., Advances in corrosion Science and Technology, Vol. 3, Plenum Press NY (1973), pp. 1-66.
23. Kraev, N.D., Zotov, V.V., Starkov, O.V., Zhuravleva, T.S. Mochalova,G.N., Luk'yanova, I.N., and Solove'v, V.A., Kernenergie, 21 (1978) 244.
24. Kolster, B.H., Proc., 7 th Int. Conf. on Metallic Corrosion, Rio de Janeiro, Brazil (1978).
25. Noden, J.D., J. Brit. Nucl. Energy Soc., 12 (1973) 32.
26. Weeks, J. R., and Isaacs, H.S., Proc. Symp. on Chemical Aspects of Corrosion and Mass Transfer in Liquid Sodium, (Ed) Jansson, S.A., Metallurgical Soc., (1971) pp. 207-222.
27. Kolster, B.H., Bos, L., Proc. Third Int. Conf. on Liquid Metal Engineering and Technology, Oxford, BNES, 1984, Vol.1, p. 235.

SOME ASPECTS OF CARBON BEARING SPECIES BEHAVIOUR IN A SYSTEM SODIUM - STEELS - PROTECTIVE GAS - MINERAL OIL

Yu.I. Zagorulko, Yu.I. Kovalov, F.A. Koslov, O.V. Starkov

Institute of Physics and Power Engineering,
249020 Obninsk, Kaluga Region, Russia

1. INTRODUCTION

The mineral oil is one of the most important source of Fast Reactor sodium coolant contamination by the carbon bearing and hydrogen impurities. Oil can penetrate into sodium coolant at reactor normal operation as a result of transfer of its vapours and aerosols from the circulation pumps systems. Accidental ingresses of oil are also possible due to the faults of personnel or equipment malfunctions.

Taking into account the complexity of the reactions and the processes determining the behaviour of oil - sodium system, it is not clear up to - date the possible consequences of the oil ingresses into sodium coolant for reactor long - term operation.

Therefore, there exists an essential necessity in detail studies of physical and chemical properties of the system "mineral oil - sodium - structural materials - protective gas" and their technological applications for Fast Reactor operational conditions.

The basic purposes of oil - in - sodium behaviour studies are as follows:

to identify the most stable gaseous oil pyrolysis products enabling the gas blanket of reactor to be monitored for oil ingresses as in normal so in accidental operation conditions;

to determine content of carbon bearing species in sodium and the distribution of solid phase oil pyrolysis products in sodium circulating loop in order to estimate the possible contamination of surfaces of equipment in contact with oil and formation of depositions of carbon bearing species in the loop zones with the flow stagnant regimes;

to determine hydrodynamic consequences for the sodium loop flow regimes due to the formation of particulate phase after oil ingresses;

to predict the influence of oil ingresses on the sodium carbonisation potential in respect to structural materials of equipment and alterations of their mechanical characteristics as in short - term so in long - term aspects.

In this report are presented some results of the experimental studies of the above mentioned problems, carried out in IPPE, Obninsk.

Liquid Metal Systems, Edited by H.U. Borgstedt
and G. Frees, Plenum Press, New York, 1995

2. EXPERIMENTAL INVESTIGATIONS OF PHYSICAL AND CHEMICAL PROPERTIES OF THE SODIUM - MINERAL OIL SYSTEM

2.1 Experimental part

Investigations were performed using the oils, applied on the Nuclear Power Installations (NPI): vacuum oil VMG and turbine oil TP-22.

Kinetics of the volatile oil - in - sodium pyrolysis products contents changes was studied using the ampoules, made of stainless steel X18H10T, with a volume of 100 cm^3 and cross - section of 10 cm^2. Quantities of sodium charged into ampoules were taken to be of the order of 5 g and oil of about 5 mg. (The abbrevations of steel qualities are as in the original paper.)

Experimental determinations of the condensed oil - in - sodium pyrolysis products contents were carried out in cylindrical ampoules with diameter values of 6 - 7 mm and the wall thickness of 0.35 - 0.5 mm. As materials for ampoules stainless steel X16H15M3B (fuel pins cladding material), stainless steel X18H10T (structural material of the main equipment of reactor primary circuit), nickel and iron were used (for the purposes of comparison). The volume of ampoules was equal to approximately 1 cm^3 and oil quantities of about 3 mg.

Morphological properties of the oil - in - sodium pyrolysis products particulate phase, found in the sodium flow, and depositions on the surfaces in the gas plena of circuit were studied in an experimental installation with reaction vessel having the volume of 21 l, as well as on the sodium circulating loops (with sodium inventories of hundreds litter). In the experiments carried out on the installation in the reaction vessel was supplied the flow of argon, saturated with oil vapour at temperature of 100 °C. In the experiments performed at the circulating circuits oil was introduced into loop in quantities of several hundreds grams.

2.2 Experimental results and their discussion

Content of oil pyrolysis products. Chromatographic analysis has shown that in the system sodium - mineral moil are formed more than 20 hydrocarbon volatile compounds and hydrogen. In the Figs. 1, 2 are given kinetic curves, characterizing time dependence of the contents of H_2 and hydrocarbons C_1 - C_4 at temperatures of 350 and 550 °C.

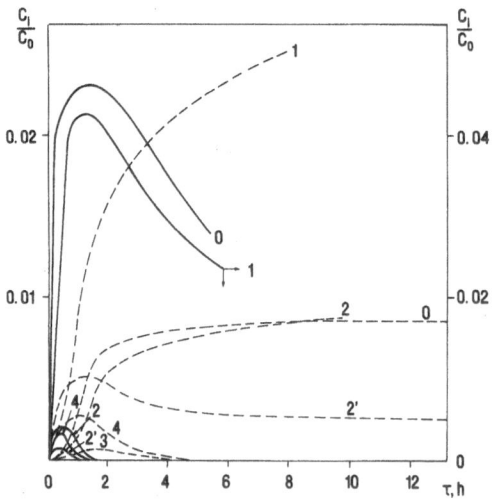

Fig. 1. Kinetic curves of the formation of H_2 and hydrocarbons in Na at 350 °C

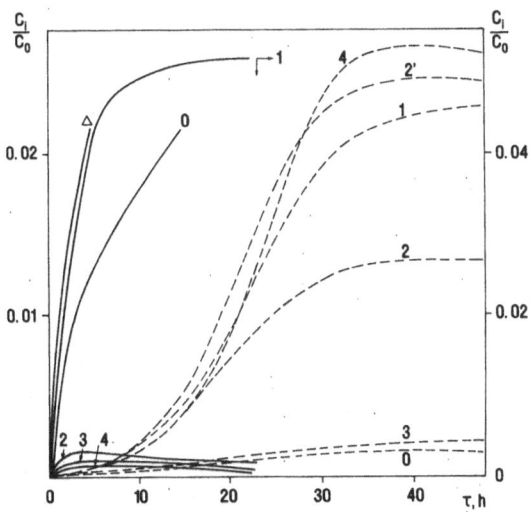

Fig. 2. Kinetic curves of the formation of H_2 and hydrocarbons in Na at 550 °C

On the ordinate the relations of the particular compound content to the initial quantity of vacuum oil are plotted. Kinetic curve for Cl (at 350°C) obtained in the check experiment, where the oil batch was dropped in the preliminarily heated ampoule, is approximately equidistant to the similar curve, obtained in the experiment, where oil was heated together with sodium (time difference corresponds to the duration of heating). From these figures one may conclude that formation of oil pyrolysis products in temperature interval of 350 - 550 °C begins with a big rate immediately after contact of oil with pyrolysis there is certain time delay in the formation of C_1 - C_4 and H_2. These results have shown also, that in the presence of sodium, hydrogen and methane are formed in the larger quantities, but output of the other hydrocarbons, particularly of unsaturated ones, was diminished.

Estimations of the maximum output of hydrogen and different forms of carbon into gas phase are given in the Table 1.

Table 1. Results of estimations of the output of reaction products

Temp. (°C)	Time		without sodium				with sodium			
	with Na	without Na	Hydr ogen alltog.	as H_2	Car bon alltog.	as CH_4	Hydr ogen alltog.	as H_2	Car bon alltog.	as CH_4
350	> 30	7	13	1	10	2,5	23*	14*	5	4,7
550**	8	< 1	13	6,6	3,5	2	24	16	4,1	4,0

*For the time indicated here the hydrogen content continued to increase.
**Results underestimation is possible for the ampoule with sodium due to the high rate of volative products formation and results sensitivity to the error in estimation of ampoule heating time.

Initial rates of volatile products formation were also estimated and corresponding data are given in the Table 2.

Table 2. Initial rates of formation of volatile reaction products in sodium

Volatile oil pyrolysis products	$(C_i/C_0)'\tau =0;10^{-8}c^{-1}$			
	without sodium		with sodium	
	350 °C	550 °C	350 °C	550 °C
Methane C_4	5	320	360	7800
Ethane C_2	3	80	30	3100
Propane C_3	1	14	9	150
Buthane C_4	4	160	38	2800
Ethylene C_2'	4	500	33	3000
Hydrogen	1	110	150	5200

From these data it follows that presence of sodium increases by 1 - 2 orders of magnitude the initial rates of volatile oil pyrolysis products formation.

Along with volatile products during the oil pyrolysis in sodium are formed also acetylide and oil - like (coke - like) carbon bearing compounds. Content of the oil - like, chloroform - soluble compound is diminished in time.

At 550 °C acetylide content for the time of several tens of seconds reaches its maximum values, which remain unaltered for several tens of minutes (during homogeneous oil pyrolysis acetylide is not formed at all). Averaged values of acetylide carbon, corresponding to the plateau region of the curve (referred to the carbon present in the oil batch) are equal to 0.06 for turbine oil and 0,04 for vacuum oil in the ampoules made of steel X16H15M3B and 0.08 - 0.05 for the same kinds of oil in the ampoules made of steel X18H10T. Acetylide content in the ampoules made of nickel was 3 - 5 times more.

The next stage of pyrolysis process has shown relatively slow decrease of acetylide content. At that, the rate of acetylide content decrease depends on the ampoules material. Chloroform - soluble forms of carbon content was decreased down to 0.06 - 0.08 during the time intervals of several minutes. Further on the rate of this decrease is considerably slow down.

During the time of several tens minutes the content of carbon condensed forms increased up to the value of 0.5 and further it changed insignificantly (in nickel ampoules it grew up to 0.8 - 0.9).

It should be noted that during the washing of the condensed carbon forms from the ampoules, using alcohol and chloroform solutions, about the third of the whole content has remained on the walls. In the case of homogeneous pyrolysis the whole mass of oil pyrolysis products is strongly adherent to the ampoules walls. The latter fact suggests about certain protective role, played by sodium on the coke - type depositions formation on the walls in contact with oil - sodium system.

The sum of carbon forms, determined by analysis in the ampoules made steels X16H15M3 and X18H10T, was equal, correspondingly, to 0.7 and 0.5.

During the oil homogeneous pyrolysis the balance on the carbon is fulfilled. The same situation took place during the oil pyrolysis in sodium in the case of nickel ampoules. From this it follows that during interaction of oil with sodium at higher temperature in stainless steel ampoules was present some carbon bearing compound that was not determined by the methods used in analysis procedure. Presumably this compound can be a complex iron and chromium carbide, found also in similar conditions by A. Thorley et al [1]. The latter suggestion is supported by the fact that there is no carbon mass unbalance in the experiments performed in nickel ampoules.

Experimental results obtained for the ampoules, made of nickel and iron, at the temperature of 550 °C have shown that in time interval of 4 - 8 s the acetylide output was higher than that in the ampoules made of steels. The condensed forms of carbon are equal in both cases. Acetylide and condensed carbon forms formation mean rates for initial pyrolysis stage are plotted in the Fig. 3 and 4.

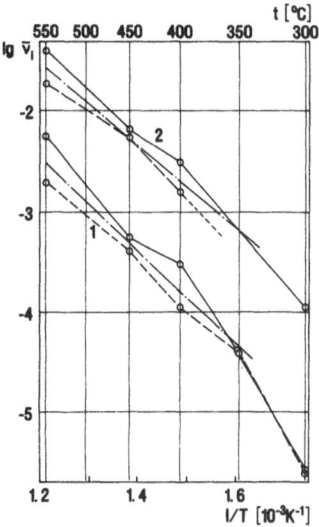 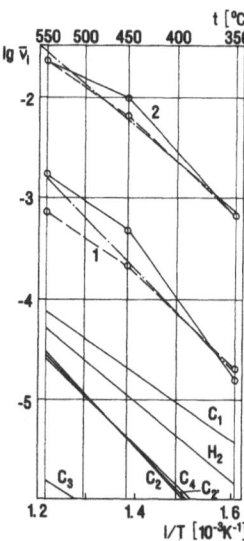

Fig.3: Mean initial rates of the oil-in-sodium pyrolysis products formation in dependence on the temperature.(Ampoules made of steel X16H15M3?): ---- - oil Tp -22; ---- - oil VGM; ---- - approximation curves; 1,2 - acetylide and condensed carbon forms

Fig.4: Mean initial rates of the oil-in-sodium pyrolysis products formation in dependence on temperature (Ampoules made of steel X18H10T): ---- - oil p - 22; ---oil VGM;approximation curves; 1,2 - acetylide and condensed carbon-forms; H_2 - hydrogen; C_1 - C_4- hydrocarbons

It may be seen that the ratios of rates of acetylide and condensed carbon compounds formation have changed considerably with temperature, not depending on the ampoules made of steel X16H15M3? this ratio was equal to 12 - 14 and for ampoules made of X16H10T - to 30 - 20 (for vacuum and turbine oil, correspondingly). Volatile oil pyrolysis products were formed with still lesser rates.

Estimations of activation energies of the particular carbon forms formation during oil - sodium pyrolysis for investigated temperature range have given the values of the order of 80 - 90 kJ/mol.

2.3 Dispersive characteristics of oil - in sodium pyrolysis products.

Condensed coke - like oil pyrolysis products form in sodium system a dispersion phase, which partially is present as a particulate phase in the flowing sodium and mainly is deposited on the equipment surfaces in contact with sodium and in the circuit plena. Depositions on the surfaces in gas plena have a dendrite - like structure impregnated with sodium condensate Particulate phase in sodium flow is formed of fine particles and aggregates of different dimensions. They originated essentially from the fragments of the hydrocarbon film that is formed during the first stage of sodium - oil interaction as an interphase surface between gaseous phase (mixture of oil vapours and volatile oil pyrolysis products) and sodium.

Dispersive analysis of the oil pyrolysis products depositions substrate and of sodium particulate phase samples has shown the following:

The complex aggregates found in the samples of oil pyrolysis products depositions substrate are easily disintegrated into small sizes particles during their treatment with ethanol - water solutions.

The whole set of particles by their characteristic sizes can be subdivided for some arrays, depending on the time passed after oil injection into sodium (for particulate phase in sodium) or locality of the depositions from which substrate samples were taken.

Size distributions parameters is described approximately by the normal - logarithmic law. Particles distributions parameters (median diameter and standard deviation) are given in Table 3.

Table 3. Parameters of the distribution of particle sizes

Particles array number	Characteristics of samples (sodium and depositions substrate)	D_o, cm	σ	S/V, $10^6 m^{-1}$
1	Particulate phase in sodium Samples were taken from sodium rig in time interval of 126 h after oil injection (Temperature 400 - 500 °C)	0,6	3,5	0,55
2	The same as for array 1 Time interval of 456 h.	0,8	2,4	1,1
3	Deposition substrate Samples were taken from the reaction vessel, of experimental installation. (Temperature of sodium 550°C); temperature in gas plenum 300°C)	5,4	3,8	0,72
4	Samples were taken from the surface of the cap of sodium rig pump tank Temperature of sodium 430°C)	5,4	5,5	0,0007
5	Samples were taken from the wall of the pu mp tank on the level of sodium mirror (Sodium temperature 430°C)	2,1	3,9	0,027

It should be noted that the particles of larger dimensions in array 1-3 were characterised by two comparable sizes and the third one with much smaller value. Dimensions of particles in the array 4 and 5, and cylindrical shape for the array 1 - 3.

The smaller values of median diameter for the sodium particulate phase can be explained by aggregates disintegration in the sodium turbulent flow. Small values of specific surface for the particles found in the substrate samples taken from the surfaces of the gas plenum inside of circuit pump may be attributed to the fact that a fine structure of particles was not accounted for in spherical estimation of their shapes.

Total size distribution functions for the particles arrays 1 - 5 are given in the Fig. 5.

3. EXPERIMENTAL INVESTIGATIONS OF STEEL CARBURIZATION IN A SODIUM RIG POLLUTED BY MINERAL OIL

3.1 Experimental part

Experiments were performed on the sodium rig made of steel X20H14C2. Layout of the rig is given in Fig. 6.

Sodium inventory during the experiments was about 200 kg. Mineral oil (turbine oil of the type 22 L) was introduced in two ways, by pouring it onto the sodium surface in the pump tank and by injecting it through a hole cut away in the rig pipe.

The first method was used during the initial experimental run with whole duration of 2000 h. Oil batches of volume 50 ml during this time were supplied into pump tank three times. The quantity of oil, introduced into the rig hot leg with temperature of 700 °C, was about 70 ml.

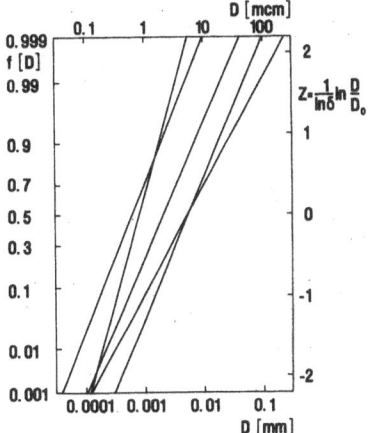

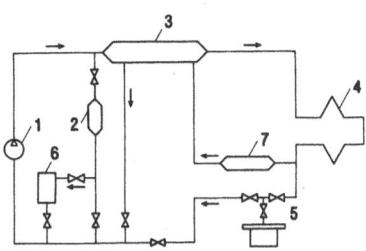

Fig.5. Total functions of the particles size distributions for the oil-in-sodium pyrolysis products (Digits for the curve are given according to particles arrays numbers in the Table 3.

Fig.6. Block diagram of experimental rig for study of carbon migration under non-isothermal conditions: 1-electromagnetic pump, 2-cold trap, 3-heat exchanger, 4-heater, 5-sampler-distiller, 6-compensation tank, 7-experimental section

Different steel specimens (see Table 4 for their chemical compositions) were placed in the rig test section held at a temperature of 700 °C, to study their carburization and variation in mechanical properties.

Table 4. Chemical composition of steels

Steel	Element concent mas (%)								
	C	Mn	Si	Cr	Ni	Nb	Ti	V	Mo
IX18H10T	0,1	1,32	0,36	17,0	10,8	-	0,61	-	-
X16H11M3	0,06	1,25	0,30	15,6	10,98	-	0,05	-	2,1
X16H15M3B									
(EI-847)	0,08	0,40	0,20	16,1	15,6	0,8	-	-	2,8
IX2M	0,1	1,27	0,34	2,16	-	-	-	-	0,66
EI-513	0,06	0,82	0,61	2,5	0,06	0,38	-	0,25	0,72

In some time intervals specimens were taken off to perform their analysis and mechanical tests.

The sodium temperature in the rig during experiments was 300 °C. Samples of sodium were taken and analysed by distillation and combustion of the residue.

3.2 Experimental Results

The main results obtained in the experiments on the carbon content in sodium and in steels after testing in a flow of sodium at 700 °C are presented in the Tables 5 and 6. Results of the tests of mechanical properties are given in the Tables 7 and 8 and Fig. 7.

Table 5. Carbon content in sodium samples after introduction of oil into circulation circuit

Run No.	Time of adding oil and taken sample from start of operation of rig	Quantity of oil added (ml)	Carbon content in sodium (wt. %)	Place of addition of oil
1	0	50	-	pump tank
	24	-	4,5	-"-
	276	50	-	-"-
	300	-	5,5	-
	576	50	-	pump tank
	600	-	3,8	-
	1800	-	1,9	-
	2000	-	-	-
2	2024	70	5,3	rig pipe
	2100	-	3,1	-
	3700	-	3,7	-
3	4100	-	3,2	-
4	5060	-	3,7	-
	5800	-	-	-
5	7300	-	4,9	-

Table 6: Carbon content in specimens of austenitic chromium-nickel steel after testing in a flow of sodium at 700 °C

Type of Steel	Initial	1 Run 2000 h		2 Run 1800		1 + 2 Run 3800 h	
		cal.	exp.	cal.	exp.	cal.	exp.
1	0,1	0,17	0,14	0,7	0,88	0,68	0,89
2	0.06	-	0,14	-	0,62	-	0,62
3	0.07	0,12	0,12	0,64	0,66	0,65	0,65

Type of Steel	3 Run 300 h		1+2+3 Run 4100 h		4 Run 1700 h		6 Run 1500 h	
	cal.	exp.	cal.	exp.	cal.	exp.	cal.	exp.
1	0,88	0,52	0,75	0,98	0,30	0,39	0,18	0,20
2	-	0,48	-	0,82	-	-	-	-
3	0,65	0,54	0,94	0,83	0,44	0,35	0.30	0,27

Explanation of steel types:
1 12Cr18Ni10T   0,12%C, 18 %Cr, 10 %Ni, 1 %Ti
2 Cr16Ni11M3   16 %Cr, 11 %Ni, 3 %Mo
3 Cr16Ni15M3B   16 %Cr, 15 %Ni, 3 %Mo, 1 %N

Table 7:Mechanical properties (at 20°C) of austenitic steels specimens after their test in the sodium flow at 700 °C

Steel	after	heat	treatment	after	treatment	Ar 700°C	1	Run	2000 h
	σ_B	$\sigma_{0,2}$	δ_{10}	σ_B	$\sigma_{0,2}$	δ_{10}	σ_B	$\sigma_{0,2}$	δ_{10}
	kg/mm 2	kg/mm²	%	kg/mm²	kg/mm²	%	kg/mm²	kg/mm²	%
1	65,8	29,6	47,8	67.0	24,8	40,4	71,0	-	39,4
2	57,5	30,5	55,3	67,6	29,0	41,8	67,0	37,7	37,0
3	62,0	33,0	38,8	67,1	30,4	28,8	73,8	53,6	26,8

Steel	2	Run	1800 h	1+2	Run	3800 h	3	Run	300 h
	σ_B	$\sigma_{0,2}$	δ_{10}	σ_B	$\sigma_{0,2}$	δ_{10}	σ_B	$\sigma_{0,2}$	δ_{10}
	kg/mm²	kg/mm²	%	kg/mm²	kg/mm²	%	kg/mm²	kg/mm²	%
1	108	86,3	5,9	115,5	85,6	7,0	84,5	84,5	5,9
2	100,6	82,0	6,5	103,5	76,5	8,5	88,0	88,0	6,5
3	99,8	83,5	7,9	95,6	74,5	10,7	86,0	69,1	11,5

Steel	1+2+3	Run	4100 h	4	Run	1700 h	5	Run	1500 h
	σ_B	$\sigma_{0,2}$	δ_{10}	σ_B	$\sigma_{0,2}$	δ_{10}	σ_B	$\sigma_{0,2}$	δ_{10}
	kg/mm²	kg/mm²	%	kg/mm²	kg/mm²	%	kg/mm²	kg/mm²	%
1	115	115	4	81,5	49	27,3	-	-	-
2	98,8	98,8	2,6	-	-	-	-	-	-
3	97,3	84,5	10,9	78,5	53,3	23,1	-	-	28,0

Steel	6	Run	1500 h
	σ_B	$\sigma_{0,2}$	δ_{10}
	kg/mm²	kg/mm²	%
1	73,0	30	37,6
2	-	-	-
3	73,0	31,8	28,4

Explanation of steel types:
1   12Cr18Ni10T   0,12%C, 18 %Cr, 10 %Ni, 1 %Ti
2   Cr16Ni11M3   16 %Cr, 11 %Ni, 3 %Mo
3   Cr16Ni15M3B   16 %Cr, 15 %Ni, 3 %Mo, 1 %N

Analysis of these data shows that sodium carburisation potential was dependent on the method of oil introduction into the rig. With pouring of 150 ml of oil on the sodium surface in the pump tank at temperature of 300 °C the rate of carburization of austenitic steels specimens was calculated to be $3.5 \cdot 10^{-5}$ mg/cm^2 .h. At the same time, injection of only 70 ml oil directly into test section with subsequent rise of sodium temperature up to 700 °C (before test section was pumped), has led to high carburization of steel specimens with the rate of 2.4 . 10^{-3} mg/cm^2 . h.

Low alloyed perlitic steels 10X2M and EI-531, have not exhibited practically notable alterations in their carbon content and degradation of their mechanical properties due to carbon transfer.

Table 8: Variation of mechanical properties (at 20°C) of perlitic steels specimens after their test in the flowing sodium at 700 °C

Steel	Heat treatm.			Ar 700 °C 2000 h			1 Run 2000 h		
	σ_B	$\sigma_{0,2}$	δ_{10}	σ_B	$\sigma_{0,2}$	δ_{10}	σ_B	$\sigma_{0,2}$	δ_{10}
	kg/mm^2	kg/mm^2	%	kg/mm^2	kg/mm^2	%	kg/mm^2	kg/mm^2	%
10X2M	60,5	42,0	22,7	40,7	25,0	31,8	34,3	23,8	28,7
EI-531	62,1	43,0	24,0	44,5	32,2	27,9	34,0	23,5	18,7

Steel	2 Run 1800 h			1 + 2 Run 3800 h			3 Run 300 h		
	σ_B	$\sigma_{0,2}$	δ_{10}	σ_B	$\sigma_{0,2}$	δ_{10}	σ_B	$\sigma_{0,2}$	δ_{10}
	kg/mm^2	kg/mm^2	%	kg/mm^2	kg/mm^2	%	kg/mm^2	kg/mm^2	%
10X2M	35,3	22,1	27,7	32,1	20,8	26,6	37,5	27,1	32,6
EI-531	38,3	21,8	24,6	33,8	20,3	19,9	44,5	31,5	28,0

Steel	1 + 2 +3 Run 4100 h		
	σ_B	$\sigma_{0,2}$	δ_{10}
	kg/mm^2	kg/mm^2	%
10X2M	31,5	19,5	28,1
EI-531	33,0	23,0	15,0

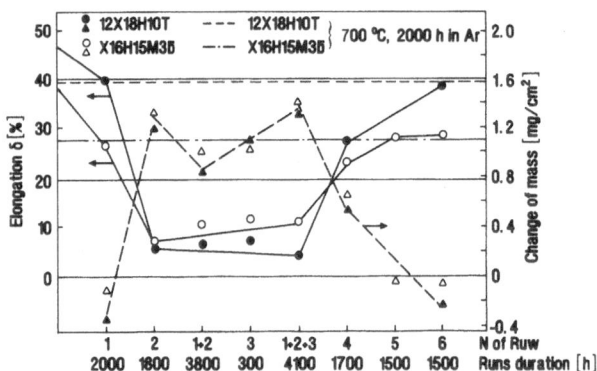

Fig.7. Diagram of steels specimens elongation and mass variations after test in sodium flow at 700 °C

As well there was not established any carburization potential of sodium, that was estimated on the basis of austenitic steels mechanical properties alterations.

Reference

1. A.W. Thorley, A.Blundell, J.Prescott, J. Hilditch, in :Proc. 4 th Intern. Conf. on Liquid Metal Engineering and Technology, Soc. Franc. Energie Atom., Paris 1988, Vol.2, p.507, 1-12

PHYSICAL CHEMISTRY OF SODIUM-CAESIUM-GRAPHITE SYSTEMS

A.I. Lastov, E.E. Konovalov, O.V. Starkov

Institute of Physics and Power Engineering
249020 Obninsk, Russia

1. INTRODUCTION

The sodium coolant of a fast reactor is contaminated with different radionuclides during its operation. These nuclides include activation and fission products as well as nuclear fuel components, if failures of fuel element clads may occur. The main radioactive impurities of sodium coolant are sodium-24, sodium-22, and caesium-137. The last one becomes predominant after a period of eight to ten days after shut-down of the reactor. The trapping of caesium radionuclides out of the sodium coolant is, therefore, required.

The unique property of carbon-graphite materials to interact with caesium can be used for purifying sodium coolant of caesium radionuclides. At present, accumulated domestic and foreign experience is available for the use of graphite based traps to decrease the radioactivity of primary circuit sodium. The PGI-graphite was used as the active component of a trap in the BR-10 reactor [4]; GMZ-mark graphite was used in the BOR-60 reactor [6]; RVC-type graphite was used in American work [6,7]. The effectiveness of caesium removal by all types of graphite is high enough, but the possibility of carbon transfer into the sodium circuit has to be considered.

2. EXPERIMENTAL RESULTS

The results of the research on the interactions of different carbon-graphite materials with sodium containing radioactive caesium are presented in this paper. The research work was performed under static and dynamic conditions of isothermal and non-isothermal tests. The character and degree of the graphite-sodium interaction depends on the type of graphite, the temperature and the availability of thermocycles during the tests. The oxygen content of sodium is also influencing the reactions (Table 1).

The results of the research on the interactions of different carbon-graphite materials with sodium containing radioactive caesium are presented in this paper. The research work was performed under static and dynamic conditions of isothermal and non-isothermal tests. The character and degree of the graphite-sodium interaction depends on the type of graphite, the

Liquid Metal Systems, Edited by H.U. Borgstedt
and G. Frees, Plenum Press, New York, 1995

temperature and the availability of thermocycles during the tests. The oxygen content of sodium is also influencing the reactions (Table 1).

The saturation process occurs at the expense of pore packing in graphite as well as chemical interaction of graphite and its impurities with sodium (Table 1).

Table 1: Material properties after the test

No.	Exposure conditions		Oxygen concentr.	Size of specimens	$\Delta M/M_{uc}$	Observations and results
	t (°C)	τ(h)	mass (%)	mm		
A	GMZ					
1	500	1000-	$4 \cdot 10^{-3}$	φ 40x60	0,10-0.15	Saved
2	550	3000	$4 \cdot 10^{-3}$	φ 40x60	0,10-0,16	Saved
3	600		$4 \cdot 10^{-3}$	φ 40x60	0,20-0,22	Saved
4	500		$4 \cdot 10^{-3}$	φ 20x40	0,12-0,20	Cracks,failing
5	550		$4 \cdot 10^{-3}$	φ 20x40	0,14-0,19	Step spallings
6	600		$4 \cdot 10^{-3}$	φ 20x40	0,14-0,16	Face cracks
7	550		$4 \cdot 10^{-3}$	φ 10x20	0,17-0,20	Step spallings
8	550		$1 \cdot 10^{-2}$	φ 10x20	0,20-0,24	Swelling,decay into pieces
9	550		$5 \cdot 10^{-2}$	φ 10x20	0,20-0,30	Face cracks
10	550		$5 \cdot 10^{-2}$	φ 10x20	0,25-0,34	Cracking
11	350	2000	$2 \cdot 10^{-4}$	φ 3	0,12	Saved
B	PK-0					
12	500	1000	$4 \cdot 10^{-3}$	φ 40x60	-	Failed in powder
C	SU 1300			SU 2000		
13	250	100	10^{-3}	2x5x6	-	Failed
D	MPG-6			VPG		
14	250-350	100	10^{-3}	φ5x2	-	Saved
15	418	100	10^{-3}	"	-	Failed
16	302	1920	10^{-3}	"	-	Failed after 38 cycl.
17	350	1350	$2 \cdot 10^{-4}$	"	0,11	Crack
E	UUK			PG-2100		
18	200-418	100	10^{-3}	2x6x7	-	Saved
19	302	1920	10^{-3}	2x6x7	-	Saved at 38 cycl.
20	350	2000	$2 \cdot 10^{-4}$	2x6x7	0,12	Separation into lay.

Contact of graphite with sodium at 500 to 600 °C and during 3000 hours results in the increase of the geometrical dimensions of the graphite specimen (swelling), crack initiation and surface layer crumbling. The diameters of PK-0 specimens increased by 25 % during 1000 hours at 500 °C, and the specimens occupied the whole volume of the test vessel. The increased oxygen content of more than 10^{-2} mass % and 5 - 20 thermocycles between 550 and 250 °C enhanced the failure of the GMZ specimens (Table 2).

The glass-carbons SU-1300 and SU-2000 fail completely to particles of a size of 20 μm. The MPG-6, VPG and GMZ graphites either crack into a small number of parts or keep their initial form which increases in the dimensions. In the pyrolytic graphite PG-2100, which was tested at 300 - 350 °C, a separation of oriented planes is observed, while the specimens do not fail. The carbon-carbonic composite (CCC-) keeps its integrity at 300 - 350 °C but shows swelling. The MPG-6 and GMZ graphites fail after 18 - 38 thermal cycles, in which they were heated to 300 °C and roughly cooled down to room temperature by means of immersion into water.

Table 2:Effect of thermocycling on graphite GMZ in sodium

Conditions of exposure				Oxygen concentr.	Specimen size	$\Delta M/M_{ish}$	Observations and results
t (°C)	t (°C)	τ (H)	number of cycles		mm		
550	250	1000	5	$4 \cdot 10^{-3}$	ϕ 20x40	0,14	Spallings
			10			0,13	Spallings
			20			0,15	Cracking
		200	5	$4 \cdot 10^{-3}$	ϕ 20x40	0,14	
			10			0,16	Cracking
			20			0,18	
		3000	5	$4 \cdot 10^{-3}$	ϕ 20x20	0,14	
			10			0,12	Cracking
			20			0,15	
302	20	1920	18	$1 \cdot 10^{-3}$	ϕ 3	-	Falure
		200	30	$1 \cdot 10^{-3}$	ϕ 3	-	

3. PHYSICO-CHEMICAL-EVALUATION

The kinetic dependence of the sorption process of caesium-137 is evaluated of the half time $\tau_{1/2}$, for which the Arrhenius relation is determined:

$$\log \tau_{1/2} = 607/T - 0.164 \tag{1}$$

with $\tau_{1/2}$ as the period of time in which one half of caesium-137 is taken out of the melt (h); T as the absolute temperature (K).

The most significant characteristic of the sorption process is the coefficient of the distribution of caesium-137 between carbon-graphite material and the molten sodium,

$$K_M = A_C / A_M \tag{2}$$

K_M is the coefficient of the distribution of caesium-137 (kg Na/kg graphite); A_C is the specific activity of caesium-137 in graphite (Bk/kg graphite); A_M is the specific activity of caesium-137 in sodium (Bk/kg sodium).

The quantities of K_M have maximum values at about 250 to 300 °C within a small scatter band for all types of carbon-graphite material (Fig. 1).

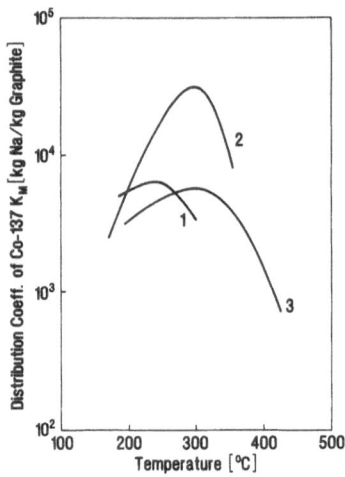

Fig. 1. Distribution coefficients K_M of caesium-137 activity for three different types of carbon-graphite

The cleaning of the sodium coolant in working reactors is usually performed by means of the single loading of carbon-graphite materials into the primary circuit. The attained cleaning degree n is connected with the relationship of the weight of graphite and sodium, G_C and G_M, as shown in equation (3):

$$n = K_M \cdot G_C / G_M + 1 \tag{3}$$

Taking $G_C / G_M = m$, we get

$$n = m \cdot K_M + 1 \tag{4}$$

The equation (3) allows to evaluate the amount of the loaded sorbent, which is sufficient to maintain the desired degree of clearing of sodium at a certain value of caesium-137 distribution coefficient, which is only dependent on temperature.

At the same time it appears to be possible to attain a high degree of clearing by means of the use of a much smaller portion of graphite, if the operation may be repeated many times. If i is the ordinal number of the injection operation of the given amount of graphite (G_C) into sodium (G_M), equation (5) is valid according to the definition:

$$K_M = (A_c^{i+1} \cdot G_H) / (A_H^{i+1} \cdot G_C) = (A_H^i - A_H^{i+1}) / (A^{i+1}) \cdot (G_H/G_C) \tag{5}$$

Transforming (5) we receive

$$A_H^{i+1} = A_H^i [1 / /K_M \cdot G_C)/(G_H+1)] \tag{6}$$

The following operation results in

$$A_H^{i+2} = A_H^{i+1}[1/(1+(K_M \cdot G_C)/G_H)] = A_H^i \cdot [1/(1+(K_M \cdot G_C)/G_H)^2] \tag{7}$$

In consequence of S operations:

$$A_H^{i+s} = A_H^i \cdot [1/(1+(K_M \cdot G_C)/G_H)]^s \tag{8}$$

For i = 0

$$A_H^{CT} = A_H^0 \cdot [1/(1+(K_M \cdot G_C)/G_H)]^s \tag{9}$$

According to equation for with $G_C/G_M = m$, we finally receive

$$A_H^0 / A_H^{CT} = (1+m \cdot K_M)^s \tag{10}$$

For the case of a single loading of graphite in an amount S, the analogous expression has the form

$$A_H^0 / A_H^{0G} = 1 + S \cdot m K_M \qquad (11)$$

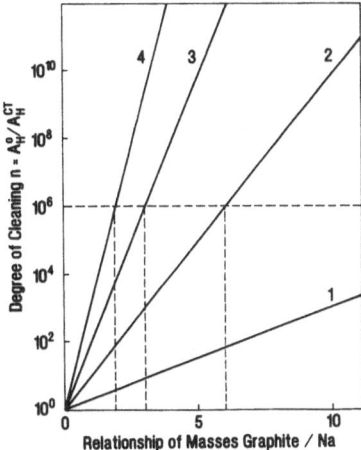

Fig. 2. Degree of sodium clearance in relation to the mass relationships of graphite and sodium

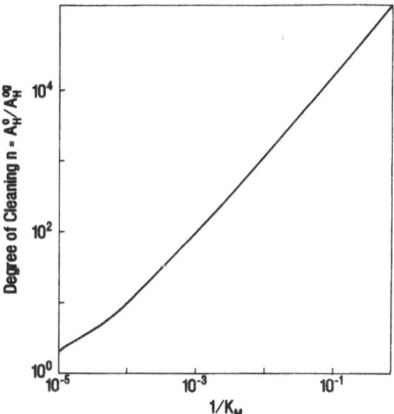

Fig.3. Degree of sodium clearance in relation to the mass relationships of graphite and sodium

4. CONCLUSIONS

The comparison of equations (10) and (11) shows that the clearing in steps is more effective than the single loading of the whole amount of adsorbent, if the product of $m K_M >$ 0.2-0.3. The plots constructed on the basis of equations (10) and (11) (Figs. 2,3) allow to optimize the clearing technology of sodium in respect to caesium-137. K_M changes the amount of graphite at a given distribution coefficient, as can be seen from the lines 1,2,3, and 4 on Fig. 2. The clearing coefficient of sodium in respect to caesium remains constant (n), if the operation numbers (s) change.

SOLUBILITY OF METALS IN THE LIQUID ALKALI METALS; THE SOLUBILITY DATA PROGRAMME OF THE IUPAC

Hans Ulrich Borgstedt and Cezary Guminski *

Karlsruhe Nuclear Research Center, IMF III,
P.O. Box 3640, W-7500 Karlsruhe, Germany
*Department of Chemistry, University of Warsaw,
Pasteura 1, 02093 Warsaw, Poland

1. INTRODUCTION

Activities of the Solubility Data Commission of the International Union of Pure and Applied Chemistry are briefly described. Samples of material, collected during the preparation of the monograph on the solubility of metals in the liquid alkali metals, are presented in form of tables and figures. General dependencies, ruling the solubilities of metals in low melting metals, are outlined.

IUPAC is the International Union of Pure and Applied Chemistry. This organisation is composed of several divisions, interested in typical chemical branches as physical, analytical, inorganic, etc. chemistry. Every division is composed of several specialising commissions. Actions of the IUPAC, making assessments in many divisions of chemistry, are very useful. Scientific and engineering experiments produce enormous amount of data, which ought to be systematised, elaborated, compared and finally selected for recommendation. There is practically no place for such treatment in encyclopaedias, handbooks, advisers, in which one meets numbers with a meagre comment, if any. A complete analysis of available data is seldom presented in monographs and authors of regular papers in journals make such exhaustive analysis even rarer. It seems to be quite typical that many unassessed data from monographs or handbooks are quoted in such circumstances. Reliability of values in experimental reference sources, especially when they are further used for technological calculations and scientific conclusions, is therefore very important. Everyone likes to minimise the printing errors but they were, are and probably will be in the scientific and technical literature. However, we should eliminate some incorrect values originated from erroneous methods, calculations and interpretations, or impure substances being used in experiments. After critical evaluation of all available references one may recommend a set of data.

One of the commissions, within the Analytical Chemistry Division, is the Solubility Data Commission, V.8., Which primarily was a section in the Equilibrium Data Commission. Due to importance of the solubility phenomena in which a large number of members is involved, a separate Solubility Data Commission was established at the late seventies.

Liquid Metal Systems, Edited by H.U. Borgstedt
and G. Frees, Plenum Press, New York, 1995

There are three main activities of the Commission: (i) publication of completed and evaluated solubility data on selected subjects in a series of volumes, (ii) preparation of a book on experimental methods of the solubility determinations and (iii) organisation of symposia on solubility phenomena every even year. The Solubility Data Commission is composed of three subcommittees: on the solubilities of gases, liquids and solids. Except the solid-solid combination (for which no specialists were recruited) all other combinations of state, including polymers and liquid crystals, are of the commission's interest. The activity of the group is not only relevant to chemistry but also to material science and engineering, biology, medicine, pharmacology, geology and environmental science.

As one may see from a leaflet showing the publication activity of the IUPAC Solubility Data Commission, quite various systems were elaborated and many others are under preparation to be published. A publication contract with Pergamon Press for about 50 volumes is finished and the Commission looks for a new publisher now. There are two volumes published in 1986 and 1992 on the solubility of metals in mercury and solubility of intermetallic compounds in mercury, respectively. Both were prepared by Guminski and Galus [1, 2].

The volume on solubility of metals in the liquid alkali metals by Borgstedt and Guminski [3] is completed. The starting point of the project was to collect the corresponding literature which has been well spread over books, journals, popular and very rare reports published in various countries. For this reason, the Chemical, Metals and Nuclear Science Abstracts, and Atom Index were systematically inspected. Moreover, the well known monographs on constitution of binary alloys [4, 5, 6] served as a very informative source. A part of material was discovered during literature studies on the solubility and corrosion phenomena. Each of the information was inspected in original, and if not otherwise stated, no secondary sources were used in our work. Statistically, a use of a secondary source decreases reliability of an information for several percent.

Let us look at the way of presentation of material in our volume. Such arrangement is kept in the whole series of the Solubility Data Series. Every one original, experimental paper is presented in a form of, so named, data sheet. Here we have one example. All essential information from a paper is sorted out into the boxes for easier orientation. "Components" contain the chemical names, their symbols and numbers according to the Chemical Abstracts. The first is always a solute, the second is a solvent metal.

"Original Measurements" contain full reference to an original paper. "Variables" inform us about conditions that were changed in specified ranges during the solubility measurements (temperature, pressure, concentration of important impurities and additives). All data sheets in this volume were prepared by the authors but if one would acquire a service of someone`s help in translation or preparation one should inform about this in the proper place. "Experimental values" contain presentation of all data from a paper in the numerical form. If the original contains only a figure, the figure is not depicted but the data are read out from such figure. The accuracy of this procedure is mentioned in the "Estimated error". A presence of non-metallic contaminants in both solute and solvent may increase the solubility determined, especially in case of the transition metals, and therefore the non-metallic concentrations are also specified in separate columns.

"Method/Apparatus/Procedure" contains details connected with the heading, from a sample preparation to the end of its analysis. "Source and Purity of Materials" contains all details connected with the heading. A method of further purification, if such was additionally applied, may be added. Concentrations of the contaminating elements (O, N, H, C) are always specified if corresponding information is available.

"Estimated Error" of solubility, temperature or pressure is either taken from the original paper or estimated by the compilers, if it is possible. One should also specify in this place what kind of error is estimated: accuracy, precision, standard deviation, detectivity level of analysis or accuracy of the reading-out procedure. "References" contain information about other literature sources used for preparation of this sheet.

The compilers tried to be essential, as much as possible, in filling up the boxes, that a reader should not be forced to study the original paper again. The complete set of the data sheets for a chosen solute in a chosen solvent starts with the "Critical Evaluation" of such system. Here again, we have the box of "Components" and the names and affiliations of the evaluators. In the "Critical Evaluation" text, there is a brief mentioning of all works connected with the subject and comparison of the solubility data. Additional data, originated from papers with undescribed experimental details, are quoted in this text. Explicit mistakes are pointed out and very deviating results are classified for rejection, although they are cited for completion. These erroneous or incompletely described works have no corresponding data sheets. Additionally, some qualitative information on compatibility and corrosion tests performed for a system are also enclosed for comparison with the quantitative solubility data, however, literature on this subject is not complete.

If possible, a solubility fitting equation for a specified temperature range is either quoted (after testing its validity) or calculated by the evaluators. Suggested values of the solubility at selected temperatures are listed in the table. These data belong to three classes: recommended, tentative and doubtful. If agreement of at least two independent studies is within their experimental errors, the solubility values are assigned to the recommended category. If one reliable result is reported or if the mean value of two or more reliable studies is outside the error limits, the solubility values are assigned to tentative category. The doubtful category is for one, not convincing result, or mean value of several quite spread out results. In the tabulation, three, two or one digit number is assigned for respective precision that are better than $\pm 1\%$, 10 % and worse than $\pm 0\%$ 10 %.

For the majority of systems, the most contemporary phase diagrams are enclosed. Few diagrams published recently in the reference source [7] were even further corrected or supplemented. Insertion of the phase diagram should help a reader to identify composition of a corresponding equilibrium solid phase, that could be a practically pure solid metal, a solid solution or an intermetallic compound, and to support understanding the selected data and conclusions. If schematic phase diagrams, as for example W-Li, W-Na. W-K, W-Rb and W-Cs, are essentially of the same type, only one diagram is presented (mostly for a solute metal with Li) and for the rest of systems a reader is guided to the one diagram shown.

In order to illustrate complexity of critical evaluations of some systems, the evaluators constructed two diagrams in the form of logarithm of the solubility versus reciprocal temperature. These figures, constructed for the Fe-Na and Cr-Li solubility systems, show somewhat large scatter of data from all sources, in which the formation of ternary compounds Fe-Na-O and Cr-Li-N interferes with the dissolution of these metals. For the former system, at lower temperatures and high O concentrations the most probable equilibrium phase is $NaFeO_2$. Above 760 K the Na_4FeO_3 is the equilibrium phase and the pure Fe may be in equilibrium at very low O concentrations and higher temperatures. The diagram for the latter system, Cr-Li, demonstrates that the more pure Li is used for equilibration, the nearer is position of the experimental results to the theoretical solubility dependence.

Since the alkali metals have rather low temperatures of boiling, many investigations had to be performed at a constrain pressure. This condition was pointed out but the effect of elevated pressure seems to be rather negligible because the pressures in the experimental conditions were not much higher than normal, and as we know the phase diagrams begin to change upon pressure above 0.1 GPa.

No detailed material was collected on mutual solubilities in the family of alkali metal systems because they are characterised by either complete miscibility or extensive immiscibility in liquid state. These ten systems are presented at the beginning of this volume in form of assessed phase diagrams. Another reason of such presentation is a page limit of our volume, which should not overrun 500 pages, and the alkali metal binary combinations have to be presented on more than 50 pages.

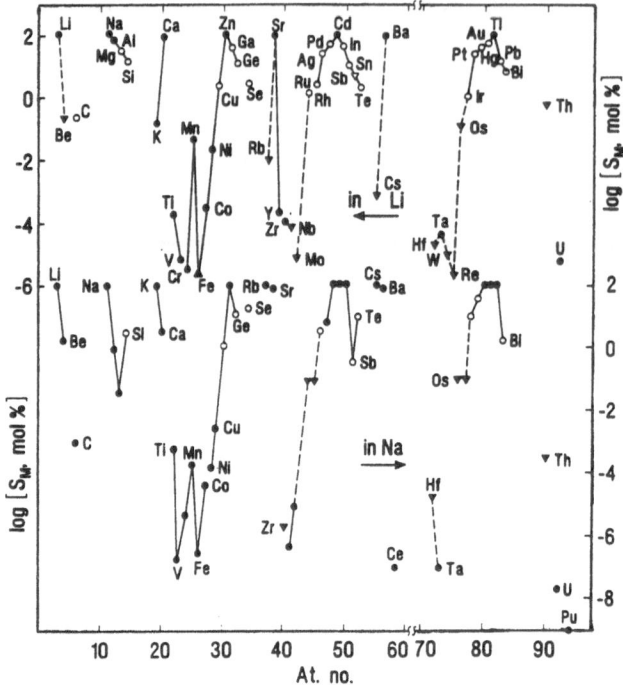

Fig. 1. Comparison of the solubility of metals in Li and Na at 873 K

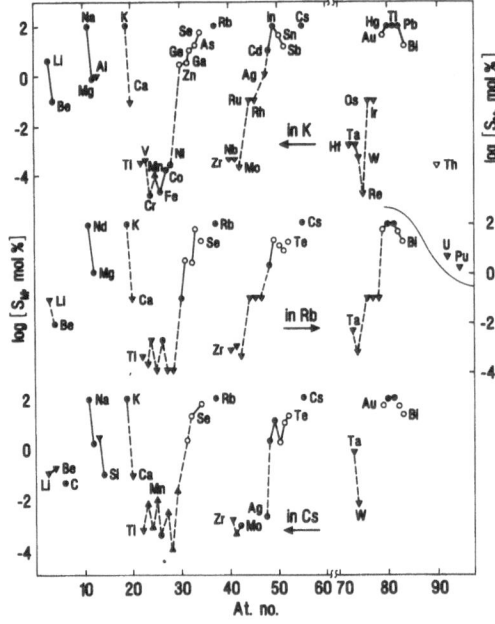

Fig. 2. Comparison of the solubility of metals in K, Rb and Cs at 873 K

The critical evaluations are always finished with complete list of references. The oldest works being used in our volume are from 1889 (by Heycock and Neville, and Tammann) and the most contemporary are from 1992. It was observed that the most intensive investigations of the solubility of metals in the liquid alkali metals were performed in sixties, what was directly connected with the intensive development of the peaceful uses of atomic energy. Various methods were used for the solubility determinations in the metallic systems, which fact is reflected in the table. This subject will be separately presented in another monograph [8] prepared by the Commission. (Sample sheets concerning the Au-Na system are attached.)

As it was shown earlier [9], the solubility of metals and metalloids in the liquid low melting metals change along the periodic table when one orders them according to their atomic numbers. In liquid Hg, Ga, In, Sn, Pb, and Bi solvents, observed jumps in the solubility behaviour of the solute metals occur, when the configuration of valence electrons in the solutes is drastically changed; for example, going from Xe to Cs in liquid Hg (from 10^{-7} to 4.4 mol % at 298 K). Eu and Yb, being two valent lanthanides, are distinctly better soluble than their three valent neighbours. Solubilities of other lanthanides as well as actinides are similar. Mn is distinctly better soluble than other transition metals. Such solubility behaviour is at best correlated with the reciprocal hardness of the metals but considerably worse with typical thermodynamic parameters of melting, boiling or sublimation. After all, the dissolution process is composed of melting and then solvation, this second stage must not be neglected.

If one looks for a comparison of the solubility of metals in the liquid alkali metals at 873 K (Figs. 1 and 2), one will easily observe that such dependencies, although more fragmentary, are almost parallel. Again the high melting metals show the lowest solubility, the low melting metals are relatively well soluble, and the anomalous behaviour of Mn is always observed. The general conclusion may be repeated that the solubility of metals in the low melting metals is mainly related to properties of a solute and in small extent to a solvent.

In case of the alkali metals, one should remember that in fact one measures the apparent solubilities at certain levels of impurities, and we still do not know for each of the systems a degree of the over-estimation of the solubility due to the presence of non-metallic traces in the systems being investigated. We still do not know whether we would observe the solubilities, as predicted by thermodynamic models [9], if we would use perfectly pure metals.

It is therefore much to do to complete and verify the dependencies shown in the figures, especially in liquid Rb and Cs solvents. A subsequent project on solubilities of non-metallic substances in the liquid alkali metals for the IUPAC is just started and this will complete our conclusions concerning solubilities of all elements in the liquid alkali metals.

REFERENCES

1. C. Guminski, Z. Galus, "Metals in Mercury", Solubility Data Series, C. Hirayama,Ed., Pergamon, Oxford, 1986, 450 pp.
2. C. Guminski, Z. Galus, Intermetallic Compounds in Mercury, Solubility Data Series, J. Osteryoung, M. Schreiner, Eds., Pergamon, Oxford, 1992, 250 pp.
3. H.U. Borgstedt, C. Guminski, Metals in the Liquid Alkali Metals, Solubility Data Series, in press
4. M. Hansen, K. Anderko, Constitution of Binary Alloys, McGraw-Hill, N.Y., 1958; I supplement by R.P. Elliott, 1965; II supplement by F.A. Shunk, 1969.
5. W.G. Moffatt, Handbook of Binary Phase Diagrams, Genium, Schenectady, 1978-1991.
6. A.E. Vol, I.K. Kagan, Stroeniye i Svoistva Dvoinykh Metallicheskikh Sistem, Fizmatgiz, Moskva, 1959, 1962, 1976 and 1979.
7. T.B. Massalski, H. Okamoto, P.R. Subramanian, L. Kacprzak,□Binary Alloy Phase Diagrams, Eds., Am. Soc. Mater., Materials Park,
8. C. Guminski, H.U. Borgstedt, in Experimental Methods of Determination of Solubility, (to be published).
9. C. Guminski, Z. Metallk., 81, 1990, 105.

COMPONENTS:	ORIGINAL MEASUREMENTS:
(1) Gold; Au; [7440-57-5]	Kienast, G.; Verma, J.; Klemm, W.
(2) Sodium; Na; [7440-23-5]	Z. Anorg. Chem., 1961, 310, 143-169.
VARIABLES:	PREPARED BY:
Temperature: 355 - 1278 K	H.U. Borgstedt and C. Guminski

EXPERIMENTAL VALUES:

The liquidus points of the Au-Na system were determined.

t/°C	soly/mol % Au	t/°C	soly/mol % Au
94	0.61	359	31.0
85	1.28	372	33.8
82	2.02	440	34.6
83	2.83	545	38.7
90	3.48	648	42.2
108	6.37	742	44.7
110	7.73	870	50.0
122	11.2	940	54.0
140	12.8	981	60.0
163	15.3	985	61.0
220	21.6	1005	67.0
305	28.7	970	73.0

AUXILIARY INFORMATION

METHOD/APPARATUS/PROCEDURE:

Both the metals were weighed and placed in a corundum crucible inserted in a quartz tube filled with Ar (1). The corundum as well as the quartz were slightly attacked by the molten alloy. The composition of the alloy was chemically analyzed. The Al content of the melt was 0.05 mass % after 60 hours of equilibration at 1073 K. Cooling curves of the molten alloys were recorded by means of a Pt/Pt-Rh thermocouple.

SOURCE AND PURITY OF MATERIALS:

Au: nothing specified.
Na: nothing specified.
Ar: from Linde, further purified by means for reaction with Ti at higher temperature.

ESTIMATED ERROR:
Solubility: nothing specified.
Temperature: precision ± 2 K, according to (1),
accuracy not better than ± 10 K (compilers).

REFERENCES:
(1) Dorn, F.W.; Klemm, W. Z. Anorg. Chem., 1961,
309, 189-203.

Components:	Evaluator:
(1) Gold; Au; [7440-57-5]	H.U. Borgstedt, KfK, Karlsruhe, Germany
	C. Guminski, Dept. of Chemistry, Univ. of Warsaw,
(2) Sodium; Na; [7440-23-5]	Poland
	November 1989

Critical Evaluation:

The solubility pf Au in liquid Na is quite large. Additions of up to 3.5 mol % Au decrease the melting point of Na to 354 K. The early publications (1-3) show good agreement in this range of concentrations. The results of (4) differ from these data by more than the experimental error. The reliability of the data of (4) seems, therefpore, to be questionable even in the range of high temperatures. The solubility values from (3) and (4) are in a band of the width ± 20 K. The single determinations of (5) and (6) indicated a lower solubility of Au than recorded in (4) in the temperature range 650 - 1050 K. The result of (3) at 975 K may be rejected, since it shows internal inconsistency and deviate from other data (4-6).

The results presented in (6) are not compiled, since experimental details were not reported. A solubility of 45 mol % Au at 1048 K was established by means of titration using a solid electrolyte.

The saturated solutions of Au in liquid Na are in equilibrium with $AuNa_2$ (below 536 K), AuNa (below 645 K) and Au_2Na up to 1275 K, as is indicated in the Au - Na phase diagram which was reported in (7). The existence of a compound Au5Na has to be confirmed.

The tentative (t) and doubtful (d) values of the solubility of Au in liquid Na

T/K	$soly$/mol % Au	source
354	3.5 (t) (eutectic)	(1,3), mean value
373	4.5 (t)	(3)
436	15 (d) (peritectic)	(4)
473	20 (d)	(4), interpolated
573	28 (d)	(4)
645	33 (d) (peritectic)	(4)
673	34 (d)	(4), interpolated
773	36 (d)	(4), interpolated
873	40 (d)	(4), interpolated
973	43 (d)	(4), interpolated
1073	46 (t)	(4,6), interpolated
1173	50 (t)	(3,5)
1275	67 (d)	(4)

References

1. Heycock, C.T.; Neville, F.H. *J. Chem. Soc.* 1889, *55*, 666.
2. Tammann, G. *Z. Phys. Chem.* 1889, *3*, 441.
3. Mathewson, C.H. *Intern. Z. Metallogr.* 1911, *1*, 81.
4. Kienast, G.; Verma, J.; Klemm, W. *Z. Anorg. Chem.* 1961, *310*, 143.
5. Nicoloso, N.; Schmutzler, R.W.; Hensel, F. *Ber. Bunsenges. Phys. Chem.* 1978, *82*, 621.
6. Egan, J.; Algasmi, R., Brookhaven National Laboratory, 1984, private communication to A.D. Pelton, as reported in 7.
7. Pelton, A.D. *Bull. Alloy Phase Diagr.* 1986, *7*, 136.

APPLICATION OF SOLUTION MODELS FOR THE PREDICTION OF CORROSION PHENOMENA IN LIQUID METALS

V.P. Krasin

Moscow Motor-Car Construction Institute
16, Avtozavodskay ul.
Moscow 109280, Russia

1. INTRODUCTION

It is known that the compatibility of construction materials with a metallic melt depends on the equilibrium solubility of the material in the liquid phase. In case of some binary melts such as Na-K and Pb-Bi of eutectic composition, there are experimental data which allows the estimation of their compatibility with constructional materials of different types. For other two-component metallic melts, however, only limited data on the solubility of metals are published [1]. Additional difficulties are encountered, if the melts contain impurities like oxygen, nitrogen, carbon and hydrogen, which have a significant influence on the compatibility of solid and liquid metals. A preliminary estimation of the compatibility of constructional materials with multi-component melts is proposed which is based on a procedure to calculate the solubility of solid metals in pure two-component melts and also in melts containing non-metallic impurities that would be possible.

2. SOLUTION MODELS

It is assumed that a metal A_3 which is a component of a constructional material is dissolved in a melt of two low-melting metals, A_1 and A_2. According to the model of regular solution [2] for a three-component system one may write.

$$\ln \gamma_3 = [E_{13}^1 x_1^2 + E_{23}^1 x_2^2 + x_1 x_2 (E_{13}^1 + E_{23}^1 - E_{12}^1)] / RT \qquad (1)$$

where γ_3 is the activity coefficient of A_3 in the ternary liquid metal system A_1-A_2-A_3; x_1 and x_2 the mole fractions of the components A_1 and A_2 in the melt; E_{ij}^1 the parameter of the pair interaction of the components in the liquid phase; R is the universal gas constant and T the absolute temperature.

In the case of a two-component regular solution, the interaction parameter is related to the activity coefficient of the second component as follows:

$$E_{12} = [R T \ln \gamma_{2 (1)}] / (1 - x_2)^2 \qquad (2)$$

Liquid Metal Systems, Edited by H.U. Borgstedt
and G. Frees, Plenum Press, New York, 1995

where $\gamma_{2(1)}$ is the activity coefficient of the second component in the solution, the concentration of which is x_2 (in mole fraction).

If the melt contains a fourth, non-metallic component A_4 as an impurity, the activity coefficient of the metal A_3 in this melt can be obtained by means of integrating the Gibbs-Duhem equation, as in an earlier publication [3]:

$$\ln \gamma_3 \text{ (IV)} = \ln \gamma_3 \text{ (III)} + \int_0^{y4} [\partial \ln \gamma_3 / \partial y_4] \cdot d y_4 \qquad (3)$$

where γ_3 (III) and γ_3 (IV) are the activity coefficients of A_3 in the three-component melt A_1-A_2-A_3 and in the four-component melt A_1-A_2-A_3-A_4, respectively; y_4 is the concentration of the element A_4 in the melt, expressed as the ratio of the number of moles of A_4 to the number of moles of the element A_1. Subsequently, it is necessary to change from mole fractions to concentrations of the components of the melts, expressed as follows:

$$y_1 = 1;$$
$$y_2 = x_2/(1-x_2-x_3-x_4);$$
$$y_3 = x_3/(1-x_2-x_3-x_4);$$
$$y_4 = x_4/(1-x_2-x_3-x_4);$$

If the treatment is restricted to low concentrations of the impurity ($y_4 \ll 1$), the Eq. (3) can be written as:

$$\ln \gamma_3 \text{ (IV)} \cong \ln \gamma_3 \text{ (III)} + \sigma_3^4 \cdot y_4, \qquad (4)$$

where $\sigma_3^4 = \partial \ln \gamma_3 / \partial y_4$ (for $y_2 = y_2$, $y_3 = y_3$ and $y_4 \rightarrow 0$) is called the specific interaction parameter. The following equation is applicable [3] for any composition of the multi-component solution:

$$\sigma_3^4 = \sigma_4^3 = \partial \ln \gamma_4/\partial y_3 \qquad (5)$$

It should be noted that the corresponding equations do not always apply, if the concentrations of the components are expressed in mole fractions.

The specific interaction parameters are calculated in our earlier paper [4] using the ideal solvent model of Kapoor [5]. In order to obtain a more exact value of σ_3^4 in the present work, we develop the coordination cluster theory of Blander et al. [6] for four-component systems. The interesting features of this model are that it takes into account the non-ideal behaviour of the solvent as well as the changes in interactions between the solvent atoms which are neighbours of solute atoms.

The basic assumptions of the coordination cluster theory are:
1) the concentration of the non-metallic solute in the solution is low enough that the Sievert's law is obeyed;
2) the dissolved solute atoms occupy quasi-interstitial sites due to the high diffusivities of oxygen, nitrogen and carbon in molten metals;
3) a binominal distribution of complexes, which consist of a non-metallic solute atom surrounded by three atoms of the metallic elements with varying amounts of these three elements, is valid.

Thus, it is assumed that the atoms of the non-metallic element A_4 in the melt of three metals, A_1, A_2 and A_3, occupy interstitial positions with a coordination number z. Each A_4 atom in the melt has j nearest neighbours of A_1 atoms, k of A_2 and l of A_3 atoms (l = z - j - k). The melt contains [z (z+1)/2] kinds of these configurations, which are called clusters and are denoted as $A_4[(A_1)_j, (A_2)_k, (A_3)_l]$.

The thermodynamic characteristics of the solution depends on the distribution of A_4 atoms among the clusters and is determined by the affinity of the metal atoms of each type to the element A_4. The following equation was obtained for the calculation of the activity coefficient of A_4 in the melt.

$$\gamma_4^{-1} = \sum_{j=0}^{z} \sum_{k=0}^{(z-j)} \{C_z^j \ C_{z-j}^k \ [x_1 \ \gamma'_{1(1-2-3)} / \gamma_{4(1)}^{1/z}] j \cdot [x_2 \ \gamma'_{2(1-2-3)} / \gamma_{4(2)}^{1/z}] k \cdot$$

$$[x_3 \ \gamma'_{3(1-2-3)} / \gamma_{4(3)}^{1/z}] \ l \cdot \exp \left[(jkh_{12} + klh_{23} + ljh_{31})/2RT\right] \tag{6}$$

where $\gamma_4(1), \gamma_4(2)$ and $\gamma_4(3)$ are the activity coefficients of A_4 in the binary melts A_1-A_4, A_2-A_4, and A_3-A_4, respectively; C_z^j means the number of combinations of j from z elements; h_{12}, h_{23} and h_{31} are the energy parameters, which are constants for the ternary systems A_1-A_2-A_4, A_2-A_3-A_4 and A_1-A_3-A_4 at each temperature, t is a geometric parameter characterizing the relative decrease in the strength of metallic bonds formed between the atoms A_1 and A_2 in the cluster $A_4[(A_1)_j,(A_2)_k,(A_3)_l]$ in vicinity of the solute atom A_4.

The activity coefficients can readily be calculated from the solubility data of the non-metallic elements in the liquid metals. The values of $\gamma_{1(1-2-3)}, \gamma_{2(1-2-3)}, \gamma_{3(1-2-3)}$ may be calculated using Eq. (1). The following expression was obtained [7] for calculating the energy parameters:

$$h_{mn} = 2t \cdot E_{mn}/(z-1) - E_{mn} \left[0.09 \ (V_n / V_m)^2 + 2/z^2\right] - 0.04 \ RT \ | \ln \ (\gamma_{4(m)} / \gamma_{4(n)}) \ |$$
$$m = 1,2,3, \qquad n = 1,2,3, \qquad .m \neq n$$

where V_m and V_n are the metallic valences given by Pauling [8] of A_m and A_n, respectively.

The best agreement between calculation and experiment for the three component system is observed for the values $z = 6$ and $t = 0.33$ [6]. We shall use these values in our calculations.

The values of σ_4^3 for the four-component system A_1-A_2-A_3-A_4 can be obtained by means of numerical differentiation of Eq. (6).

The corrosion of the solid metal A_3 in a melt of the two low-melting metals A_1 and A_2 containing the non-metallic impurity A_4 can now be considered. The solid metal A_3 is obviously in contact with the four-component melt A_1-A_2-A_3-A_4. For systems, in which the components A_1 and A_2 do not form solid solutions in A_3, the activity of the component A_3 in the saturated solution is equal to the activity of the pure solid metal.

$$a_3(IV) = \gamma_3 \ (IV) \ x_3^0 \ (IV) = a_3^* = 1 \tag{7}$$

where a_3 (IV) and γ_3 (IV) are the thermodynamic activity and the activity coefficient of A_3 in the system A_1-A_2-A_3-A_4; a_3^* the activity of the pure component A_3 in the solid state; and x_3^0 (IV) the solubility of A_3 in the four-component melt. Repeating the arguments used for the three-component system A_1-A_2-A_3 one obtains:

$$a_3 \ (III) = \gamma \ (III) \ x_3^0 \ (III) = 1 \tag{8}$$

where a_3 (III), γ_3 (III) and x_3^0 (III) are the same quantities as in Eq. (7),but for the system A_1-A_2-A_3. Eqs. (7) and (8) can be rewritten as follows:

$$\gamma_3 \ (IV) = 1 / x_3^0 \ (IV) \ ;; \ \gamma_3 \ (III) = 1/x_3^0 \ (III) \tag{9}$$

Combining the Eqs. (4), (5) and (9) one obtains an equation for the calculation of the solubility of a component of a constructional material in a melt of two metals containing a non-metallic impurity:

$$x_3^0 \ (IV) = x_3^0 \ (III) \ \exp \left(- \sigma_4^3 \ y_4\right) \tag{10}$$

The applicability of the cluster model for the prediction of the corrosion behaviour of materials in binary melts has been shown in earlier publications [9,10].

The equations which were developed on the basis of the cluster model were used to calculate the solubility of iron, chromium, nickel in liquid lead-lithium alloy of the eutectic composition Pb-17Li. The effect of oxygen on the solubility of components of structural materials was evaluated using the coordination cluster theory (CCT) equations. These systems were chosen, since the Pb-17Li melt is considered as an attractive breeding material for future fusion reactors. The initial data for the calculations are listed in Table 1. The binary constitution data of the Pb-Li system, which were used in the calculations, are taken from Saboungi et al. [15].

The results of the calculations (Figs. 1-3) show that the pure molten Pb-17Li alloy is less aggressive towards constructional materials than pure lead. The solubility of nickel in Pb-17Li which does not contain any non-metallic impurities is higher than the solubility of iron and chromium. The concentration of oxygen has no effect on the solubility of nickel. The solubility of chromium increases, if the molten alloy contains oxygen. The effect of the oxygen content on the solubility of chromium is stronger at low temperatures. The calculations by means of Eqs. (5) and (6) show that $\sigma^O_{Fe} \ll \sigma^O_{Cr}$, and the content of oxygen as impurity has only a weak effect on the solubility of iron.

Table 1: Parameters of the temperature dependence $\gamma = \exp(C+D/T)$ of the activity coefficients of elements in dilute binary melts in the range 550 - 900 K.

Element	System	C	0.001·D	Reference
Ni	Li-Ni	-0.55	8.32	[11]
Fe	Li-Fe	-8.0	24.6	[11]
Cr	Li-Cr	-6.48	20.95	[11]
Ni	Pb-Ni	0.62	2.96	[1]
Fe	Pb-Fe	4.05	5.64	[12]
Cr	Pb-Cr	-3.20	12.3	[11]
O	Li-O	15.97	-67.04	[13]
O	Pb-O	6.13	-14.22	[14]
O	Ni-O	3.69	-9.07	[14]
O	Fe-O	3.07	-14.02	[14]
O	Cr-O	9.53	-35.78	[4]

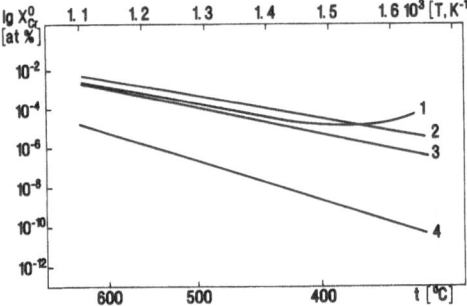

Fig. 1. The solubility of Fe in Li (3), in Pb (1) and in the Pb-17Li alloy (2)

The results of the calculations are in satisfactory agreement with the experimental mass transfer data by Barker et al. [16]. The corrosion data of Barker et al. show that the oxygen impurity in the Pb-17Li alloy increases the mass transfer rate of chromium, but has no influence on the corrosion behaviour of nickel and iron.

Thus, the proposed method to estimate the solubility of the components of structural materials in multi-component melts allow the prediction of the corrosion behaviour of constructional materials in such liquid metal systems.

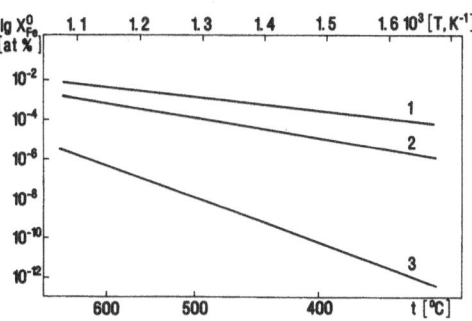

Fig. 2. The solubility of Cr in Li (4), in Pb (2) and in the pure Pb-17Li alloy (3) and in the alloy containing $3 \cdot 10^{-3}$ at % oxygen (1)

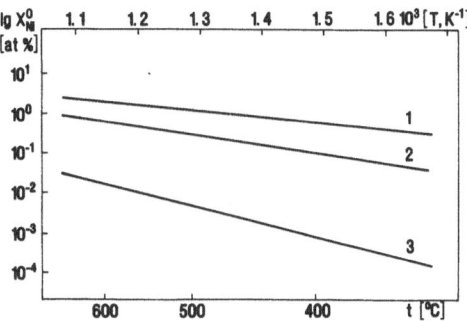

Fig. 3. The solubility of Ni in Li (3), in Pb (1) and in the Pb-17Li alloy (2)

REFERENCES

1. M.G. Barker, T. Sample, Fusion Engng. Design 14 [1991] 219.
2. L. Kaufmann, H. Bernstein, Computer Calculation of Phase Diagrams with Special Referencee to Refractory Metals (Refractory Materials, Vol. 4) Russian tranlation, Izdatelstvo Mir, Moskva, 1972.
3. R. Schuhmann, Metall. Trans., B 16 [1985] 807.
4. V.P. Krasin, I.E. Lyublinskii, Yu.V. Mitin, Russ. J. Phys. Chem. 64 [1990] 660.
5. M.L. Kapoor, Scr. Metall. 10 [1976] 323.
6. M. Blander, M. Saboungi, P. Cerisier, Metall. Trans., B 10 [1979] 613.
7. T. Chiang, Y.A. Chang, Metall. Trans., B 7 [1976] 453.
8. L. Pauling, The Nature of the Chemical Bond, 3rd. ed., Cornell University Press, Ithaka, 1970.
9. N.M. Beskorovainyi, V.P. Krasin, Metally i Splavy dlya Atomnoi Tekhniki, Energoatomizdat, Moskva, 1985, 30.
10. V.P. Krasin, Yu.V. Mitin, V.B. Kirilov, Russ. J. Phys. Chem. 64 [1990] 1490.
11. G.M. Gryaznov, V.A. Evtikhin, L.P. Zavyalskii, A.Ya. Kosukhin, I.E. Lyublinskii, Materialovedenie Zhidkometallicheskhikh Sistem Termoyadernykh Reaktorov, Energoatomizdat, Moskva, 1989.
12. I. Ali Khan, in: Material Behaviour and Physical Chemistry in Liquid Metal Systems, Ed. H.U. Borgstedt, Plenum Press, N.Y. 1982, 237.
13. K. Natesan, J. Nucl. Mater. 115 [1983] 251.
14. O. Kubaschewski, C.B. Alcock, Metallurgical Thermochemistry, 5th ed., Pergamon Press, Oxford, 1979.
15. M. Saboungi, J. Marr, M. Blander, J. Chem. Phys. 68 [1978] 1375.
16. M.G. Barker, J.A. Lees, T. Sample, P. Hubberstey, J. Nucl. Mater. 179-181 [1991] 599.

SOLUBILITY OF ZIRCONIUM IN LIQUID SODIUM

K. Künstler, H. Heyne

KAI e.V./WIP
in the Forschungszentrum Rossendorf e.V.,
P.O. Box 51019, D-01314 Dresden, Germany

1. INTRODUCTION

The radioactivity of the primary system is one of the special problems of the operation of sodium-cooled fast breeder reactors. It is caused by activation of the coolant and by radioactive corrosion and fission products. The interaction of many parameters in technical plants, the difficulties of examination important values without injury at the same time the operation of the plant make it necessary, to examine individual fission products in laboratory scale. $^{95}Zr/^{95}Nb$ belong to these fission products, which must be expected to enter into the sodium system in case of more serious fuel element damages.

The solubility of zirconium was determined in dependence on the
-equilibration time
-temperature
-chemical form of zirconium (metallic, ZrO_2, Na_2ZrO_3 and $BaZrO_3$)
- and the impurities in the sodium.

2. DETERMINATION OF THE SOLUBILITY

The applied procedures for the determination of the solubility cause a number of problems in the liquid metals. The solubility of a metal or a compound in a liquid metal can be influenced by subsequent chemical reactions or by exchange of impurities (oxygen, carbon, nitrogen, hydrogen) between the liquid metal and the solid metal or the compound.

Also reactions are possible to ternary compounds so that it is hardly to say in which form the solute exists. There is an additional source of error. Suspended particles pretend a larger solubility in the liquid phase. This can be prevented if the solution equilibrium is maintained at the wanted temperature for longer time. In this process a grain growth occurs and the larger particles settle.

Following conditions must be fulfilled for the exact solubility measurements:
-a solid phase as pure phase of known composition
-the solid phase is in contact with a larger amount of pure solvent

-the solid phase and the dissolved component are no subjects of chemical reactions leading to the formation of new compounds.

-after the evaporation of the equilibrium solution the solid phase is formed in the same composition as before, that is, the process of dissolution is reversible.

It is advantageous to perform the determination of the solubility by different methods and to incorporate the expected stable compounds into the consideration of the solubility in order to eliminate the error source at the determination of the solubility. The obtained solubility values are always to regard under the condition of the test procedure (impurities in the sodium, above all of the oxygen content in the sodium, crucible material, purity of the dissolving compound).

3. STATE OF KNOWLEDGE

Investigations on the solubility of zirconium in sodium are sporadically described in the literature [1,2,3]. Systematic investigations of various influences on the solubility do not exist. The measured values for the solubility have been summarised in the Table 1. The solubility values very heavily scatter and cannot be compared with each other, since the experimental assumptions for the determination of zirconium in sodium are different.

Table 1. Test Systems for the Solubility Studies

Temp. °C	Oxygen content µg/g	Solid phase which was equilibrated	Container material	Solubility µg/g	References
800		Zr/Nb-1	Mo	1.7;0.8:<1	
1000		"	"	<1	
1010		"	"	0.6	
1180		"	"	1.0	
1185	no data	"	"	2.4	[1]
1200		"	"	<1	
1380		"	"	3.6	
1500		"	"	4	
600		Zr	Zr	34	
750		"	"	77	[2]
900		"	"	370	
722	no data	"	Stainless steel	0.09	[3]

4. EXPERIMENTAL

4.1 Materials

Stainless steel of the X8CrNiTi18.10 type was used for the capsule experiments. The analysis of the tubes and plugs is shown in the Table 2. Before use the capsules were mechanically cleaned, then treated with methanol in a Soxhleth apparatus for two hours, etched in a moving solution at 40 °C for 10 minutes (etching solution 25 vol% of HNO_3, 70 %, 4 vol% of HF, 60 %, in water), rinsed with tap water for 30 minutes, then with distilled water, dried at 200 °C for 24 hours and finally stored in an exsiccator or in an inert gas box [4].

Table 2. Chemical composition of stainless steel Type X8CrNiTi 18.10

Element	Tube material (wt %)	plate material (wt %)
C	0.09	0.08
S	0.015	0.013
P	0.015	0.017
Si	0.50	0.40
Mn	1.67	1.27
Cr	17.10	17.10
Ni	10.30	10.20
Ti	0.64	0.56
	balance Fe	balance Fe

The contents of impurities of zirconium and zirconium dioxide were determined by emission spectrometry and are presented in Table 3.

Table 3. Content of impurities in Zirconium and Zirconium Dioxide (The determination of contents followed by atom emission spectrometry, semiquantitative)

Element	Zirconium µg/g	Zirconium dioxide µg/g
Al	< 70	< 70
Be	< 10	~ 10
Ca	< 50	< 50
Cr	~ 30	< 30
Cu	~ 40	~ 40
Fe	< 200	~200
Hf	~ 1	~ 1
Li	-	< 5
Mg	< 1	< 5
Mn	< 5	-
Mo	< 7	< 100
Ni	-	< 100
Pb	-	< 90
Si	< 50	~ 50
Ti	< 50	

Zirconium sheets were also cleaned before use in the same manner as the steel capsule, but the etching solution was of different composition (40 vol% of HNO_3, 70 %, 5 vol% of HF, 60 %, dissolved in water). The etching period was one minute. The oxygen content of the zirconium sheets was 20 to 40 µg/g. All chemicals for the preparation of $BaZrO_3$ and

Na_2ZrO_3, as there are $BaCl_2$, NaOH and the sodium metal are from Merck and were of p.a. (analytical) quality. The main impurities in the sodium were (in µg/g) Fe < 10, Ca < 500, K < 100 and heavy metals (as Pb) < 5. The filling of sodium was carried out in an inert gas box. The contents of impurities in argon were (in cm^3/m^3) oxygen < 2, water vapour < 6 and carbon dioxide < 1. The sodium was melted in a vessel over a glass frit (G4, pore size >3 to 10 µm) and was held liquid at 120 °C for five hours. It was sucked by vacuum into the receiver. The sodium filled under these conditions had contents (in µg/g) of oxygen 17, carbon 1 and hydrogen < 0.3.

4.2 Procedure

Steel capsules (130 mm long, 25 mm or 20 mm OD, and 22 mm or 16 mm ID) were used for the solubility investigations in different ways (Fig. 1).

The radioactively labelled zirconium sheet or zirconium compound or a zirconium steel crucible system was placed on the bottom of the capsule (Fig. 1a to 1d). The capsules or crucibles were filled with sodium, sealed by a steel plug, welded together and maintained at a given temperature (500 °C to 900°C) up to 25 days. Afterwards, the capsules were quenched in liquid nitrogen, disassembled and the sodium samples taken according to the scheme given in Fig. 2.

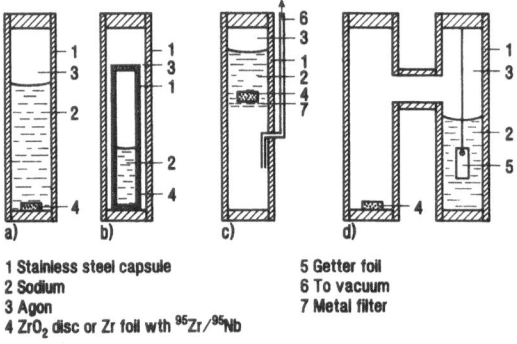

1 Stainless steel capsule
2 Sodium
3 Agon
4 ZrO_2 disc or Zr foil wth $^{95}Zr/^{95}Nb$

5 Getter foil
6 To vacuum
7 Metal filter

Fig. 1. Test system for the studies

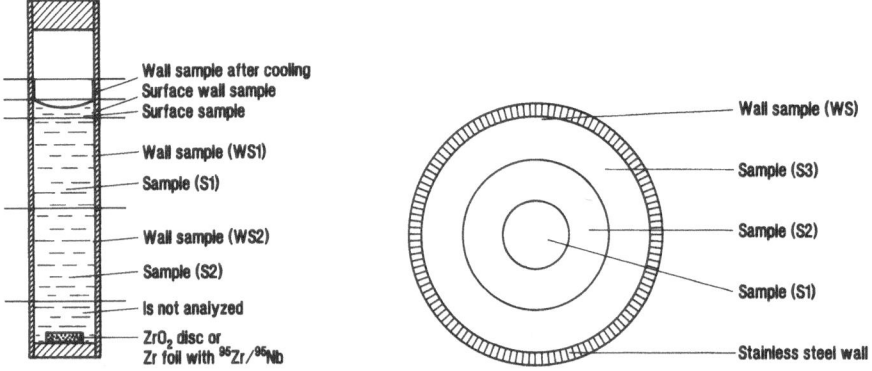

Fig. 2. Sectional drawing for the sampling procedure   Fig. 3. Horizontal section through the sample

The disassembling according to Fig. 3 was also possible. Quenching in nitrogen was necessary, in order to maintain the solubility equilibrium which was attained at higher temperatures.

The results from the quenching experiments were completed by other procedures for the determination of the solubility. Thus, the sodium samples could be filtered (Fig. 1c) or decanted (Fig. 1d). Several samples were also centrifuged at the test temperature and afterwards quenched. In order to obtain more results on the precipitates at the steel wall tilting experiments with steel sheets were carried out. This technique is illustrated in Fig. 4. In this way sodium and the steel wall could be separated and a normal cooling of the solution was possible.

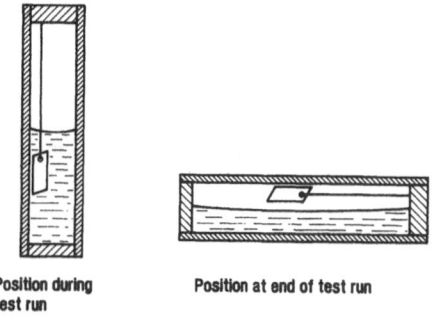

Position during
test run

Position at end of test run

Fig. 4. Test run for the tilting experiments

4.3. Preparation

The zirconium sheets and crucibles used for the investigations were activated in a thermal reactor. For the preparation of the zirconium dioxide $ZrOCl_2$ $8H_2O$ was dissolved in 0.2 N HCl, added to a solution of $^{95}Zr/^{95}Nb$ and zirconium was precipitated by adding a 16% mandelic acid solution. The zirconium tetra mandelate was then decomposed to ZrO_2 at 1000°C, followed by comminution, pelletizing and annealing at 1000 °C for 24 hours. Na_2ZrO_3 and $BaZrO_3$ were obtained by solid state reactions between NaOH resp. $BaCO_3$ and radioactive labelled ZrO_2 [5,6]. For the solubilities these compounds were used in form of pellets, too. The existence of these compounds could be proved by X-ray analysis.

4.4 Analytical Procedures

At the parts marked in Fig. 2 the steel capsules were disassembled and analysed according to the sectional drawing. Central (S 1, S 2) and wall samples (WS 1, WS 2), respectively, were obtained by means of prickling out. Parts of the central samples were weighed and then measured on a γ-spectrometer with a Ge(Li)-detector. The sodium remaining at the wall during the pricking out was dissolved in a methanol-water-mixture (ratio 3:1).

The sodium content was determined by titration with hydrochloric acid, followed by adding a zirconium carrier solution, concentrating to dryness, and measuring the residue. By means of running a zirconium standard parallel to the sample, the zirconium could be determined. The sensitivity of the method in the range of the determined solubility values was 0.05 µg/g.

5. RESULTS AND DISCUSSION

5.1 Time dependence

The solubility data of zirconium (inserted as ZrO_2) show a dependence on the experimental time. It could be shown in preliminary tests, that a constant solubility value of zirconium at first adjusts at a temperature of 700 °C in the steel capsule after 10 days (Fig. 5).

The experimental times were extended until 25 days. No changes in the solubility values were observed. The reasons for the time dependence of the establishment of the solubility values were seen in the slow establishment of the equilibrium of the zirconium or its compounds to the gaseous phase, steel surface and boundary surface steel / gaseous phase/sodium. In an earlier work it could also be proved that the distribution of oxygen in the system sodium/steel is time and temperature dependent [7,8]. The indicated solubility data refer to a time of experiment > 10 days in this work.

5.2 Distribution

The distribution of zirconium in the steel capsule at the end of the experiment is

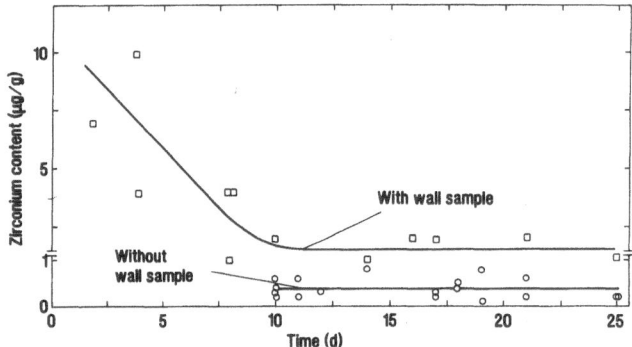

Fig. 5. Dependence of the zirconium content in the sodium on the experimental time

demonstrated in Table 4.

In all experiments similar distribution patterns were obtained:

- Enrichment of zirconium or its compounds near the interfaces sodium / steel / argon.

- In the sodium / steel interface the zirconium content was also higher than in the sodium bulk.

- By filtration or decantation of the sodium at test temperature could be proved, that this result was not caused by the quenching of the samples in liquid nitrogen. The separation of sodium and the wall area under the conditions of the experiment was also carried out by tilting of the steel capsule and letting it cool down (Fig. 4). After that the sodium adhering to the steel sheets was analysed. The zirconium contents there were always higher than in the bulk. The pricking out of sodium samples according to the sectional drawing given in Fig. 3 did not result in different zirconium contents of the samples (S1, S2, S3), too, but again the zirconium content was higher in the sodium from the wall of the steel.

Table 4. Zirconium distribution in the stainless steel capsule according to sectional drawing for the sampling (compare Fig. 2)

Analysed region	Zirconium content * $(ZrO_2$, 17 µg/g oxygen in Na, 800°C) µg/g
Wall sample after cooling	(164 ±59)
Surface sample	(0.9 ±0.6)
Surface wall sample	(21.9±6.1)
Sample (S 1)	(0.5 ±0.4)
Wall sample (WS 1)	(7.8 ±0.9)
Sample (S 2)	(0.5 ±0.4)
Wall sample (WS 2)	(8.6 ±0.4)

* Averages from 6 test values

5.3 Solubility

5.3.1 Zirconium

The results of the solubility measurements of zirconium in a zirconium crucible are demonstrated in Fig. 6. The solubility equation is

$$\log c \ (\text{µg/g Zr}) = 0.669 - 209 / T.$$

The partial enthalpy of solution was calculated to be 7.37 kJ/mol. This equation is only valid in the field of the α-Zr. At temperatures of 900 °C and 1000 °C the zirconium content in sodium was measured to be 47 µg/g and 117 µg/g, respectively. Further investigations were not carried out in this temperature region. After addition of 1000 µg/g oxygen to the sodium no changes of the solubility values could be observed. From the phase diagram zirconium-oxygen it can be seen that α-Zr is able to absorb more than 20 at% of oxygen. Filtration experiment also delivered corresponding solubility values of zirconium in sodium. The investigations showed that the real solubility of zirconium in sodium was measured. By addition of NaCN, Na_2CO_3 or NaOH higher values of zirconium solubility in sodium were obtained (NaCN additive until 7µg/g, Na_2CO_3 additive until 6 µg/g and NaOH additive until 11 µg/g as zirconium). A well defined dependence of the zirconium solubility on temperature could not be detected. Complicated interactions between impurities, zirconium and sodium must be expected, influencing the content of zirconium in sodium. In experiments with $^{95}Zr/^{95}Nb$ labelled zirconium sheet in steel crucibles (5 mm x 5 mm, 1 mm thick) solubility values were obtained like those, when ZrO_2 respectively Na_2ZrO_3 in sodium were used. The surfaces of the inserted zirconium sheets were investigated by X-ray spectrometry [9]. The analysis showed, that Na_2ZrO_3 was formed at the surface of the zirconium sheet. Furthermore a Zr-O compound could be detected which was not identical with the monoclinic or cubic ZrO_2.

5.3.2 ZrO_2, Na_2ZrO_3

The solubility values of ZrO_2 and Na_2ZrO_3 are reported in Table 5. The solubilities of the compounds weakly changed with the temperature. If oxygen was added to the sodium, the solubilities did not change. By X-ray analysis could be proved, that Na_2ZrO_3 is formed if solid ZrO_2 is present in sodium. A further reaction of the so formed Na_2ZrO_3 does not occur [9]. It cannot be decided, whether zirconium in liquid sodium exists in form of ZrO_2, Na_2ZrO_3 or in both.

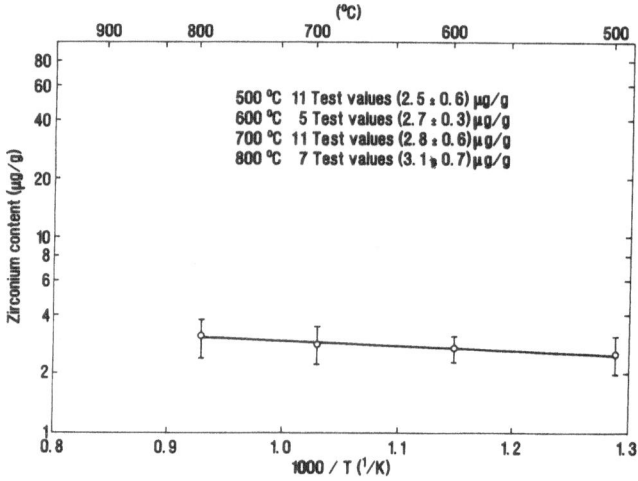

Fig. 6. Mean Values of the Solubility of Zr in Na at 500 to 800 °C

Table 5. Zirconium content in liquid sodium (solid phases were ZrO_2 and Na_2ZrO_3)

	ZrO_2			Na_2ZrO_3		
Temp. [°C]	Test values	Zr content [µg/g]	ZrO_2 calculated [µg/g]	Test values	Zr content [µg/g]	Na_2ZrO_3 calculated [µg/g]
700	20	(0.4 ± 0.2)	(0.5 ± 0.3)	10	(0.3 ± 0.2)	(0.6 ± 0.4)
800	17	(0.5 ± 0.3)	(0.7 ± 0.4)	12	(0.4 ± 0.2)	(0.8 ± 0.4)
900	10	(0.5 ± 0.3)	(0.7 ± 0.4)	12	(0.4 ± 0.2)	(0.8 ± 0.4)

5.3.3 BaZrO$_3$

The contents of zirconium and barium in dependence on temperature are demonstrated in Table 6.

These contents increase with rising temperature. The Ba to Zr ratio is always constant in sodium as it is determined by the formula. By X-ray analysis no other compound but $BaZrO_3$ could be detected. For that reason it is supposed that $BaZrO_3$ exists in this form in the liquid sodium. The solubility reads

$$\log c \ (\mu g/g \ BaZrO_3) = 2.700 - 2398 / T$$

The partial enthalpy of solution is calculated to be 44.22 kJ/mole.

6. CONCLUDING REMARKS

The results of this investigation indicated how complex the determination of the solubilities of metals or compounds in liquid metals may be. A series of suppositions must be fulfilled for the solubility measurements:

Table 6. Zirconium and barium contents in liquid sodium (solid phase was $BaZrO_3$)

Temp.	Test values	Zr content [µg/g]	Ba content [µg/g]	$BaZrO_3$ calculated [µg/g]
600	6	(0.3 ± 0.2)	(0.4 ± 0.2)	(0.9 ± 0.5)
700	6	(0.6 ± 0.3)	(0.9 ± 0.2)	(1.8 ± 0.6
800	6	(0.9 ± 0.3)	(1.4 ± 0.3)	(2.8 ± 0.7)

-establishment of the equilibrium

-knowledge about the absorption processes in the investigated system

-which atom or molecule is dissolved

-and no suspended particles of the dissolved materials in the solution.

For the thermodynamic interpretation of solubility data, it must be exactly examined whether the metal or a compound exists as the material in equilibrium with the solution.

References

1. R.L. McKisson, R.L. Eichelberger, R.C. Dahleen, J.M. Scarborough, G.R. Argue, Solubilities studies of ultra pure transition elements in ultra pure alkali metals, Report NASA-CR-610 (1966).
2. R.L. Eichelberger, R.L. McKisson, Solubility studies of Cr, Co, Mn, Mo, Ni, Nb, Ti, V and Zr in liquid sodium, Report AI-AEC-12955 (1970).
3. T.O. Claar, Reactor Technology 13 (1970) 124-146.
4. K. Künstler, K. Betzl, H.-J. Grosser, W. Furkert, O. Novotny, Aufbau und Erprobung eines Boxensystems für Arbeiten mit luftempfindlichen und radioaktiven Stoffen, Report ZfK-558 (1985).
5. V.S. Lyer, V. Venugopal, Smruti Mohapatra, Ziley Singh, K.N. Roy, R. Prasad, D.D. Sood,J. Chem. Thermodynamics 20 (1988) 781-784.
6. Smruti Dash, Ziley Singh, R. Prasad, D.D. Sood, J. Chem. Thermodynamics 22 (1990) ,557-562.
7. K. Künstler, B. Schossig, Zur Verteilung von Sauerstoff im System Natrium-Stahl, in Report ZfK-451 (1981) 133-135.
8. K. Künstler, B. Schossig, Zur Ablagerung von Sauerstoff aus flüssigem Natrium an den Stahl X8CrNiTi18.10, in Report ZfK-489 (1982) 122-124.
9. K.Künstler, H.-P. Schützler, Zum Reaktionsverhalten von Zirkonium und Niob im Natrium, in Report ZfK-312 (1976) 37-39.

SOLUBILITY MEASUREMENT OF SODIUM IODIDE IN SODIUM - STAINLESS STEEL SYSTEMS

Norihiko Sagawa *, Shinya Miyahara ** Tohru Sone **

* Department of Mechanical Engineering,
Faculty of Engineering, Ibaraki University
Hitachi, 316, Japan
** Oarai Engineering Center, Power Reactor &
Nuclear Fuel Development Corp. Ibaraki, 311-13, Japan

1. INTRODUCTION

It is highly important in the safety assessment of fast reactors that fission products released from ruptured fuel elements are to be retained in the sodium coolant. Radiological consequences of iodine isotopes are so significant that experimental and theoretical efforts have been put forward for obtaining a better understanding of iodine behavior in the sodium system [1-8]. Any iodide or free iodine released into the sodium is known to be converted to sodium iodide, and its behaviour in the sodium system is essentially controlled by the iodide solubility in sodium. The solubility was measured by Bredig [9,10] at temperatures between 550 - 900 °C and by Allan [11] between 250 - 400 °C, as shown in Fig. 1.

The extrapolations of both data are not likely to meet each other in the temperature range of 400 - 550 °C. The solubility was recently supplemented by Miyahara [12] over the temperature range of engineering importance, 350 - 800 °C. The solubility expressions given by Miyahara agree with the data of Bredig in the temperature region higher than 600 °C, whereas they indicate values 10 times as large as the data by Allan and Sagawa [13] in the lower temperature region

In order to find a reason why Miyahara's expression yields a far larger value than Allan's expression does, two possibilities were examined using capsules of different types; one possibility that sodium iodide may gather near the free surface of liquid sodium, as reported by Jordan [14], and the other possibility that the iodide may form solid particles in the sodium which undergoes temperature fluctuations during a long period of heating. Neither local enrichment nor particle formation of sodium iodide is observed in the experiments [15].

The aims of this paper are to evaluate the solubility data obtained from different methods and to discuss a reason why the solubility data are divided into the higher values and the lower ones.

Liquid Metal Systems, Edited by H.U. Borgstedt
and G. Frees, Plenum Press, New York, 1995

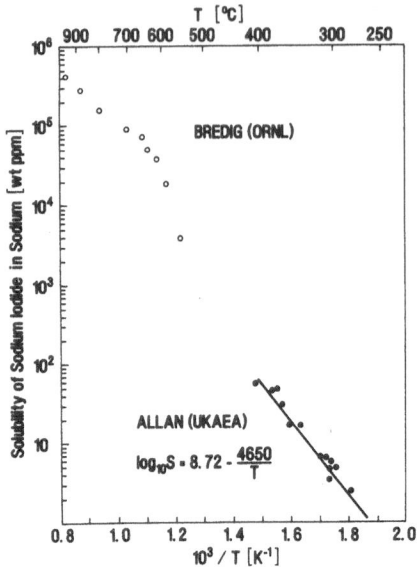

Fig. 1. Reported solubility of sodium iodide in sodium

2. EXPERIMENTAL

2.1. Reversed J Type Capsule

A capsule made of JIS type 316 stainless steel is composed of a base cell with a mouth for containing 7.5 - 8 g sodium and 2 - 3 g fragments of the iodide crystal and of a side arm cell with a tube assembly for sampling the sodium solution, as illustrated in Fig. 2. Equilibration of the iodide in solution with the iodide crystal was made in the base cell under uniform temperature distribution in an electric resistance furnace. Details of the measurements were described in a previous publication [12].

After a given period of heating, a portion of the equilibrated solution was decanted into the side arm cell by tilting the capsule about 110° with the furnace capable of rotating round its horizontal axis. The side arm cell was sectioned at position 40 mm apart from its bottom, and the sodium in the cell was transferred to a beaker by heating the sectioned cell in an argon atmosphere glove box. The sodium was dissolved in distilled water. Iodine in the solution was determined by oxidation-reduction titration with a standard solution of potassium permanganate.

2.2. Short Capsule

A short capsule was made of JIS Type 304 stainless steel tube 10 mm in inner diameter, 70 mm in length, welded at the bottom with an end cap and the top with a nozzle. The capsule was filled with 1 g fragments of sodium iodide crystal and 2 - 3.5 g sodium of reactor grade. The capsule was set on a space held with a pair of stainless steel holder to be heated in the central portion in a vertical electric resistance furnace, as illustrated in Fig. 3. Upper and lower ceramic fibre insulators were used to make temperature distribution uniform along the capsule. The short capsule experiments were reported in detail in a preceding paper [15].

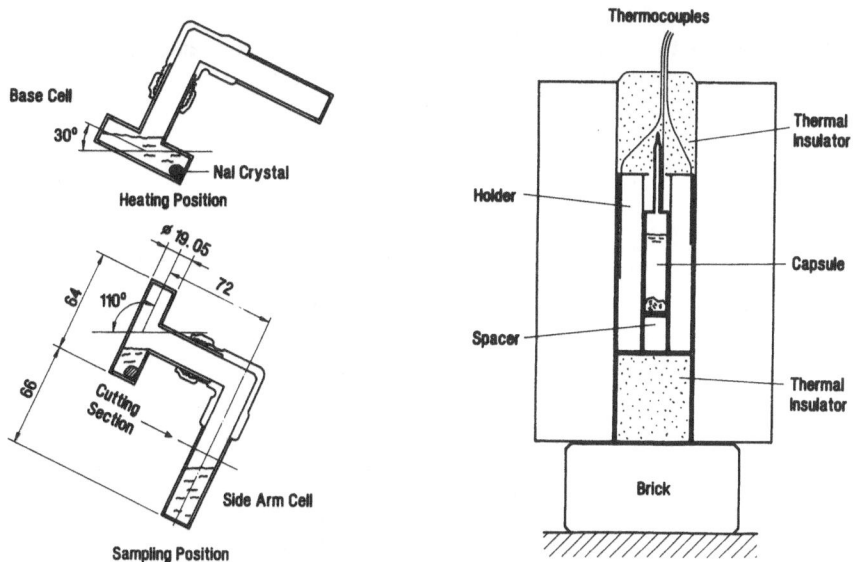

Fig. 2. Capsule used for equilibration between liquid sodium and solid sodium iodide

Fig. 3. Arrangement of the heating device for the short capsule

The capsule was heated in the furnace for 380 - 1,000 h depending on the heating temperature for establishing equilibration between the solution and the crystal. After the given period of heating, the capsule was pulled up from the furnace and cooled quickly in quenching oil for fixing the iodide in solidified sodium.

X-ray photographs were used to find the cut positions so that the top sample involved entirely the portion of free surface in the sectioned samples 10 mm long. The sodium in the samples was shared into a core sodium extruded from the capsule tube and a periphery sodium remaining on the tube. These portions of sodium were dissolved in distilled water and iodine in the solution were determined by an ion chromatograph with an electrochemical detector and a conductivity meter.

2.3. Long Capsule

A long capsule was composed of JIS Type 304 stainless steel tube, 10 mm in inner diameter, 180 mm in effective length, sealed at the bottom and a vessel provided with two nozzles. The capsule was loaded with 20 g sodium of reactor grade and 0.1 - 0.3 g powders of sodium iodide, and was installed vertically with the upper part inserted into an electric resistance furnace and with the tube bottom placed on brass plates, as illustrated in Fig. 4. The long capsule experiments were reported elsewhere [13].

Surface treatments were adopted to stabilize the iodide deposition in the capsule prior to the long term of heating for the measurement, that is, surface cleaning was done for removing the oxide film on stainless steel with the loaded sodium, and the iodide dissolving for coating the sites of adsorption with iodide dissolved in the sodium. The heating was continued for longer than 700 h to let the iodide have settled on the bottom of the capsule by falling and diffusion of the iodide through the sodium. The capsule was quickly cooled in quenching oil to fix the iodide in solidified sodium and sectioned into samples 10 mm long for chemical analyses similar to ones applied to the short capsule.

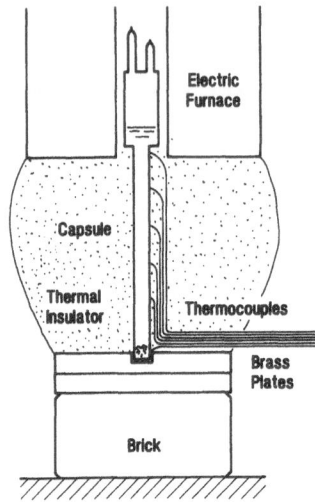

Fig. 4. Arrangement of the heating device for the long capsule

3. RESULTS

3.1. Solubility Data in the Higher Temperature Region

Results measured from the reversed J type capsule are plotted in Fig. 5 with hollow square marks. These marks can be seen to lie on straight lines, which cross at a temperature close to the melting point of the iodide, in the forms

$$\log\ S = 8.46 - 3440\ /\ T \quad \text{above } 660\ °C$$
$$\log\ S = 9.65 - 4550\ /\ T \quad \text{below } 660\ °C,$$

where S is the solubility in the unit of ppm by weight and T the temperature in the unit of Kelvin.

In the temperature region above the melting point, the straight line shows a gentler slope than that given in the region below the point. It is observed in Fig. 5 that the data determined by Bredig [9,10] fit well with the above expressions in the higher temperature region, whereas they fall steeply with decreasing temperature.

3.2. Solubility Data in the Lower Temperature Region

Representative concentrations of sodium iodide obtained from the short capsule are shown in Fig. 6 as a function of the sample position, the numbers of which decreases from the sample involving the free surface of sodium toward the bottom. The results indicate that the concentration in the core sodium decreases gradually with raising to the elevated sample positions towards the free surface. The horizontal, full line denotes the solubility obtained from Miyahara (PNC)'s expression [12] at the temperature of this measurement. The full line is observed to be between the concentrations in the core and periphery, while they are about ten times as large as the broken line denoting Allan's expression [11] at the temperature.

The average concentration in the samples cut from the short capsule was evaluated by dividing a total amount of sodium iodide in the cores and peripheries by that of sodium.

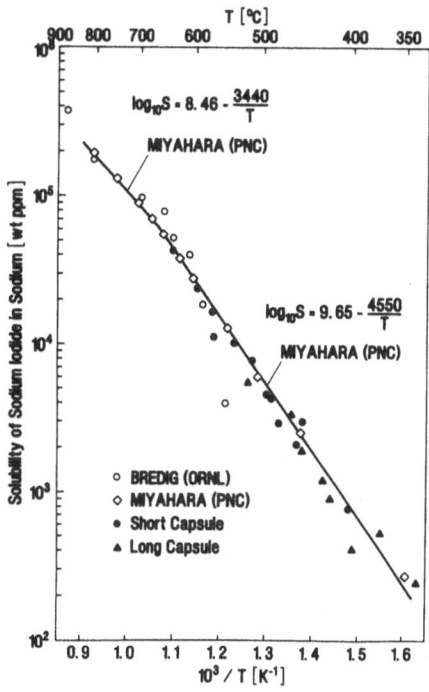

Fig. 5. Solubility data determined by reversed J type, short and long capsules

Representative concentrations of sodium iodide obtained from the long capsule are plotted in Fig. 7 as a function of the reciprocal of sample temperature measured during the run. The concentration in the core indicates a nearly constant level in the lower temperature region, while the concentration trends to fall in the higher temperature region. This fall shows an evidence happened during the quick quenching; the sodium precipitated the iodide, which formed small particles to fall down to the bottom of the capsule and was fixed at the middle stage in the solidified sodium.

Since the iodide saturated in the liquid sodium segregated in the solidified sodium during the quenching process, it was adopted for most proper evaluation of average concentration that a total amount of the iodide in all the cores and peripheries is divided by that of sodium in all the sections except the one indicating the steep jump in the concentration. The saturation temperature was read from the lowest temperature of the section except the one indicating the jump. The average concentration and the saturation temperature are shown respectively with a horizontal arrow and a vertical one drawn in Fig. 7. These concentrations are plotted in Fig. 5 against the reciprocal absolute temperature measured during the run, with thick circular marks for the short capsules and thick triangle ones for the long capsules. Both results for the short and long capsules are in good agreement with Miyahara's expression [12] of the iodide solubility over the range of 350 - 650 °C.

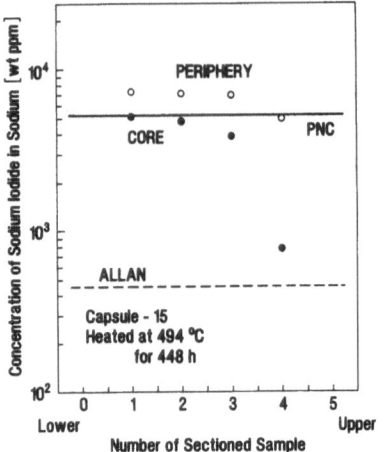

Fig. 6. Axial distribution of sodium iodide in sodium along the short capsule [15]

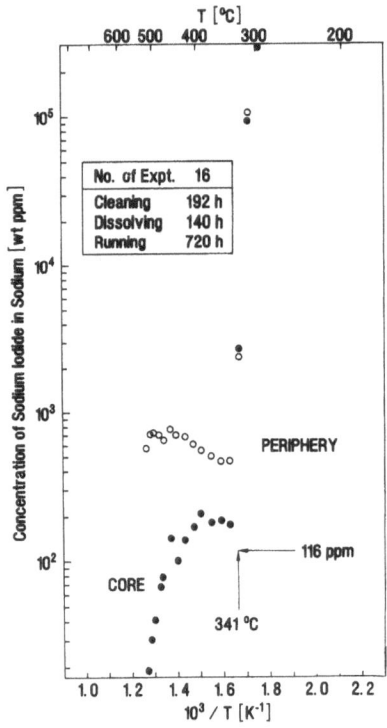

Fig. 7. Axial distribution of sodium iodide in sodium along the long capsule [15]

4. DISCUSSION

4.1. Local Enrichment at Free Surface

The distributions of iodide in sodium were analysed by a layer-by-layer method after experiments in the direction of the height level in a stainless steel vessel containing 100 g sodium, and a considerable enrichment was observed at the pool surface [14]. The experiments using the short capsules indicate no evidence of the enrichment at the free surface, as shown in Fig. 6. In view of the process that any contamination by air was minimised in loading the iodide crystal and the sodium into the short capsules, the enrichment can be attributed to the iodide gathering accompanied by other impurities, which have the nature to float on the free surface.

4.2 Particle Formation in Sodium

Temperature gradients of about 20 °C/cm were established along the long capsules descending from the top toward the bottom of the capsule. These temperature gradients suppress the mixing of sodium by natural convection in the vertical direction of the capsule and, at the same time, will accelerate falling of solid particles by thermal diffusion in the sodium along the capsule tube. Thus, the temperature gradients ensure the sodium in the tube being free of containing solid iodide particles, which may happen to form in the sodium during the heating.

Figure 5 shows that the solubility data measured using the long capsules are in good agreement with those by Miyahara by means of the reversed J type capsules. This experimental fact reveals that no significant amount of solid particles had existed in the sodium transferred to the side arm cell of the J type capsules. It is noted that the segregation seen in the results in Fig. 7 did not take place during the heating process, but did during the quenching process for fixing the iodide in solidified sodium.

4.3. Iodide Solubility in Sodium-Stainless Steel Systems

Results plotted in Fig. 5 show that the solubility data measured using the short and long capsules are in good agreement with those which were determined in the J type capsules. This agreement should strengthen the validity of the solubility expression over the range of 350 - 650 °C. However, the straight line drawn among the data is seen to be parallel with the extension of Allan's expression [11] above 400 °C.

For interpretation of this difference, another determination of solubility was adopted with the use of the long capsule; an average concentration was determined from a total amount of the iodide divided by that of the sodium in all core portions of sodium, and the saturation temperature was read from the lowest temperature of the sectioned sample except one indicating precipitation of the iodide. Figure 8 represents the solubility obtained in this manner with thick triangle marks. The present data appear to go along Allan's expression in the temperature region below 350 °C, while they trend to leave the expression above 350 °C and to approach the Miyahara's expression. A straight line can be drawn among these data between both expressions in the form

$$\log S = 8.43 - 5360 / T.$$

The temperature coefficients of the solubility of metals in sodium are reported to be 4116 for iron, 9010 for chromium and 1230 for nickel [16-18]. The dissolution of iron and chromium may contribute to the coefficient 5360 in the above expression.

All the capsules used in the measurement were made of stainless steel, so that its constituents will be leach into the sodium during the period of equilibration. For examination of this possibility, another measurement was made with varying the surface treatment prior to the measurement. One long capsule underwent the cleaning process for removing the oxides on stainless steel surface with loaded sodium, while the long other was free of this treatment. After fixing the iodide in solidified sodium by quick quenching, the capsule were sectioned into the samples 10 mm long as above-mentioned. A Teflon extruding rod and holders, and a ceramic knife were used to minimise contamination with metal elements during the sampling.

Two samples (10 mm long x 3) were taken carefully from upper and lower parts of each capsule for vacuum distillation in the argon atmosphere glove box and for the determination of the metallic elements by the use of an inductively coupled plasma mass spectrometer. One sample (10 mm long x 5) taken from the central part of each capsule was submitted to sodium dissolution with distilled water and to iodine determination by an ion chromatograph and the mass spectrometer.

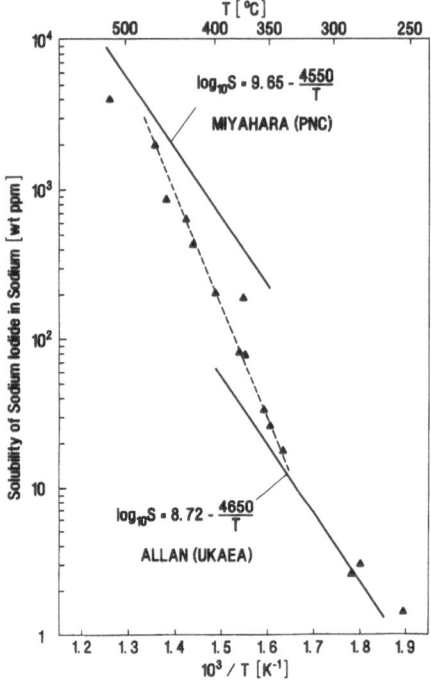

Fig. 8. Solubility data determined from average concentration of sodium iodide in sodium taken from the core of the long capsule

Table 1: Metal and iodine concentrations in sodium in the long capsule

Capsule number	29			31		
Surface cleaning	without			at 430 °C for 100 h		
Saturation temperature (°C)	400			400		
Sample position	Upper	central	lower	upper	central	lower
Sodium (g)	1.61	2.7	1.64	1.59	2.5	1.54
Iron (ppm)	1.3		2.3	-		-
Chromium (ppm)	1.5		1.0	0.3		0.2
Manganese (ppm)	0.7		0.6	0.2		0.1
	0.1		0.1	-		-
Iodine (ppm)	67			200		

The results of chemical analyses are shown in Table 1. It is evident from Table 1 that the capsule without the pre treatment indicates higher metal concentrations and a lower iodine value in comparison with those of the capsule pre-treated with loaded sodium. This striking contrast can be interpreted as that iodine shows the tendency to combine with metallic elements dissolved in the sodium and precipitate from the solution forming insoluble materials.

5. CONCLUSION

1) The solubility data measured over the temperature range of 350 - 800 °C are expressed in the form of two equations

$$\log S = 8.46 - 3440 / T \qquad \text{above 660 °C}$$

$$\log S = 9.65 - 4550 / T \qquad \text{below 660 °C,}$$

where S is the solubility in the unit of ppm by weight and T the temperature in the unit of Kelvin.

2) Good agreement with Miyahara's expression is seen in plots of an average of iodide concentrations in all sodium sampled from the short capsule against heating temperature and from the long capsule against saturation temperature.

3) Local enrichment of sodium iodide is not found at the free surface of liquid sodium in the measurement with using a short capsule heated under an uniform temperature distribution.

4) Solid particles of the iodide is not observed to form in the liquid sodium in the long capsule heated under a large temperature gradient for a period required for particle settling.

5) Plots of an average of the concentrations in the sodium, sectioned and cored from the long capsules, as a function of the saturation temperature trend to leave Allan's expression and to approach Miyahara's expression.

6) Solubility data determined by using stainless steel capsules are considerably affected by metal elements leached from the steel into the sodium.

ACKNOWLEDGEMENT

The authors wish to express their appreciation to Mr. H. Hara, Safety Engineering Division and to Dr. N. Mizoo and Mr. T. Funada, Experimental Reactor Division, Oarai Engineering Center, PNC, for their fruitful discussions given to this work and to the members of Reactor Technology Section of the latter division for their technical supports in chemical analysis of sodium samples.

REFERENCES

1. W. P. Kunkel, Report NAA-SR-11766 (1966).
2. W. S. Clough, J. Nucl. Energy, 21 (1967) 225-232.
3. A. W. Castleman and I. N. Tang, Nucl. Sci. Eng., 29 (1967) 159-164.
4. W. P. Kunkel, B. D. Pollock, J. Guon, G. B. Zwetzig, M. Silberberg, and S. Berger, Report AI-AEC-12687 (1968).
5. G. W. Kieholtz and G. C. Battle, Report ORNL-NSIC-37 (1969).
6. A. W. Castleman, Nucl. Safety, 11 (1970) 379-390.
7. H. Feuerstein, A. J. Hooper and F. A. Johnson, Atomic Energy Review, 17 (1979) 697-761.

8. A. W. Thorley, Proc. of IAEA Specialist's Mtg. on Fission and Corrosion Products Behavior in Primary Circuits of LMFBR's, Report KfK-4279 (1987) 433-468.

9. M. A. Bredig, J. W. Johnson and W. T. Smith, J. Amer. Chem. Soc., 77 (1955) 307-312.

10. M. A. Bredig and H. R. Bronstein, J. Phys. Chem., 64 (1960) 64-67.

11. C. G. Allan, Proc. of BNES Conf. on Liquid Alkali Metals, Nottingham (1973) 159-164.

12. S. Miyahara, N. Sagawa, T. Sone, T. Arakawa and H. Hara, J. Nucl. Sci. Technol., 29 (1992) 351-357.

13. N. Sagawa, J. Nucl. Sci. Technol., 28 (1991) 305-313.

14. N. Jordan, Proc. of IAEA Specialist's Mtg. on Sodium Fires and Prevention (1978) 208-210.

15. N. Sagawa and S. Miyahara, J. Nucl. Sci. Technol., 29 (1992) 427-435.

16. R. M. Singer, A. H. Fleitman, J. R. Weeks and H. S. Isaacs, Corrosion by Liquid Metals, Plenum Press (1970) 561-576.

17. W. P. Stanaway and R. Tompson, Proc. of 2nd Int. Conf. on Liquid Met. Eng. and Tech., CONF 800401-P2 (1980) 18.54-18.61.

18. C. R. Pellet and R. Tompson, Proc. of 3rd Int. Conf. on Liquid Met. Eng. and Tech., Vol. 3 (1984) 43-48.

SOLUTIONS OF NITROGEN IN LIQUID LITHIUM CONTAINING DISSOLVED ALKALINE EARTH METALS: SOLUBILITIES, PRECIPITATING SPECIES AND PHASE RELATIONSHIPS

P. Hubberstey and P. G. Roberts

Chemistry Department,
University of Nottingham,
Nottingham, NG7 2RD, England

1. INTRODUCTION

Several years ago we reported the results of a detailed study of the reaction of nitrogen with solutions of alkaline earth metals in liquid sodium [1-4]. For the Na-Ba-N system, resistivity [1] and solubility [2] studies of the solutions coupled with X-ray powder diffraction studies of the solid products [3] indicated that an initial solution process is followed by the precipitation of Ba_2N which, in the presence of excess nitrogen, reacts further to form Ba_3N_2. Since the extent of the solution process is determined solely by the amount of barium dissolved in the liquid sodium, precipitation commencing at a Ba:N ratio of 4:1, it has been suggested [5] that the nitrogen, which is effectively insoluble in liquid sodium [6], is solvated by the barium forming a soluble "Ba_4N" species. The reaction sequence, which can be summarised as follows:

$$8\,Ba_{(Na)} + N_{2(g)} \rightarrow 2\,Ba_4N_{(Na)}$$
$$2\,Ba_4N_{(Na)} + N_{2(g)} \rightarrow 4\,Ba_2N_{(s)}$$
$$6\,Ba_2N_{(s)} + N_{2(g)} \rightarrow 4\,Ba_3N_{2(s)}$$

has also been shown [3] to be compatible with a simple Na-Ba-N ternary phase diagram dominated by a very steep Ba_2N precipitation face with an effectively constant BaN molar ratio of 4:1.

Analogous experiments [4] on the Na-Sr-N system showed a similar behaviour pattern, the only difference being a more limited solution process, precipitation of Sr_2N starting at a Sr:N molar ratio of 17:1.

Extension of our interests into liquid lithium systems has shown that its solvent behaviour is quite different from that of liquid sodium [7]. Of particular significance in the present context is the fact that lithium reacts with and dissolves nitrogen [8,9]. This difference in behaviour, which can be traced to the lattice enthalpies of Li_3N and Na_3N [10], presents an opportunity to study the role of a potentially active alkali metal solvent for the reaction of nitrogen with dissolved alkaline earth metals; feasible products are M_2N or M_3N_2 (M = Ca, Sr, Ba), the lithium acting as an inert solvent, or LiMN or Li_3N, the lithium adopting an active role.

Liquid Metal Systems, Edited by H.U. Borgstedt
and G. Frees, Plenum Press, New York, 1995

In this paper, we report the results of a comprehensive study of the Li-M-N (M=Ca, Sr, Ba) systems; resistivity measurements have been used to determine solubilities and precipitating phases, X-ray powder diffraction techniques to confirm the identity of the products, and thermal analysis methods to determine phase relationships.

2. SOLUBILITIES AND PRECIPITATING PHASES

Analysis of equilibrium resistivity-composition data gives an indication of both the solubility of nitrogen in the liquid metal mixture and the identity of the precipitating phase. Such data were obtained for the addition of nitrogen to both Li-Ba (initial x_{Ba} = 2.59 at %; 38 aliquots) and Li-Sr (initial x_{Sr} = 9.78 at %; 45 aliquots) mixtures under isothermal conditions (673 K). The results for the Li-Ba-N system are summarised in Figure 1, where they are compared to the corresponding data obtained for the addition of nitrogen to pure lithium. As the nitrogen is added, the resistivity exhibits an initial increase followed by an effectively constant section. The initial increase is attributed to a solution process, the subsequent constant section to a precipitation process. For the ternary system the increase can be split into two sections. For dilute solutions, the increase is small and non-linear with an increasing gradient (section AB); it is only for more concentrated nitrogen solutions that a linear increase in resistivity is observed. The gradient of this latter section (section BC) is considerably smaller ($5.6 \cdot 10^{-8}$ Ωm (at %)$^{-1}$) than that (section AX) for the binary system ($7.0 \cdot 10^{-8}$ Ωm (at %)$^{-1}$), which is linear throughout, inferring that the solution process involves not only dissolution of the nitrogen but also interaction with dissolved barium. The marked change at point C, when the resistivity becomes constant (section CD), represents a saturation point in the system. The nitrogen concentration at saturation in the Li-Ba-N system point C; x_{Ba} = 2.51 at %; x_N = 3.06 at %) is much higher than in the Li-N system (point X; x_N = 1.45 at %), again suggesting that dissolved barium plays a significant role in the solution process.

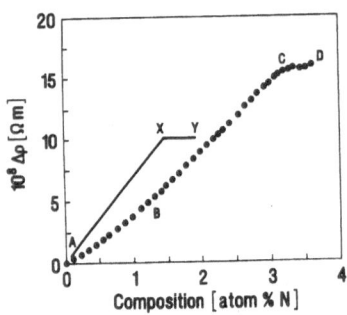

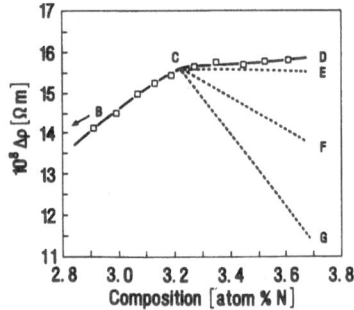

Fig. 1. Equilibrium resistivity-composition data on addition of nitrogen to a Li- Ba alloy (2.59 at % Ba) at 673 K.

Figure 2. Detail of the equilibrium resistivity-composition data observed on addition of nitrogen to a Li-Ba alloy (2.59 at % Ba) at 673 K

The product which precipitates from the binary system has previously been shown to be Li_3N [8,9]; that which precipitates from the ternary system could be one of three phases, Li_3N, LiBaN or Ba_2N. By comparison of the experimental resistivity data obtained after saturation with those predicted for saturation of these three phases the identity of this material can be ascertained. The analysis is shown in Figure 2. If Li_3N is precipitated from solution the resistivity change in the system may be predicted to increase along line CE. This slight increase results from the increase in the barium concentration of the solution owing to the reduction in the amount of lithium present as Li_3N is precipitated. Resistivity decreases are predicted to occur when the precipitating compound contains barium; that expected for LiBaN is shown by line CF, that for Ba_2N by line CG. Comparison of the experimental results with the three possibilities suggests that Li_3N is the precipitating phase.

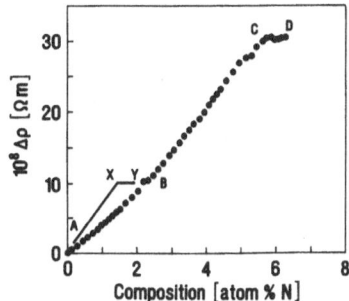

Fig. 3. Equilibrium resistivity-composition data on addition of nitrogen to a Li-Sr alloy (9.78 at % Sr) at 673 K.

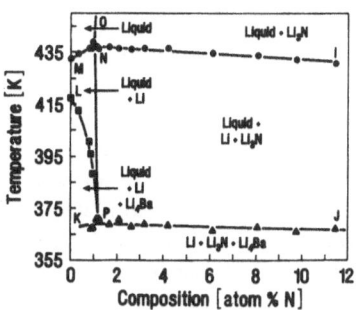

Fig. 4. Phase equilibria derived from thermal analysis data for solutions of nitrogen in a hypoeutectic Li-Ba solution (4.8 at % Ba)

The corresponding data for the Li-Sr-N system are shown in Figure 3. They are very similar to those for the Li-Ba-N system. The gradient of the linear section of the resistivity increase (section BC) is $5.5 \cdot 10^{-8}$ Ωm (at %)$^{-1}$. The similar gradients may be highly significant. If the dissolution process results in a random arrangement, the resistivity change might be predicted to be independent of alkaline earth metal. However, assuming the formation of a soluble species in which the nitrogen is solvated by a given number of alkaline earth, or indeed lithium, atoms the change in resistivity must result from a combination of an increase due to nitrogen dissolution and solvation and either a decrease due to loss of alkaline earth metal or an increase due to loss of lithium. Under these latter circumstances, similar resistivity changes are most unlikely. Hence, the similar resistivity increases suggest that the solution is random. Clearly, many more data are required before this conclusion can be confirmed. The concentration at saturation in the Li-Sr-N system (point C; $x_{Sr} = 9.27$ at%; $x_N = 5.21$ at%) cannot be compared directly with that reported above for the Li-Ba-N system. However, if as a first approximation, a linear relationship is assumed between x_N and x_{Sr} at saturation, the value of x_N is given as a function of x_{Sr} by the expression:

$$x_N = 1.45 + (5.21 - 1.45)x_{Sr}/9.27,$$

similarly for Li-Ba-N solutions:

$$x_N = 1.45 + (3.06 - 1.45)x_{Ba}/2.51,$$

Thus, at $x_M = 5$ at %, $x_N = 3.48$ at % in the Li-Sr-N system and $x_N = 4.66$ at % in the Li-Ba-N system. Preliminary results for the Li-Ca-N system give a corresponding x_N value of 2.49 at %. These results, although preliminary, clearly indicate that the enhancement of the solubility of nitrogen in liquid lithium increases from calcium through strontium to barium.

333

3. PHASE RELATIONSHIPS

Phase relationships can be extracted from thermal analysis data for the ternary systems. Such data have been obtained for the Li-Ba-N system only. The reaction between Li-Ba solutions initially containing 4.75, 7.51, 13.2, and 13.91 at % Ba has been monitored by performing a thermal analysis after each nitrogen aliquot. The initial compositions were chosen to straddle the eutectic of the Li-Ba system (x_{Ba} = 10.5 at %; T = 416 K); the amount of nitrogen added in each aliquot was chosen to increase the nitrogen content of the system by $\approx$ 0.05 at % N. Typical results for the addition of nitrogen (15 aliquots) to a hypoeutectic Li-Ba solution (x_{Ba} = 4.75 at %) are shown in Fig. 4. The cooling curve for the initial solution exhibits two thermal effects (Figure 4). The first is due to the precipitation of lithium from the solution as the system passes through the hypereutectic liquidus (432 K); the second is due to the solidification of the eutectic mixture of lithium and Li_4Ba (417 K). The cooling curves for the dilute solutions exhibited three thermal effects (Figure 4). The first two may be attributed to the same phenomena as those for the pure Li-Ba solution. The temperatures at which they occur diverge with increasing nitrogen content. The first effect slowly increases eventually to reach a maximum at point N, x_N = 1.2 at %, T = 439 K; the second rapidly falls to attain the temperature of the third effect at point P, x_N = 1.2 at %, T = 368 K. The third thermal arrest is independent of composition, invariably occurring at 368 K; it is attributed to the freezing of the ternary eutectic mixture of lithium, Li_4Ba and Li_3N. For more concentrated nitrogen solutions only two thermal effects are observed. The first, which slowly decreases in temperature with increasing nitrogen content, is attributed to the solidification of the binary Li-Li_3N eutectic; the second, which is essentially independent of concentration corresponds to the ternary eutectic. The logic behind these conclusions is discussed later. The experiment was terminated when the nitrogen content reached point J, x_N = 11.5 at %.

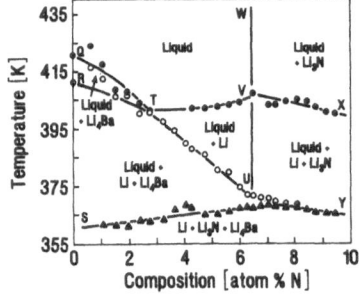

Fig. 5. Phase equilibria derived from thermal analysis data for solutions of nitrogen in a hypereutectic Li-Ba solution (13.2 at % Ba)

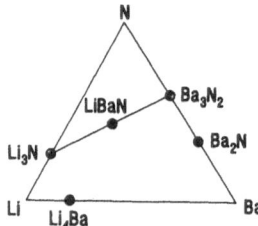

Fig. 6. Li-Ba-N ternary phase diagram

Typical results for the addition of nitrogen (26 aliquots) to a hypereutectic Li-Ba solution (x_{Ba} = 13.20) are shown in Figure 5. The cooling curve for the initial solution exhibits two thermal effects (Figure 5). The first is due to the precipitation of Li_4Ba from the solution as the system passes through the hypereutectic liquidus (419 K); the second is due to the solidification of the eutectic mixture of lithium and Li_4Ba (413 K). The cooling curves for the dilute solutions exhibited three thermal effects (Figure 5). The first two may be attributed to the same phenomena as those for the pure Li-Ba solution. The temperatures at which they occur decrease with increasing nitrogen content, eventually converging at point T, x_N = 2.8

at %, T = 400 K. With increasing composition, the third thermal arrest increases slowly to a maximum at 368 K; it is attributed to the freezing of the ternary eutectic mixture of lithium, Li_4Ba and Li_3N. The first effect was susceptible to supercooling; consequently, the temperature of the phase change can only be defined to ± 2 K. At higher nitrogen contents, three thermal effects are again observed but in this case the first two diverge, the first increasing slightly to a maximum at point V, x_N = 6.8 at %, T = 405 K, while the second falls quite rapidly to the temperature of the third invariant effect at point U, x_N = 6.8 at %, T = 368 K. In this part of the experiment, the first thermal effect is attributed to the precipitation of pure lithium from the solution and the second to the solidification of the binary eutectic mixture (lithium and Li_4Ba). The logic behind these conclusions is discussed later. At even higher nitrogen concentrations only two thermal effects are observed. The first, which slowly decreases in temperature with increasing nitrogen content, is attributed to the solidification of the binary $Li-Li_3N$ eutectic; the second, which is essentially independent of concentration is related to the ternary eutectic. The experiment was terminated when the nitrogen content reached point Y, x_N = 9.45 at%.

4. THE Li-Ba-N TERNARY PHASE DIAGRAM

The results from the thermal analysis studies can be considered as sections through the Li-Ba-N ternary system (Figure 6); it is easier, however, to consider them as sections of the $Ba-Ba_3N_2-Li_3N-Li$ pseudo-quaternary system (Figure 7).

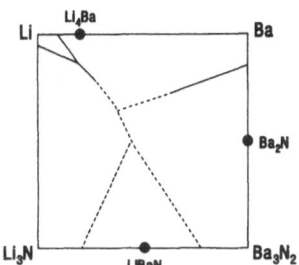

Fig. 7. $Li-Ba-Ba_3N_2-Li_3N$ pseudo-quaternary diagram

Data on the constituent binary systems are sparse. The Li-Ba binary system is that reported by Keller et al [11] modified to incorporate our recent results [12]. One intermetallic phase, Li_4Ba, which decomposes in a peritectic reaction at 429 K, exists in the system which is a simple eutectic. The hypoeutectic liquidus falls from the melting point of pure lithium (453.5 K) to the eutectic between lithium and Li_4Ba (x_{Ba} = 10.5 at %, T = 416 K); the liquidus then rises gently to the peritectic reaction (x_{Ba} = 18.4 at %, T = 429 K) before rising more steeply to the melting point of barium (998 K).

The $Li-Li_3N$ phase diagram is a simple eutectic system [13,14]; the eutectic lies close to the lithium axis (x_N = 0.068 at %, 453.24 K) [15] and the hypereutectic rises steeply from the eutectic temperature to the melting point of Li_3N at 1086 K.

The $Ba-Ba_3N_2$ phase diagram has been studied upto a maximum nitrogen concentration of x_N = 25 at % [16]; it also is based on a simple eutectic (x_N = 14.4 at %, T = 781 K).

Little is known of Ba_3N_2-Li_3N phase relationships, except for the synthesis and structural characterisation of the ternary compound, LiBaN [17].

Since the present study is concerned solely with dilute solutions of both barium and nitrogen, it is the lithium corner of the pseudo-quaternary phase diagram which is of interest. A projection of a three-dimensional representation of the lithium corner of the pseudo-quaternary Ba-Ba_3N_2-Li_3N-Li phase diagram is shown schematically in Figures 8 and 9. This projection, which can be considered as a series of divariant precipitation faces, intersecting to form a series of univariant valleys which meet at an invariant point, can be used to interpret the phase changes observed on the thermal analysis curves.

The locations of the sections, which are effectively of constant Li:Ba molar ratio, are shown by lines α - β and γ - δ in Figures 8 and 9. Considering firstly the Li-4.8 Ba solution (Figures 4 and 8). Solutions containing very small quantities of nitrogen are typified by a composition corresponding to point A on Figure 8. As the single phase solution is cooled, the first thermal inflection corresponds to the precipitation of lithium, since the solution is on the lithium-rich side of the binary Li-Ba eutectic. In this respect the behaviour of the solution is similar to that of the initial Li-Ba solution. This phase change is represented as a point on line MN of the phase diagram (Figure 4). Upon further cooling, the solution is depleted in lithium, its concentration following the line A-A'. When the line joining the Li-Ba binary eutectic with the ternary eutectic (T_1T_3 ; Figure 8) is reached, Li_4Ba starts to precipitate together with lithium, and the second thermal arrest is observed. This phase change is represented as a point on line LP on the phase diagram (Figure 4). As the temperature is lowered further, the composition of the solution follows the Li-Ba binary eutectic -- ternary eutectic line (T_1T_3 , Figure 8) until the ternary eutectic is reached whereupon the remaining liquid solidifies, and the third effectively invariant phase change is observed. This phase change is represented as a point on the line KP on the phase diagram (Figure 4).

On increasing the nitrogen content of the system, the liquidus (i.e., the first observed thermal arrest) will increase in temperature as the lithium precipitation face is traversed. However, the eutectic line T_1T_3 (and hence the second thermal arrest) is reached at correspondingly lower temperatures.

As the nitrogen content of the system is increased even further, the Li-Li_3N binary eutectic -- ternary eutectic line (T_2T_3 , Figure 8) is crossed at point B. Since the Li-Li_3N hypereutectic liquidus is quite steep, it may be assumed that the Li_3N precipitation face also rises steeply from the Li-Li_3N binary eutectic -- ternary eutectic line. Consequently, the solubility limit for Li_3N is probably exceeded immediately after crossing this line. This precipitation face is represented by a vertical line (NO) in Figure 4.

Since the solubility is exceeded on crossing the precipitation face the corresponding phase boundary will not be detectable in the thermal analysis experiment. Thus at the relatively low temperature at which the nitrogen was added (473 K), the system is no longer a single (liquid) phase but a two (liquid + Li_3N) phase system. Thus on cooling solutions with these higher nitrogen concentrations, which are typified by point C (Figure 8), the first recorded thermal arrest is due to the precipitation of the Li-Li_3N binary eutectic mixture at point C' (Figure 8) and corresponds to a point along line NI (Figure 4). As the temperature is lowered further, the composition of the solution follows the Li-Li_3N binary eutectic -- ternary eutectic line (T_2T_3; Figure 8) until the ternary eutectic is reached when the remaining liquid solidifies. This phase change is represented as a point on line PJ of the phase diagram (Figure 4).

We will now consider the Li-13.2 Ba solution (Figures 5 and 9). Solutions containing very small quantities of nitrogen are typified by a composition corresponding to point D on Figure 9. As the single phase solution is cooled, the first thermal inflexion corresponds to the precipitation of Li_4Ba, since the solution is on the barium-rich side of the binary Li-Ba eutectic. In this respect the behaviour of the solution is similar to that of the initial Li-Ba solution. This phase change is represented as a point on line QT of the phase diagram (Figure 5). Upon further cooling, the solution is depleted in Li_4Ba, its concentration following the

336

line D-D′. When the line joining the Li-Ba binary eutectic with the ternary eutectic (T_1T_3; Figure 9) is reached, lithium starts to precipitate together with Li_4Ba, and the second thermal arrest is observed. This phase change is represented as a point on line RT on the phase diagram (Figure 5). As the temperature is lowered further, the composition of the solution follows the Li-Ba binary eutectic -- ternary eutectic line (T_1T_3, Figure 9) until the ternary eutectic is reached whereupon the remaining liquid solidifies, and the third effectively invariant phase change is represented as a point on the line SU on the phase diagram. (Figure 5)

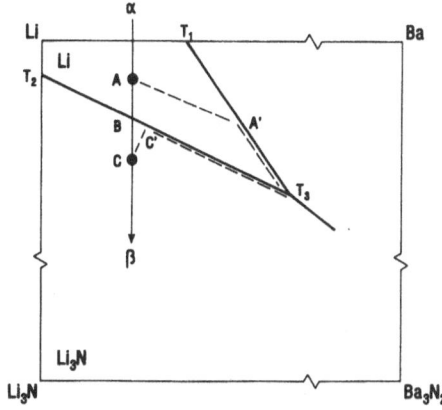 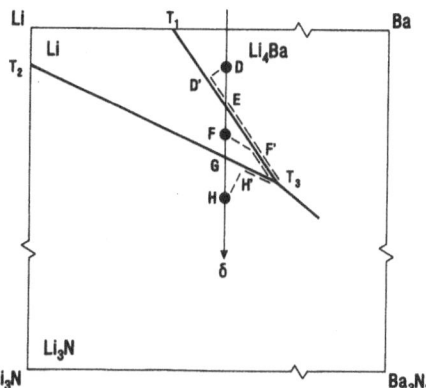

Fig. 8. Proposed quaternary phase diagram for the Li-Ba-Ba_3N_2-Li_3N system (cross-section starting at 4.8 at % Ba)

Fig. 9. Proposed quaternary phase diagram for the Li-Ba-Ba_3N_2-Li_3N system (cross-section starting at 13.2 at % Ba)

On increasing the nitrogen content of the system, it follows that the liquidus will decrease in temperature as the Li_4Ba precipitation face is traversed. Similarly, the eutectic line T_1T_3 (Figure 5) is reached at a correspondingly lower temperature. The temperature of the first two phase changes should converge and eventually become coincident. This behaviour will be shown by a solution with a composition corresponding to point E (Figure 9) when the eutectic line is reached directly on cooling; it corresponds to point T (x_N = 2.8 at %) in Figure 5.

The behaviour of more concentrated solutions will be very similar to that observed for the Li-4.8 Ba solutions. Thus, when a solution at point F (Figure 9) is cooled, lithium is precipitated initially from solution, followed by a lithium + Li_4Ba mixture and is completed by the ternary eutectic mixture, the composition of the solution following line FF′T_3. The corresponding phase changes are represented by points on lines TV, TU, and SU on the phase diagram (Figure 5). Similarly for more concentrated solutions, represented by point H (Figure 9). The two recorded thermal arrests are due to the precipitation of the Li + Li_3N eutectic mixture and the ternary eutectic, the composition of the solution following line HH′T_3 (Figure 9). The corresponding phase changes are represented by points on lines VX and UY on the phase diagram (Figure 5).

5. CONCLUSIONS

The reaction of nitrogen with liquid lithium containing dissolved alkaline earth metals (calcium, strontium and barium) follows a similar behaviour pattern. Initially, nitrogen dissolves in the liquid metal; subsequently, Li_3N is precipitated.

Nitrogen solubility increases with increasing alkaline earth metal content. Thus, at 473 K the phase boundary representing the solubility of Li_3N in the Li-Ba-N system is dependent on the barium content of the solution, rising from $x_N \approx 0.07$ at % for pure lithium through $x_N \approx 1.5$ at % for $x_{Ba} = 4.75$ at % and $x_N \approx 2.2$ at % for $x_{Ba} = 7.50$ at % to $x_N \approx 6.3$ at % for $x_{Ba} = 13.2$ at %.

Of the alkaline earth metals, barium is the most effective in solubilising nitrogen and calcium least effective. Thus, at $x_M = 5.0$ at %, $x_N = 4.66$ at % in the Li-Ba-N system, $x_N = 3.48$ at % in the Li-Sr-N system and $x_N = 2.49$ at % in the Li-Ca-N system.

This behaviour pattern is different from that in the corresponding sodium systems where the solubility is directly proportional to the alkaline earth metal content of the solution and the precipitating phase is the alkaline earth metal subnitride, M_2N (M = Ca, Sr, Ba)

Phase relationships in the Li - Ba - N system ($x_{Li} > 85$ at %) were studied using thermal analytical methods. In brief, three crystallisation fields (of Li, Li_4Ba and Li_3N) are separated by eutectic valleys which fall from the binary eutectics (Li - Ba: $x_{Ba} = 10.5$ at % and 416 K; Li - N: degenerate at 453.5 K) to the ternary eutectic of the Li-Ba-N system (368 K).

Acknowledgements

We would like to thank the SERC for financial support (to PGR).

References

[1] C.C. Addison, G.K. Creffield, P. Hubberstey and R.J. Pulham, J. Chem.Soc. Dalton Trans., (1976) 1105.
[2] C.C. Addison, R.J. Pulham and E.A. Trevillion, J. Chem.Soc. Dalton Trans., (1975) 2082.
[3] P. Hubberstey, Proc. Int. Conf. "Liquid Alkali Metals", BNES, London 1973, p. 15-19.
[4] P. Hubberstey and P. R. Bussey, Proc. Int. Conf. "Liquid Metal Engineering and Technology", BNES London, 3 (1984) 143.
[5] C. C. Addison, Sci. Progr. Oxf., 60 (1972) 385.
[6] E. Veleckis, K. E. Anderson, F. A. Cafasso and H. M. Feder, USAEC Rept., ANL-7520 (part I) (1968) 295.
[7] R. J. Pulham and P. Hubberstey, J. Nucl. Mater., 115 (1983) 239.
[8] P. Hubberstey, A. T. Dadd and P. G. Roberts, in "Material Behaviour and Physical Chemistry in Liquid Metal Systems", Ed. H U Borgstedt, Plenum Press, New York, 1982, 445
[9] P. Hubberstey, in Proc. Int. Conf. "Liquid Metal Engineering and Technology", Oxford, 1983, BNES, Vol 3 (1984) 85.
[10] G. J. Moody and J. D. R. Thomas, J. Chem. Educ., 43 (1966) 205
[11] D. V. Keller, F. A. Kanda and A. J. King, J. Phys. Chem., 62 (1958) 732.
[12] P. Hubberstey and P. G. Roberts, unpublished results.
[13] P. F. Adams, M. G. Down, P. Hubberstey and R. J. Pulham, J. Less Common Metals, 42 (1975) 325.
[14] R. M. Yonco, E. Veleckis and V. A. Maroni, J. Nucl. Mater., 57 (1975) 317.
[15] P. Hubberstey, R. J. Pulham and A. E. Thunder, J. Chem. Soc., Faraday Trans. I, 72 (1976) 431.
[16] V. A. Russell, M Sc Thesis, Univ. of Syracuse, New York, 1949.
[17] J. F. Brice and J. Aubry, Compt. Rend. Ser. C. (1970) 825.

PHYSICAL - CHEMICAL PRINCIPLES OF LEAD - BISMUTH COOLANT TECHNOLOGY

B. F. Gromov, Yu. I. Orlov, P. N. Martynov,
K. D. Ivanov, V. A. Gulevsky

The Institute of Physics and Power Engineering
Bondarenko Sq. 1, 249020 Obninsk, Kaluga Region, Russia

1. INTRODUCTION

Lead - based melts, like melts based on other metals (Na, Na-K), have been treated as possible coolants to cool nuclear reactors for a long time. Recently, however, the interest to the lead - containing melts has grown, particularly, in relation to the increased requirements of nuclear power safety and the activation in the development of electro - nuclear units for the transmutation of radioactive wastes. This is due to the fact that these coolants are characterized by a number of unique properties.

It is sufficient to note that they have low chemical reactivity, particularly, if they interact with water, water vapour and oxygen. This fact significantly reduces the risk of their application as compared to alkali metals. From the standpoint of using the above mentioned coolants in electro - nuclear units, the highest efficiency can be expected in the case of utilizing them as liquid targets, in which their thermophysical and nuclear - physical properties can be harnessed to a maximum degree.

Among the Pb - containing melts, the Pb - Bi eutectic holds a special place because of its low melting temperature (123,5 °C), by which many technical difficulties can be avoided. This circumstance stimulated the commercial application of this coolant to a great extent. One of the key problems to be solved was the problem of ensuring a long - term safe and reliable operation of the circulating heat - transfer loop. Applied to any heat - transfer loops, the solution of this problem requires that two conditions have be fulfilled: the provision for corrosion and erosion resistant construction materials used in contact with the coolant and the provision for an adequate purity of the coolant itself as well as of the inside surfaces of pipes and other equipment of the loop. The practical implementation of these conditions is achieved by special arrangements of the whole complex with reference to the coolant under consideration, which is called "the Pb - Bi Coolant Technology ".

2. CORROSION OF CONSTRUCTION MATERIALS

Provision for corrosion - resistant materials contacting with the Pb - Bi melt is a complicated problem which, depending on the purpose of using the circulation loop, involves the solution of such issues as the selection of available steels and the development

Liquid Metal Systems, Edited by H.U. Borgstedt
and G. Frees, Plenum Press, New York, 1995

of the new ones, their alloying, the formation of protective covers on their surfaces, etc.. The Pb - Bi coolant technology covers a more narrow scope of problems only concerned with checking and correcting the qualitative composition of the coolant during the operation of the liquid metal loop. The possibility of providing the materials being used with the resistivity to corrosion by means of this method is based on the accepted technique of the protection of constructional steels, which implies the creation of a diffusion barrier preventing the formation of protective covers on their surface. As far as these covers essentially consist of oxide compounds of steel components, their stability is governed by the activity of oxygen and the relevant metal components in the Pb - Bi melt to a great degree. For example, consider the thermodynamic stability of the magnetite (Fe_3O_4) forming the base of the oxide cover of stainless steels.

The reaction of the formation and the dissociation of this oxide in the Pb - Bi melt can be expressed as:

$$4\,O + 3\,Fe \leftrightarrow Fe_3O_4 \tag{1}$$

with the equilibrium constant

$$K = a_{[Fe3O4]} / a_O{}^4 \cdot a_{Fe}{}^3 \tag{2}$$

where Fe, O and Fe_3O_4 are iron, oxygen and iron oxide, respectively, dissolved in the melt; a is the thermodynamic activity of the substances present in the solution.

With an excess solid phase of magnetite, $a_{[Fe3O4]} = 1$ can be put in Eq. 2: the correlation between the iron and oxygen activities will then have quite a definite nature. Thus, with a temperature, $T = 400°C$ and an iron activity value, $a = 1$, the equilibrium activity of oxygen will amount to $1 \cdot 10^{-6}$, corresponding to an oxygen concentration, $C_{[O]} = 1 \cdot 10^{-10}$ mass %. With a lower activity of iron, the corresponding equilibrium value of the oxygen activity in the melt will be higher.

Fig. 1 shows the results of calculations of the equilibrium oxygen activity values as a function of the reciprocal temperature 1/T for constant concentrations of oxygen (dotted lines) and for the conditions of the magnetite dissolution at $a_{[Fe]} = 1$ (solid lines). Setting now the temperature

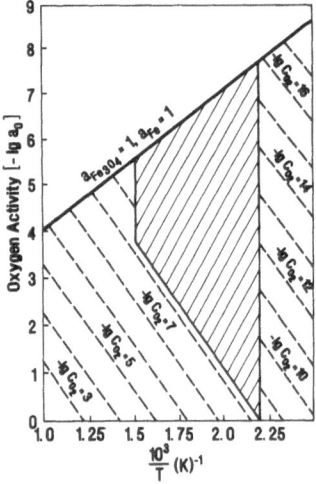

Fig. 1: Assessment of the range of permissible values of the oxygen activity in Pb - Bi melts.

interval corresponding to the max. and min. temperature of the melt in the circuit and restricting the oxygen concentration values to the saturation concentration at the minimum temperature of the melt, $C_{[O]} = C_s$ [at $T = T_{min}$), the domain of oxygen activity values can be obtained which, as a first approximation, can be considered a domain of permissible values of oxygen activity. From above, the domain is restricted by the magnetite dissociation line, and from below, by the line representing the crystallization of the excess solid at a decreasing temperature of the melt.

If the coolant may be operated under the conditions of a real non - isothermal loop, a process can develop, which yield the oxygen activity values to fall beyond the limits of permissible values. Thus, the deoxidizing action of impurities, which are related to oxygen more closely than iron, decreases the oxygen activity in the melt. On the contrary, the penetration of oxygen in the circulation loop due to leaks contributes to the increase of its activity. Considering the importance of control of this parameter, an instrument for the measurement of oxygen activity, called " the oxygen activity measuring cell", was developed and constructed in the Institute of Physics and Power Engineering on the basis of the galvanic concentration cell with solid electrolite. Wide use of this measuring cell allows not only the on - line control of the Pb - Bi coolant state but also the development of a number of devices for the correction of this state immediately in the course of the operation of the heat - transfer loop. These devices function on the basis of the following to influence the coolant:

The method involving the reaction of the gaseous oxygen interaction with the Pb - Bi melt. The rise of oxygen activity in the melt is achieved due to the successive proceeding of reactions yielding lead oxide

$$2 Pb(l) + O_2(g) \rightarrow 2 PbO(s) \tag{3}$$

and its dissolution in the coolant

$$PbO(s) \rightarrow [O]_{diss} + Pb \tag{4}$$

The method based on interactions of the melt with solid oxides of the coolant, that are used as charge in various mass - transfer apparatus. In the course of their operation the oxygen activity in the melt increases due to reaction (4).

The method based on the interaction of the Pb - Bi melt with reducing-oxydizing gas mixtures. The reaction of the melt with the hydrogen and water vapour mixture occurs as in Eq. (5):

$$Pb + H_2O \leftrightarrow PbO + H_2 \tag{5}$$

In view of the low partial pressure of oxygen at the dissociation of water vapour, the process of promoting the oxygen activity proceeds without the formation of solid lead oxide. Beside, in contrast to the first two methods, the present technique allows the opportunity that occurring processes might be changed through the variation of the partial pressures of water vapour and hydrogen. An increase of the hydrogen content in the mixtures uses rises the role of the redox process. On the contrast, the relative increase of the water vapour content is able to enhance the oxidizing component to affect the coolant.

The set of the methods considered above allows one to develop the best - adequate conditions of applying these means for resistance of constructive materials to corrosion at all stages of the heat - transfer loop operation.

The second principal problem of the coolant technology is associated with the purity of the circulation loop and of the coolant itself, which are directed in one sense only, helping to keep the thermal - hydraulic performances of the circulation loop constant and to maintain reliable operation of the equipment. This problem is related to the fact that a number of impurities are accumulated in the coolant during the course of a long - term operation of the loop. Achieving the saturation concentration, they form an excess solid which deposits on the inside surfaces or circulates with the coolant in the circuit, and can cause negative consequences. The composition of deposits being formed in the loop depends on the set of constructive materials used, constructive features of the loop, its operation regimes, service peculiarities, and it can vary in a wide range.

Table 1 presents the results of composition analyses for slag deposits built - up when exploiting some research loops as an example.

Table 1. Results of chemical analyses of samples of slag deposits

Number of slag sample	Content of elements in slag, mass %					
	Fe	Cr	Ni	O	Pb	Bi
1.	$4.0 \cdot 10^{-4}$	$3.0 \cdot 10^{-4}$	$3.0 \cdot 10^{-4}$	2.4	51.6	46.35
2.	$4.0 \cdot 10^{-4}$	$2.3 \cdot 10^{-4}$	$2.3 \cdot 10^{-4}$	2.4	51.6	46.3
3.	$1.3 \cdot 10^{-2}$	$2.6 \cdot 10^{-3}$	$1.1 \cdot 10^{-3}$	1.3	52.85	45.35
4.	$2.6 \cdot 10^{-2}$	$2.4 \cdot 10^{-2}$	$1.2 \cdot 10^{-3}$	3.9	51.9	42.1
5.	$2.6 \cdot 10^{-2}$	$1.2 \cdot 10^{-2}$	$1.0 \cdot 10^{-2}$	6.5	76.0	14.2
6.	$8.0 \cdot 10^{-2}$	$1.0 \cdot 10^{-2}$	$5.0 \cdot 10^{-2}$	2.8	74.0	21.1
7.	$4.2 \cdot 10^{-1}$	$1.5 \cdot 10^{-1}$	$9.5 \cdot 10^{-2}$	3.5	55.9	39.5
8.	$5.1 \cdot 10^{-1}$	$4.1 \cdot 10^{-2}$	$9.0 \cdot 10^{-2}$	4.0	47.5	47.5
9.	0.6	$1.0 \cdot 10^{-2}$	$5.0 \cdot 10^{-2}$	2.6	75.3	18.6
10.	1.7	$8.6 \cdot 10^{-1}$	$7.0 \cdot 10^{-2}$	1.4	52.0	38.0
11.	5.3	1.3	0.1	-	-	-
12.	6.1	1.7	0.2	1.73	51.4	35.8
13.	6.9	1.2	$8.0 \cdot 10^{-2}$	0.95	53.3	32.3

The experimental results are given in the order of increasing contents of metallic impurities, thus, exhibiting the presence and the rate of the corrosion processes in the loop. It is evident from this table that as the number of the sample increases, the content of iron and chromium impurities increases more than by four orders of magnitude, whereas the nickel content varies almost by a factor of 10^3. Simultaneously, the variation of the content of oxygen impurities in slags is not definite in the same way. This is explained by the fact that the slags being formed are capable to adsorb free oxygen. In the course of taking and storing samples their initial composition in respect to a particular impurity can vary. The metal impurities detected in the deposits on the inside surfaces of the loop after discharging the coolant or on the interfaces melt / gaseous protective atmosphere primarily consist of the mixture of oxide compounds with coolant inclusions. In rare cases typical of highly deoxigenated melts, deposits consisting solely of metal phases were established.

In order to provide the coolant with necessary purity and to prevent the formation of slag deposits, filtration of the coolant is widely used.

The designs of filters and the principles of their function are rather multiple. Thus magnetic, adsorption and mechanic filters found use in different research rigs. The filter location in the loop and the flow rate of the coolant being pumped through are dependent on the constructive features of the loop to a great extent and are chosen starting from the condition of the filter competitiveness as compared to the filtrating performance of the loop itself. The matter is that the heat - transfer loop usually incorporates equipment with a developed inside surface of contact with coolant (steam generators, reactor core and other heat exchangers) which itself possesses rather good filtrating properties.

To prevent blocking of the gas pipeline of the liquid metal loop as well as to clean the coolant surface at the points of contact with the gaseous atmosphere, filtration of the gaseous phase is employed. By promoting the flow rate, directional circulation of the shielding gas is maintained, that, due to dynamic action of the gas flow, allows dust-like impurities to be separated from the coolant surface to the filter; the gas lines are simultaneously cleaned.

It is not reasonable to use filtration - based methods of purification of the coolant for the removal of lead and bismuth oxides, which are the base of slug deposits, from heat transfer loops. It seems to be more promising to reduce them immediately to lead and bismuth in the circulating loop. With this goal in view a method called " the hydrogen regeneration " has been developed, which is used to clean heat - transfer loops. The chemical reaction of the interaction of oxides with hydrogen is according to Eq. (6).

$$Me_xO_{y(s)} + y\,H_{2(g)} \leftrightarrow x\,Me_{(l)} + y\,H_2O_{(g)} \qquad (6)$$

where Me is lead or bismuth.

The conditions to conduct this reaction have been chosen in a way that any possibility of the reduction of oxides forming the structure of protective covers on constructive materials is excluded during its proceeding. Simultaneously, the variation of the slag deposit composition observed during the hydrogen regeneration and the subsequent degradation of the mechanical properties of deposits promote their destruction and the cleaning of the inside loop surfaces. Water vapour being formed during the reaction condense in a special cooler and can be easily removed from the loop. The efficiency of the hydrogen regeneration depends on the temperature conditions as well as on an efficient feed of hydrogen to slag deposit locations. Slags located on the free surface of the coolant can be easily provided with hydrogen through the gas lines of the loop, whereas special systems were developed to introduce the gas phase into the coolant in order to supply hydrogen to deposits located below the melt level.

To summarize the brief description of processes involved in the Pb - Bi coolant technology, it should be noted that this technology found its application and support while operating various industrial units.

DIFFUSION COEFFICIENTS IN LIQUID METALS AT HIGH DILUTION

Cezary Guminski

Department of Chemistry ,University of Warsaw
Pasteura 1, 02093 Warszawa, Poland

1. INTRODUCTION

The knowledge of experimental diffusion coefficients of metals and non-metals in liquid metals is essential for both, practical and theoretical reasons. These data are very important in metallurgy, corrosion investigations and electrochemistry. An interest of theoretically orientated scientists in physics, material science and chemistry is to verify several models proposed as well as to withdraw some information about the structure of liquid alloys. As physicists are mostly interested in theory and metallurgists in the determination of diffusion coefficients, an opinion of chemists on diffusion in metallic systems is seldom expressed.

2. EXPERIMENTAL METHODS

There are some excellent reviews [1,2,3,4,8a] concerning the experimental methods of the determination of diffusion coefficients. Employing radioactive or stable isotopes one may use methods as capillary-reservoir, long capillary, shear cell and diaphragm cell techniques. In some cases the concentration changes may be traced by chemical analysis, resistivity, electromotive force measurements as well as investigations of electrotransport of a solute metal forced by a current flow. The precision of these methods is estimated at the best at several %.

If the dissolution of a solid in a liquid metal is diffusion controlled, one may use the rotating disc technique where the rate of the process is determined by analysis of liquid metal fractions or a mass loss of the solid disc. The precision of this method is claimed to be better than 10 %.

Mass spectrometry and nuclear magnetic resonance techniques have found rather limited use to determine the diffusion isotopic effect. A method of scatter of slow neutrons by liquid metals give information on the structure of the medium as well as on dynamic motion of the atoms.

Electrochemical methods, which are based on various kinds of electrolyses, are not presented in the mentioned reviews. However, a major part of the diffusion data for Hg as a solvent was exactly obtained from electroanalytical experiments and the results obtained

Liquid Metal Systems, Edited by H.U. Borgstedt
and G. Frees, Plenum Press, New York, 1995

were precise and reliable. If a metal dissolved in Hg is oxidized from a stationary drop electrode in diffusion controlled chronovoltammetric, chronoamperometric or chronopotentiometric conditions, the variables of such electrolysis may be analyzed in the general form [5]:

$$iX^{1/2} = K'D_M^{1/2} - K''D_M X^{1/2} r_o^{-1} \qquad (1)$$

where i is the oxidation current, X is time parameter characteristic for every method (transition time in chronopotentiometry, time of electrolysis in chronoamperometry and reciprocal rate of polarization in chronovoltammetry), K' and K'' are constants dependent on the selected technique, electrode surface and metal concentration, D_M is diffusion coefficient of a metal and r_o is radius of spherical electrode. Thus, D_M may be calculated from an experiment by three methods: from the intercept of $iX^{1/2}$ - $X^{1/2}$ plot on the $iX^{1/2}$ axis, from the slope of this plot and from the ratio of the slope to the intercept. If the electrode radius is about 0.5 mm, the oxidation time should not exceed 30 s. For an experiment performed on a plane electrode the second term in the Eq.(1) is equal zero and there is only one way to calculate the diffusion coefficient.

Metals being dissolved in Hg may be also oxidized in, so named, amalgam polarography conditions, if the amalgam drops regularly from a capillary. The mean limiting current, recorded, i_g, is then proportional to the square root of the diffusion coefficient D_M

$$i_g = K'''D_M^{1/2} \qquad (2)$$

Since the oxidation of more noble metals (than Hg) cannot directly be carried out from their amalgams, one may utilize a fast subsequent reaction connected with an electrode process for tracing the diffusion. For example, the reaction of Au with Zn in Hg is stoichiometric and fast, therefore the reduction current of Zn(II) ions on an Au amalgam electrode, at specially selected potential, is dependent on diffusion of Au atoms to the electrode surface [6].

The precision of carefully performed electroanalytical determinations of D_M in Hg is between 2% (Cd) and 20% (lanthanides). However, for high metal concentrations (>10^{-2} mol %), long time of the electrolysis (>30s) and temperature gradient (>0.1 K), an independent convectional vortex may arise causing an overestimation of the results by even more than 100%.

3. EXPERIMENTAL RESULTS IN LITERATURE

The collection of experimental values of the diffusion coefficients of elements in liquid metals and metalloids has been achieved using the Diffusion and Defect Data [8] which is a quite comprehensive reference compilation for the period 1966-1992. The results published earlier are quoted from reviews of [9] and [10]. Since the results for diffusion in liquid Hg were not complete in the above mentioned sources, the reviews of [7] and [11] served as the basis for the compilation. Any additional sources for particular systems, which have been found during the study of literature, are additionally mentioned in the captions of tables and figures.

The D_M values reported from the literature fall into different classes of precision and accuracy. Although precision for various methods is frequently claimed to be only several %, it is much safer to assume that a typical statistical error is 20% for the majority of alloy systems. It should be underlined at this point that such uncertainty of quantitative results allows to confirm as well as to neglect the validity of various diffusion theories, the existence of isotopic effect, the shape of diffusing molecules and other structural interpretations of the diffusion data.

The experimental diffusion coefficients are frequently charged with a convectional overestimation which seriously complicate an assessment and the selection of the most accurate result. An underestimation is much less probable and may be connected with an influence of impurities in the system studied, or non-diffusional dissolution kinetics of solid in liquid metals (the rotating disk method).

Due to different diffusion theories the temperature dependence of D_M has been variously proposed for a presentation of experimental results. However, the most popular, as well as rational, is the following functional form:

$$\ln D_M = A + B/T \tag{3}$$

where B is generally treated as the activation energy of the diffusion process and A is a frequency factor. This dependence has served to perform the interpretation as well as the extrapolation of the collected data to selected temperatures. The D_M values from a distant extraploation are shown in the figures with arrows indicating rather probability limits of D_M than any realistic values.

4. THEORETICAL MODELS

Several models of diffusion in the liquid state were proposed [1,8a]: the hole, the free volume, the fluctuation and the itinerant oscillator theories derived from the solid state theory, the fluidity from the liquid state theory, and the dense gas formulation, the corresponding states correlation and the molecular dynamic calculations originated from kinetic theories of gases. The self diffusion coefficients calculated from both sorts of theories give rather good agreement with the experimental results [1,12]. In the case of the impurity diffusion, an agreement with theories is only observed for some systems [13,14,15,16]. Semi-empirical correlations of the diffusion coefficients with the relative valency of solute and solvent [13,14], the molar enthalpy or free energy of mixing [14] and differences in mass and size of the atoms [14] were additionally postulated, with only partial success.

5. CHEMICAL INTERACTIONS IN LIQUID ALLOYS

It must be borne in mind that quite stable molecules or associates may exist in some liquid alloys, and such molecules, rather than atoms, take part in the diffusion process. Expressing the stability of the molecules formed in an alloy with the help of the equilibrium constant:

$$K = a_{MM'} x / a_M a_{M'}^x \tag{4}$$

where a's are the corresponding activities and x is the stoichiometric coefficient. One may estimate that in a dilute solution of M in liquid M' the atoms of M are predominantly present in form of molecules MM'_x when K in Eq.(4) is higher than 100. For K values in the range 1-100 the diffusion of M in M' is realized by a movement of the MM'_x molecules as well as atoms M. For K values lower than 1 the diffusion of M in M' proceeds predominantly by the movement of M atoms, especially, if the chemical equilibrium is fast. Thus, if the partial free energy or activity coefficient of M in a dilute liquid solution in M' is known, one may estimate the stability of the MM'_x molecules and their participation in the diffusion process. There are derived relations [17] which allow to calculate the effective diffusion coefficient, if the equilibrium constant and the mobility of the molecule is known, but the kinetic aspect of such an equilibrium was not taken into account. The shape of diffusing molecules was also discussed [18] but no spectacular difference was observed for transport of spherical or disc-like molecules of the same volume.

6. COMPARISON OF EXPERIMENTAL RESULTS WITH THEORY

A comparison of the experimental diffusion data with those predicted by the simple theory was carried out using the Sutherland-Einstein equation:

$$D_M = kT/4\mu_{M'} \cdot r^d_M \pi \qquad (5)$$

where k is the Boltzman constant, r^d_M is the radius of the diffusing particle and $\mu_{M'}$ is the viscosity of the medium (for diluted alloys it is practically equal to the viscosity of pure metal solvent). The correctness of this equation was tested on self diffusion of more than 10 liquid metals [12] and on impurity diffusion of 37 solutes in Hg [7] which was encouraging for tests in other systems (see Table 1 and Fig. 1 for comparison). The Eq.(5) is in agreement with the modified hole theory of diffusion presented in [12]. Moreover, the simplicity of this equation and accessibility to necessary data, like $\mu_{M'}$ and r_M, is another advantage.

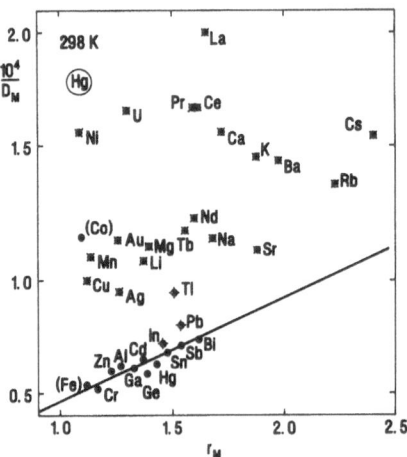

Fig. 1. Diffusion coefficients of metals in liquid Hg at 298 K;
• - no compounds in the system; ⊕ - weak interaction; ⊗ stable compound in the system.
D_M is in $cm^2 s^{-1}$, r_M is in 10^{-8} cm.

A general rule was formulated in case of amalgams that a metal forming an intermetallic compound with Hg diffuses slower than it would be predicted by Eq.(5), while the metals which do not form stable compounds with Hg diffuse as atoms, and Eq.(5) is in this case fulfilled within experimental errors of few percents.

All available D_M values in other metallic solvents were collected from the mentioned literature sources. If several different results were determined for the same system in different laboratories the lowest reliable values were generally selected because the mentioned convectional disturbance may only increase D_M. The selected D_M data of various elements in liquid Ag, Al, Bi, Cd, Cu, Fe, Ga, Ge, In, K, Li, Mg, Na, Ni, Pb, Sb, Si, Sn, Te and Zn are plotted in Figs. 2-22 in form of plot of D_M^{-1} vs. r_M. The r_M values are calculated from the formula:

$$r_M = 1/2 \ (3V_M/\pi N)^{1/3} \qquad (6)$$

where V_M is the molecular volume of the metal at given temperature and the N is the Avogadro number. The $\mu_{M'}$ and V_M data for every element were taken from [19] and [20].

In each figure the straight line corresponds to the Eq.(5), and one may easily observe that in most cases the self diffusion coefficients of the solvent metals are very close to this line. If the solute metals do not form stable intermetallic compounds with the solvent metals they

El..	D_M 10^{-5} cm^{2}s^{-1}	r_M 10^{-8} cm	r^d_M 10^{-8} cm	Diffusing entity in liquid Hg	Compound identified in liq. Hg	Compound in solid phase
Li	0.92 ±0.1	1.37	2.33	$Hg_{3(\pm1)}$	$LiHg_2$	$LiHg_3$
Na	0.84±0.15	1.68	2.55	$NaHg_{4(\pm2)}$	$NaHg_4$	$NaHg_4$
K	0.79±0.08 (293 K)	1.88	2.71	$KHg_{6(\pm2)}$	KHg_{11}^c	KHg_{9-11}
Rb	0.75±0.08	2.23	2.85	$RbHg_{4(\pm2)}$	$RbHg_{8-11}^c$	$RbHg_{11}$
Cs	0.65±0.1	2.40	3.30	$CsHg_{6(-4,+5)}$	$CsHg_{8-11}^c$	$CsHg_{12-13}$
Mg	0.90±0.1	1.40	2.38	$MgHg_{3(\pm1)}$	$MgHg_2$	$MgHg_2$
Ca	0.64±0.02 (283 K)	1.72	3.34	$CaHg_{8(\pm1)}$	+	$CaHg_{8-10}$
Sr	0.91±0.1 (293 K) ^{a}	1.88	2.33	$SrHg_{2(\pm1)}$	+	$SrHg_{11-13}$
Ba	0.70±0.07(interpol.)	1.98	3.06	$BaHg_{6(\pm2)}$	+	$BaHg_{12}$
La	0.50±0.05	1.65	4.28	$LaHg_{18(-4;+6)}$	+	$LaHg_{6\frac{1}{2}}$
Ce	0.60±0.06	1.61	3.57	$CeHg_{11(\pm3)}$	+	$CeHg_{6\frac{1}{2}}$
Pr	0.6±0.06	1.60	3.57	$PrHg_{11(\pm3)}$	+	$PrHg_{6\frac{1}{2}}$
Nd	0.78±0.08	1.60	2.74	$NdHg_{5(\pm2)}$	+	$NdHg_{6\frac{1}{2}}$
Sm	0.52±0.06	1.58	4.11	$SmHg_{16(-6;+8)}$	+	$SmHg_{6\frac{1}{2}}$
Tb	0.82±0.08	1.56	2.61	$TbHg_{4(\pm1\frac{1}{2})}$	+	$TbHg_4$
U	0.6±0.1	1.35	3.57	$UHg_{10(-4;+9)}$	+	UHg_4
Cr	≈ 2 ^{b}	1.13	1.07	Cr	-	Cr
Mn	0.90±0.08	1.14	2.38	$MnHg_{3(\pm1)}$	+	$MnHg_{2\frac{1}{2}}$
Fe	1.84±0.13 (?)	1.12	1.16	Fe	-	Fe
Co	0.84±0.04 (?)	1.10	2.55	$CoHg_4$ (?)	-	Co
Ni	0.65±0.03	1.09	3.30	$NiHg_{9(\pm1)}$	+	$NiHg_4$
Cu	1.00±0.08	1.12	2.14	$CuHg_{2(\pm1)}$	+	$CuHg_{0.84}$
Ag	1.05±0.03	1.27	2.03	$AgHg_{1\frac{1}{2}(\pm\frac{1}{4})}$	+	$AgHg_{1\frac{1}{4}}$
Au	0.85±0.04	1.26	2.52	$AuHg_{3\frac{1}{2}(\pm\frac{1}{2})}$	+	$AuHg_{\frac{1}{2}}$
Zn	1.67±0.06	1.23	1.28	Zn	-	$ZnHg_{0.33}$
Cd	1.53±0.03	1.37	1.39	Cd	-	(Cd)
Hg	1.60±0.05	1.43	1.34	Hg	-	Hg
Al	1.6±0.2 (interpol.)	1.26	1.34	Al	-	Al
Ga	1.64±0.08	1.33	1.31	Ga	-	Ga
In	1.38±0.1	1.46	1.54	In	$InHg_3$	$InHg_{0.1}$
Tl	1.05±0.05	1.51	2.03	$TlHg_{1\frac{1}{4}(\pm\frac{1}{4})}$	$TlHg_{2\frac{1}{2}}$	$TlHg_{2\frac{1}{2}}$
Ge	1.70±0.15	1.39	1.26	Ge	-	Ge
Sn	1.48±0.04	1.48	1.45	Sn	-	$SnHg_{0.12}$
Pb	1.25±0.04	1.54	1.71	$PbHg_{\frac{1}{4}(\pm0.1)}$	$PbHg_{\frac{1}{2}}$	$PbHg_{\frac{1}{2}}$
Sb	1.40±0.1	1.54	1.52	Sb	-	Sb
Bi	1.35±0.1	1.62	1.58	Bi	-	Bi
Te	1.19±0.3 (385 K)	1.59	2.95	$TeHg_{6(-4;+8)}$	+	TeHg

349

should also be placed on these straight lines. However, such agreement is not so frequently observed as it was in the case of Hg (Fig. 1).

All results in Table 1 are from C. Guminski, J. Mater. Sci., 24 (1989) 2661, except those marked with [a], which are from I.A. Makarova, A.A. Lange, S.P. Bukhman, Izv. Akad. Nauk. Kaz. SSR, Ser. Khim., 1990, no 6, 35, [b] from S. Marczak, private communication, and [c] from W. Lu, A.S. Baranski, J. Electroanal. Chem.,355 (1992) 105.

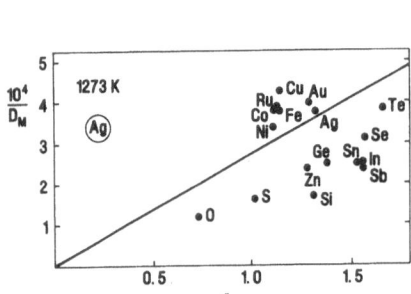

Fig. 2. Diffusion coefficients of metals in liquid Ag at 1273 K; designations as in Fig. 1.

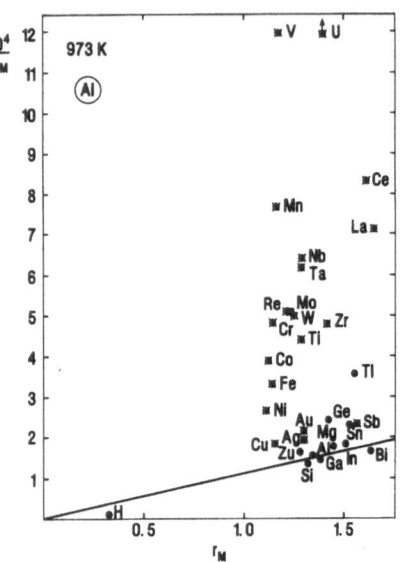

Fig. 3. Diffusion coefficients of metals in liquid Al at 973 K; D_M for Ag, Cr, Mo, Nb, Re, Sb, Sn, Ta, Ti, Tl, V, W, Zn are from Fizicheskaya Khimiya Neorganicheskich Materialov, V.N. Eremenko, Ed., Naukovaya Dumka, Kiev, 1988, p. 66,68; designations as in Fig. 1.

All metals placed below the lines diffuse faster than predicted by the Eq.(5). This fact may be interpreted that a diffusing species may be smaller than a neutral atom. However, it would be against the fundamental physical laws. Positive monoatomic ions have smaller radii than the corresponding atoms. One can imagine an easy ionization of the alkali metal atoms but it is impossible to imagine positively charged ions of N, O, S, Se in a metallic medium. Therefore, it seems probable that the majority of points placed under the lines are simply charged with the convectional overestimation of D_M.

All data points placed above the lines inform us that a diffusing species is larger than an atom; this could be an intermetallic molecule, a stable solvate or a negatively charged metallic ion, since they would have a larger radius. The latter possibility would be against the physico-chemical rules in case of a typical metallic solution. Thus the slower diffusion than predicted is certainly caused by atoms of solute and solvent bounded together, being still present in liquid alloys.

In the case of liquid Ag solvent, in which formation of intermetallics is not observed, all data are spread below as well as above the straight theoretical line. A quite good agreement with the theory is observed in Al solvent; all metals forming intermetallic compounds with Al diffuse slower than the metals which do not form intermetallics. No rational situation is observed in liquid Bi. In liquid Cd, the almost correct behaviour is observed for all except two metals. The most of the data in Cu solvent seems to be disturbed by convection, because the elements showing typical affinity to Cu diffuse faster than those which do not interact with Cu. The situation in molten Fe is not much better.

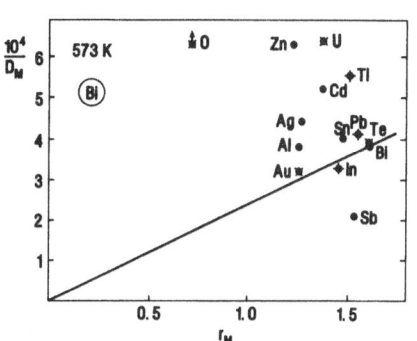

Fig. 4. Diffusion coefficients of metals in liquid Bi at 573 K; DM for O from K. Fitzner, Thermochim. Acta 35 (1980) 277; designations as in Fig. 1.

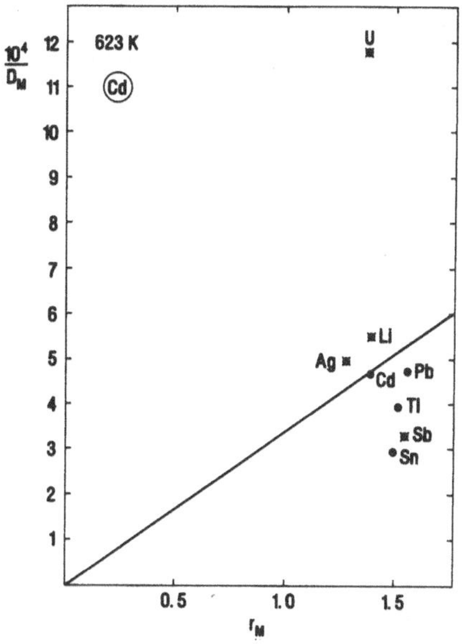

Fig. 5. Diffusion coefficients of metals in liquid Cd at 623 K, designations as in Fig. 1.

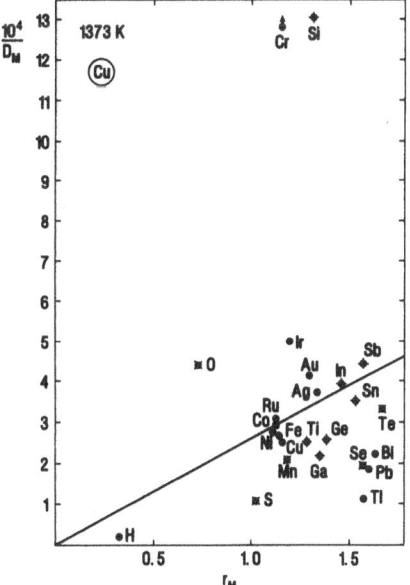

Fig. 6 Diffusion coefficients of metals in liquid Cu at 1373 K; designations as in Fig. 1.

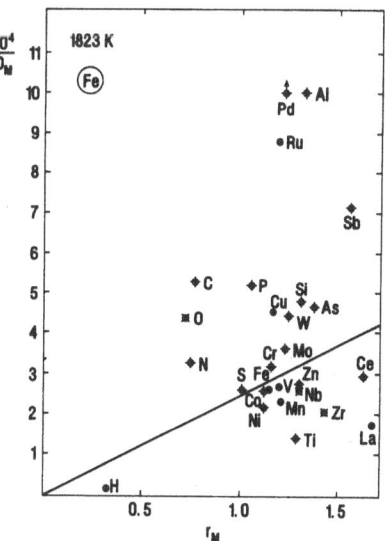

Fig. 7. Diffusion coefficients of metals in liquid Fe at 1823 K; designations as in Fig. 1

351

The diffusion data in liquid Ga show agreement with the proposed concept, however, the points for some elements indicate improbably slow diffusion of the intermetallic molecules with Au, As, Co,O). The diffusion coefficients of some elements in Ge seem to be too fast. The diffusion data in In solvent show a good agreement with the proposed model. The same observation is valid for liquid K and Li. The improbably low value of the diffusion of H but not of T in liquid Li is certainly due to incorrect interpretation of the experiments. Only a partial success is observed for the data in molten Mg. Except an unexplained behaviour of K diffusing in liquid Na, all other data seems to be at proper places.

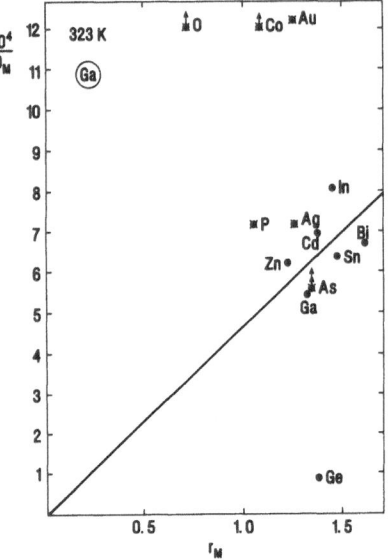

Fig.8: Diffusion coefficients of metals in liquid Ga at 323 K; designations as in Fig. 1.

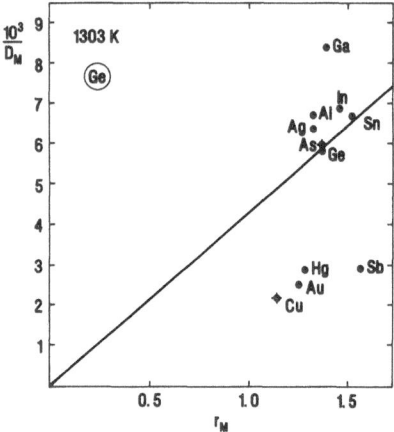

Fig. 9.Diffusion coefficients of metals in liquid Ge at 1303 K; designations as in Fig. 1.

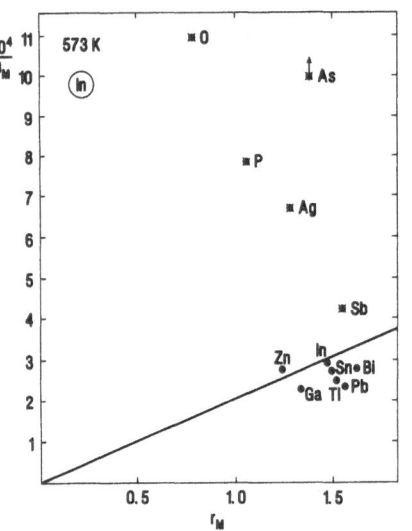

Fig. 10. Diffusion coefficients of metals in liquid In at 573 K; designations as in Fig. 1

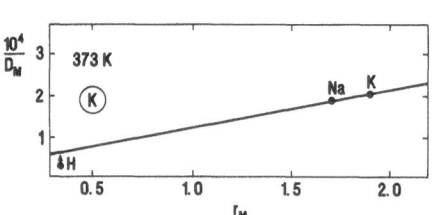

Fig. 11 Diffusion coefficients of metals in liquid K at 373 K; designations as in Fig. 1.

352

Only the diffusion of N would confirm a formation of a N-Ni aggregate in liquid Ni. The majority of diffusion coefficients in liquid Pb qualitatively confirms the concept. It seems that the singular point for diffusion of tritium in the Pb-Li eutectic suggest diffusion of T in form of T-Li molecules the same effect is, however, not observed with H. This system has to be studied further in order to find an explanation. The experimental diffusion data in liquid Si reflect the model positively but the position of Ag in liquid Sb is likely wrong. A more convincing situation is observed in liquid Sn. The excellent situation is reflected for diffusion in liquid Te, unfortunately only partly in liquid Zn.

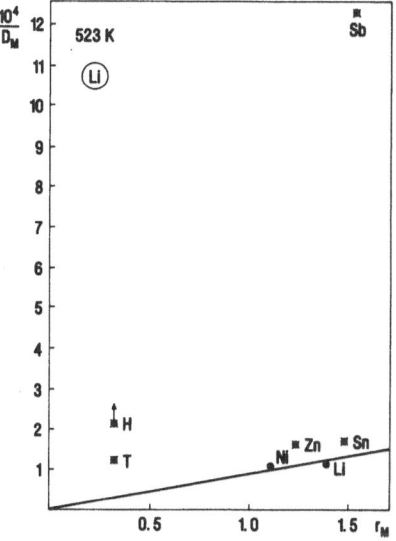

Fig. 12. Diffusion coefficients of metals in liquid Li at 523 K; DM for Ni from G.M. Gryazanov et al, [22]; designations as in Fig. 1.

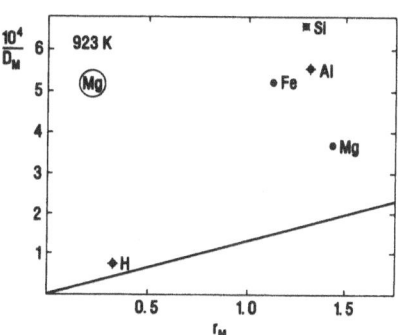

Fig. 13. Diffusion coefficients of metals in liquid Mg at 973 K; designations as in Fig. 1.

7. CONCLUSIONS

Summing up in statistical sense, a prediction of the diffusion coefficient D_M at high dilution using Eq.(5) within an accuracy of 20%, for the metals which do not form intermetallics, results in 60%. 74% of the systems, in which intermetallic compounds are formed in the solid state, have diffusion coefficients lower than it would be predicted by Eq.(5), which is in agreement with the proposed model. Totally, 66% of all investigated systems confirm the applicability of the model, with exclusion of small unbonded atoms C, H, O, N and especially H. The diffusion of these small atoms may occur by a different diffusion mechanism, the applicability of Eq. (5) might be lost, if interstitial spaces are used, fior example. An approximate composition of a diffusing molecule could be calculated from the formula [7]:

$$x = 0.74 \ (r^d_M{}^3 - r_M{}^3)/ \ r_{M'}{}^3 \tag{7}$$

where x is the number of M' atoms surrounding the M atom, r^d_M is radius of diffusing molecule.

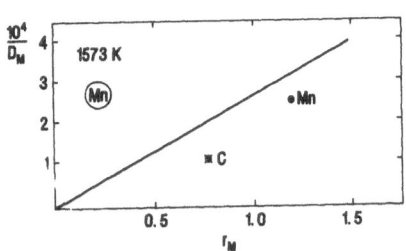

Fig. 14. Diffusion coefficients of metals in liquid Mn at 1573 K; designations as in Fig. 1.

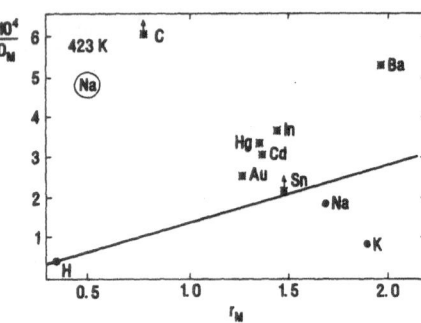

Fig. 15. Diffusion coefficients of metals in liquid Na at 423 K; D_M for H from [23,24], D_M for Xe at 773 K is $1.3 \cdot 10^{-4}$ cm^{2}s^{-1} [25]; designations as in Fig. 1.

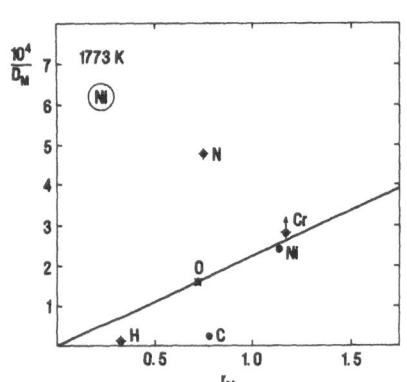

Fig. 16. Diffusion coefficients of metals in liquid Ni at 1773 K; designations as in Fig. 1

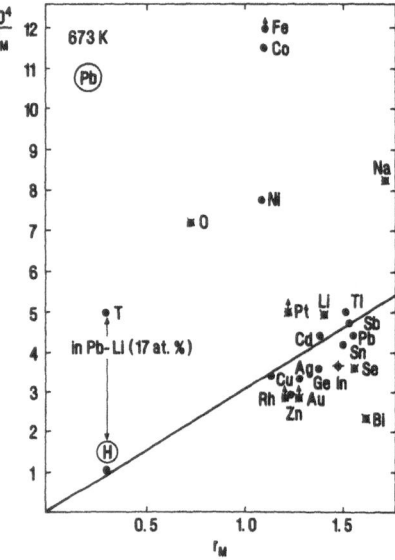

Fig. 17. Diffusion coefficients of metals in liquid Pb at 673 K; designations as in Fig. 1; D_M for H and T in liquid Pb-Li(17 at%) alloy

Only 30% of the analysed systems show the same stoichiometry of diffusing molecule in the liquid as in the solid state. The inclusion of the data in liquid Hg, in which system the model is fulfilled at best, improves the statistics.

Finally, one may withdraw a practical suggestion that an available phase diagram or thermodynamic data point on the formation of intermetallics in the investigated system, it is more probable that the diffusion process would be slower than formulated by the Eq.(5), using the radius of the atom M.

On the basis of Eq.(7) one may calculate the radius of a diffusing molecule taking the stoichiometric coefficient of a molecule of the composition of the solid phase (from the phase diagram), however, one should remember that such a treatment was experimentally confirmed only in 30% of the systems.

Experimental precision is still not sufficient to draw any more precise conclusion in subtle verification of the theories, tests of the isotopic effect for heavier atoms or projection of an alloy structure in the liquid state. Further progress is needed to increase precision of the experimental techniques. Similar scepticism was recently expressed by Iida and Guthrie [21]. Further progress is therefore needed to increase the precision of the experimental techniques which are used for the determination of diffusion coefficients in liquid metals.

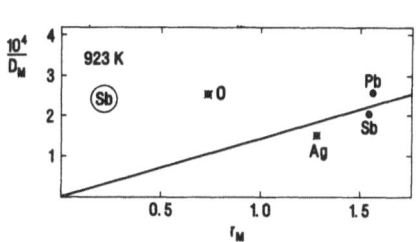

Fig. 18. Diffusion coefficients of metals in liquid Sb at 923 K; D_M for O from [26]; designations as in Fig. 1.

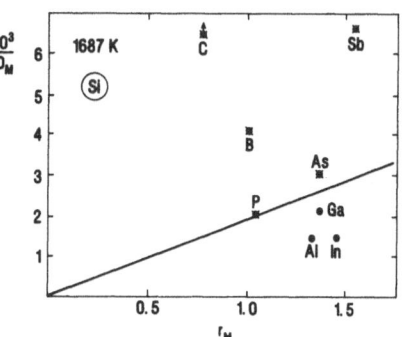

Fig. 19 Diffusion coefficients of metals in liquid Si at 1687 K; designations as in Fig. 1.

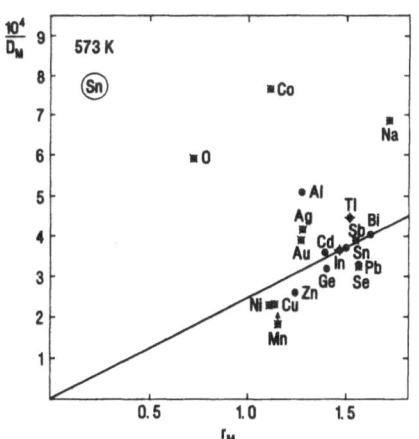

Fig. 20. Diffusion coefficients of metals in liquid Sn at 573 K; designations as in Fig. 1.

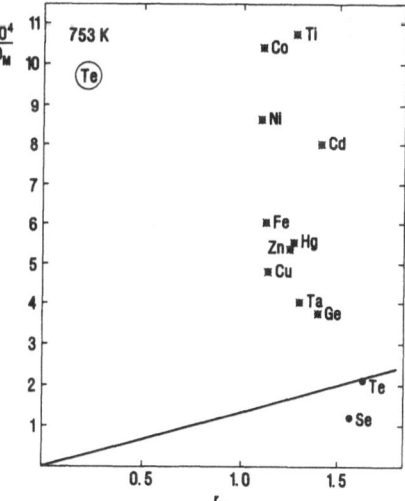

Fig. 21. Diffusion coefficients of metals in liquid Te at 753 K; designations as in Fig. 1.

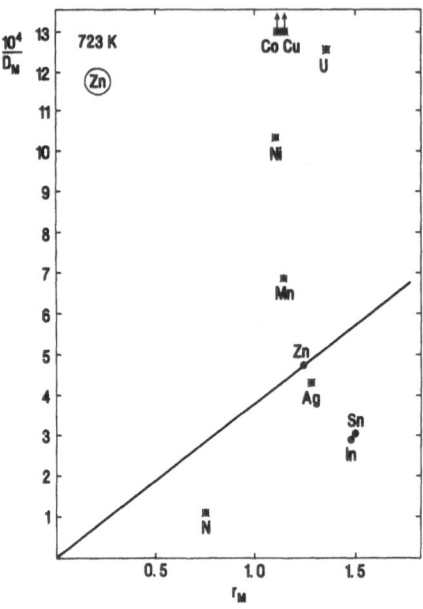

Fig. 22. Diffusion coefficients of metals in liquid Zn at 723 K.

References

1. N.H. Nachtrieb, Ber. Bunsengesel. Phys. Chem., 80 (1976), 678.
2. E.L. Cussler, "Diffusion, Mass Transfer in Fluid Systems", Cambridge Univ. Press, Cambridge, 1984, p. 132.
3. R. Alvarez, J.L. Bueno, Ingen. Quim., 15 (1983), 107.
4. V.M. Glazov, M. Vobst, V.I. Timoshenko, "Metody Issledovaniya Svoistv Zhidkikh Metallov i Poluprovodnikov", Metallurgiya, Moskva, 1989, p. 101.
5. Z. Galus, "Fundamentals of Electroanalysis", Horwood, Chichester, 1993.
6. C. Guminski, Z.Galus, J. Electroanal. Chem., 83 (1977), 139.
7. C. Guminski, J. Mater. Sci., 24 (1989), 2661.
8. Diffusion and Defect Data, 1 - 89 (1967-1992), various pages.
8a. M.Shimoji, T. Itami, Diffusion and Defect Data, 43 (1986), 1.
9. L.F. Epstein, Chem, Engin. Symp. Ser. no 20, 53 (1957), 67.
10. J.B. Edwards, E.E. Hucke, J.J. Martin, Intern. Met. Rev., 13 (1968),1
11. Z. Galus, Pure Appl. Chem., 56 (1984), 635.
12. H.A. Walls, W.H. Upthergrove, Acta Metall., 12 (1964), 461.
13. Y.P. Gupta, Adv. Phys., 16 (1967), 333.
14. T. Eijima, T. Yamamura, J. Phys., Colloque C8, 41 (1980), 345.
15. A.K. Roy, R.P. Chhabra, Metall. Trans., A, 19 (1988), 273.
16. A. Bruson, M. Gerl, Phys. Rev., B, 19 (1979), 6123.
17. H. Schoenert, Z, Phys. Chem., N.F., 119 (1980), 63.
18. H. Vogel, A. Weiss, Ber. Bunsengesel. Phys. Chem., 85 (1981), 539.
19. R.H. Lamoreaux, U.S. Dept. of Ener. Rep. LBL-4995, (1976).
20. E.E. Shpilrayn, V.A. Fomin, S.N. Skovorodko, G.F. Sokol, "Issledovaniye Vyazkosti Zhidkikh Metallov", Nauka, Moskva, (1983).
21. T. Iida, R.I.L. Guthrie, "The Physical Properties of Liquid Metals", Clarendon Press, Oxford 1993, p.199.
22. G.M. Gryazanov, V.A. Evtikhin, L.P. Zavialskii, A.Ya. Kosukhin, I.E. Lyiublinskii, Materialovedenie Zhidkometallichskikh System Termoyadernykh Reaktorov, Energoatomizdat, Moskva 1989, p. 35.
23. J. Trouvé, G. Laplanche, Liquid Metal Engineering and Technology, BNES, London 1984, vol. 1, 375.
24. F.A. Kozlov, Liquid Metal Engineering and Technology, BNES, London 1984, vol. 3, 225.
25. N. Chellew, R. Kessie, W.E. Miller, Trans. Am. Nucl. Soc. 14 (1971)626.
26. K. Fitzner, Z. Metallk. 71 (1981) 178.

BEHAVIOUR OF LITHIUM IN Pb-17Li SYSTEMS

H.Feuerstein, L.Hörner, J.Oschinski and S.Horn

Kernforschungszentrum Karlsruhe GmbH.
P.O. Box 3460, D-76021 Karlsruhe

1. INTRODUCTION

The phase diagram [1] of the lead-lithium system shows a low melting point for the eutectic mixture with 17 at.% lithium (0.68 wt.%). The molten mixture with this composition is proposed as blanket material for fusion reactors. A number of newer publications describe properties of the eutectic [2-4]. The mixture will be used in the liquid state. Generally it is assumed, that there exists one liquid phase. In this, the compound LiPb is dissolved in lead.

Partly gravitational segregation in the molten mixture was reported for static systems or during solidification. Also many experiments started with the eutectic composition and had after some time lower concentrations. Usually, this loss of Li is attributed to oxidation, but this may be not true in all cases. Recently Hubberstey [4] showed that the eutectic mixture contains only 15.7 at.% Li (0.62 wt.%). (Nevertheless we will use the writing Pb-17Li in this paper.)

The formation of a solid phase has to be avoided in a blanket system. The temperature range for blanket operation is discussed between 280 and 420°C [5]. It can be seen from the phase diagram (Fig.1) that the Li concentration must be maintained between 10 and 22 at.%. This is a wide margin for operation of a blanket, but the possibility of segregation should not be ignored.

In this paper, observations and results will be reported, connected with segregation effects in Pb-17Li systems.

2. SAMPLES FOR ANALYSIS

The most simple system is a solid sample for analysis. However, without special precaution, samples taken from a larger solid bulk material will never give reproducible results. How bad the situation may be can be seen in Figure 2. We received 11 kg bars from M,taux Sp,ciaux . There is an axial as well as a smaller radial profile for the Li concentration in the bar, ranging from 15.4 to 20.7 at.%. The average concentration of the bar was, as specified, 17 at.% . Similar effects were observed within all solidified material. Always, the Li concentration is lower in the lower part. The only way for reliable results is to take a sample from a stirred molten mixture, and analyze the whole sample. We therefore take dip samples with small stainless steel crucibles, or suck in Li-17Pb in capillaries of steel or quartz.

Liquid Metal Systems, Edited by H.U. Borgstedt
and G. Frees, Plenum Press, New York, 1995

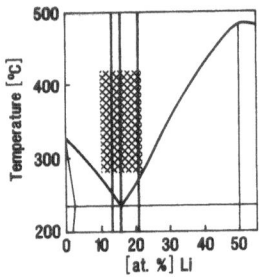

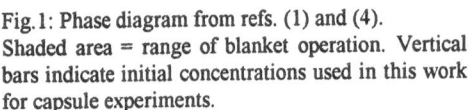

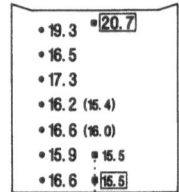

Fig.1: Phase diagram from refs. (1) and (4). Fig.2: Distribution of Li in an 11 kg bar of Pb-17Li,
Shaded area = range of blanket operation. Vertical obtained from Métaux Spéciaux, SA, France
bars indicate initial concentrations used in this work
for capsule experiments.

3. STATIC Pb-17Li

The Pb-17Li in melting pots and the drain tanks of loops is usually static. Therefore tests with the molten mixture under static conditions were performed.

Capsules of 100 mm length were heated 100 to 500 hours at 450°C in an isothermal oven. They were quenched or cooled to room temperature within 2 hours. Fig.3 shows, that segregation occurred only with slow cooling rates. The effect was most remarkable for an initial concentration of 25 at.% , it was negligible for the eutectic with 16 at.% Li. Quenched capsules showed nearly no segregation.

More dramatic was the effect if samples were heated in thermal gradients. Capsules, 430 mm long and 10 or 12 mm inner diameter, containing 340 resp. 450 g Pb-Li, were heated for 700 to 3000 hours in an oven with room temperature at the bottom and 600 or 700°C at the top. We call such capsules 'Sagawa-capsules'; Sagawa used them first for sodium studies [6].

The cooling rate - quenching or cooling within 2 hours to room temperature - had no influence on results. The found Li distribution was clearly formed during the long heating. However, the initial Li-concentration was very important for its final distribution in the capsule.

When starting with about the eutectic concentration (15.5 to 16.5 at% Li), a lead phase was formed in the lower part at temperatures around or below 250°C (Fig.4). In agreement with the phase diagram , this solid lead contained about 2 at.% Li, its melting point was between 300 and 320°C. There was about the eutectic composition in a longer range in the center part of the capsule. The Li concentration in the upper part was as high as 21 at.%.

No lead phase was observed after 900 h when starting with a hyper-eutectic mixture. The eutectic was found only near the melting point of 234°C. About two thirds of the capsule contained a mixture with 17 to 18.5 at.% Li. The concentration rose as high as 28 at.% at the upper part (Fig.5).

In other tests, capsules were heated without a solid phase in a gradient between 350 and 600°C. Initially the concentration was eutectic. The concentration profile after 750 hours is included in Fig.4. It is nearly identical to the profile in capsules with a solid phase.

In a third series of tests, capsules were heated horizontally in thermal gradients. The results were astonishing (Fig.6). When starting with the eutectic mixture, a range with low concentrations of lithium was formed as before. But this low concentration was found in the whole range where the mixture has been solid during heating.

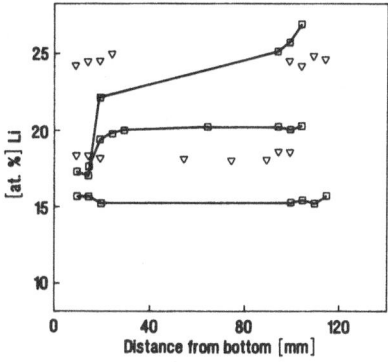

Fig.3: Distribution of Li in capsules, heated isothermally. Solid lines after slow cooling; broken lines after quenching.

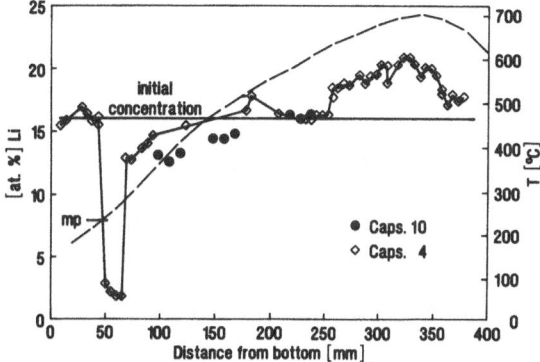

Fig.4. Distribution of Li in vertical capsules with initially eutectic composition. Capsule was heated without a solid phase. Capsule 4: 2110 hours with T_{max} =700 °C; capsule 10: 761 hours with T_{max} =600 °C.

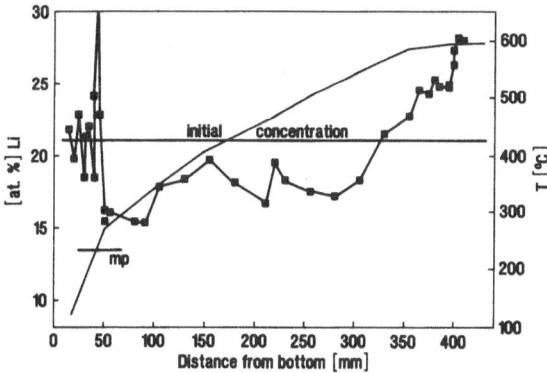

Fig. 5 Distribution of Li in a vertical capsule with initially hyper-eutectic concentration.

However, when starting with a hyper-eutectic mixture, nearly LiPb was formed over the whole solid range. No Li-profile was found within the molten range, the average concentration there was 15.6 at.% for both initial concentrations. A radial profile (gravity influence) was found only near the liquid/solid boundary in case of a hyper-eutectic initial concentration. The concentration was as high as 47 at.% at the lower and only 19 at.% at the upper part. This large difference was found over a cross section of only 12 mm !

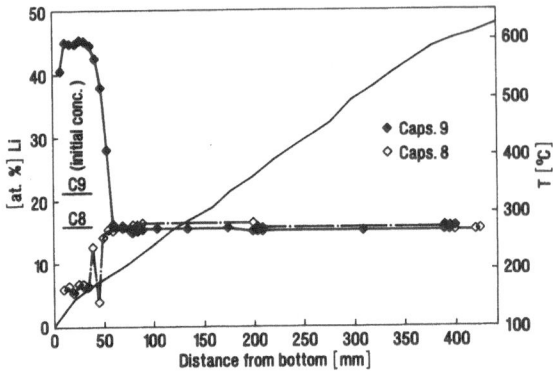

Fig. 6 Distribution of Li in a horizontal capsule with different initial concentrations. Both capsules were heated for 765 hours with Tmax 600 °C.

4. OBSERVATIONS IN A PUMPED SYSTEM

In our laboratory, the pumped loop TRITEX [7] is operated. It is filled with about 100 kg of Pb-17Li, the initial Li-concentration was 16.4 at.%. The loop was operated so far for about 8000 hours at different temperatures, with the cold trap kept between 250 and 270°C. The average Li concentration dropped during this time to 15.7 at.%, meaning a loss of 33 grams lithium.

Many samples were analysed, taken during operation and during dismantling of the loop in shutdown phases. So far only about two grams of the missing Li were found in oxide crusts, and another 2 grams in the cold trap. Here, remarkably, the Li : Pb ratio was nearly one. Unfortunately it was not possible to identify if these deposits were LiPb or oxides (compare [8]).

On the other hand, about 1 gram of small shiny beads were found in the cold trap after experimental phase 4. The diameter was up to 2 mm. The beads consisted of lead, again with the mentioned 2 at.% Li. The relatively large cooler before the cold trap was not analysed so far. Probably more lead will be found there. Smaller amounts of lead beads were found also in other parts of the loop.

5. DISCUSSION

In different experiments, partly segregation of Li and Pb in the molten eutectic mixture was observed during heating, cooling or solidification. Three main parameters influence the severity of the effect:
- the initial concentration of Li,
- the cooling rate during solidification, and
- temperature gradients in the molten mixture.

Kalinin [9] heated static Pb-17Li for 2000 hours at 350°C in an isothermal tube. There was a very long cooling time of about 10 hours. No analytical results were published, but the metallogaphic pictures show a high concentration of alpha-phase (lead) at the bottom. Also, segregation during solidification was reported before. Borgstedt [2] mentioned this effect.

Atanov [10] found values between 15.5 and 18.4 at.% Li in larger ingots. It was explained by "gravitational liquation" during solidification. Overs [11] reported concentrations in bars between 16.3 and 18 at.% Li . Always, the higher concentration is found at the top and the lower at the bottom of a sample. This is clearly an effect of gravity.

We found an even wider range of concentrations in bars : 15.4 to 20.7 at.%. Very important is the cooling rate during solidification. Samples or capsules, heated for a longer time under isothermal conditions, showed no segregation if quenched to room temperature within a few seconds. The slow cooling was probably responsible for the segregation found by Kalinin.

More important is the effect if vertical capsules are heated in thermal gradients. Segregation occurs during this heating. A lead phase is formed at the bottom of the capsules, if the initial concentration of Li was eutectic or sub-eutectic. On the other hand, concentrations up to 28 at.% Li were found at the top when using hyper-eutectic mixtures.

If capsules were heated horizontally in a thermal gradient, lithium diffuses into the solid phase in case of hyper-eutectic mixtures and out of the solid phase in case of eutectic mixtures. The diffusion rate of Li in the solid phase is obviously high. Such diffusion was observed during oxidation experiments even for room temperature [8]. In the liquid phase the mixture with 15.6 at.% Li seems to be very stable. It is found over the whole temperature range independent on the initial concentration. This value is in very good agreement with the new eutectic concentration of 15.7 at.%, given by Hubberstey [4].

Segregation effects are not limited to static systems. Even in the pumped loop TRITEX, a lead phase, as well as deposits with high Li-concentrations were found, mainly in the cold trap. So far it was not possible to identify the intermetallic compound LiPb in these deposits, but the formation of this compound can not be excluded.

Gravity is obviously responsible for the observed Li distribution in vertical capsules, caused by density differences of lead, the eutectic and LiPb (2). Pure lead may freeze out at low temperature, the remaining LiPb float to the top of the capsule. But this effect was not observed with an initial hypo-eutectic composition. That means, the explanation is not sufficient. Not clear at all is the driving force for the observed segregation in case of horizontal capsules.

Even if we do not understand so far all of the observed effects, the impact of this segregation on operation of a blanket has to be considered. For this, the following recommendations can be given :

1. Taking samples for analysis from molten or solid Pb-17Li has to be done carefully. The best way is to take small samples from stirred molten mixtures and analyzing the whole sample. Never keep static molten Pb-17Li in a thermal gradient for a longer time. Stirring of the eutectic in drain tanks should be considered.

2. Never keep static molten Pb-17Li in a thermal gradient for a longer time. Stirring of the eutectic in drain tanks should be considered.

3. To avoid the formation of a lead phase it would be better to use hyper-eutectic mixtures. This was recommended before by Kalinin [9]. However, to avoid the formation of LiPb, the concentration should not be too high.

4. During operation of a blanket, the Li concentration has to be controlled continously - e.g. with a meter. If necessary, Li has to be added.

ACKNOWLEDGEMENT

This work has been performed in the frame of the Nuclear Fusion Project of Kernforschungszentrum Karlsruhe, and was supported by European Communities within the European Fusion Technology Program.

REFERENCES

1. M. Hansen and K. Anderko, "Constitution of Binary Alloys", McGraw Hill Book Co., New York 1958.
2. H.U. Borgstedt, Report KfK 4620, Karlsruhe (1989).
3 .U. Jauch, G. Haase and B. Schulz, KfK 4144, Karlsruhe (1986).
4 .P. Hubberstey, T. Sample and M.G. Barker, J. Nucl. Mater 191-194 (1992) 283.
5. S. Malang, H. Deckers, U. Fischer, H. John, R. Meyder, P. Norajitra, J. Reimann, H. Reiser and K. Rust, Fusion Eng. Des. 14 (1991) 373.
6. N. Sagawa, H. Iba, K. Naitoh and N. Sakurama, J. Nucl. Sci. Technol. 12 (1975) 581.
7. H. Feuerstein, G. Gräbner, and G. Kieser, J. Nucl. Mater 155-157 (1988) 520.
8. H. Feuerstein, L. Hörner, S. Horn, J. Beyer, J. Oschinski, H. Gräbner, P. Welter and S. Bender, Report KfK 4927, Karlsruhe 1992.
9. G.M. Kalinin, B.S. Rodchenkov, A.V. Sidorenkov, Yu.S. Strebkov, A.S. Tanklevsky, A.G. Uchlinov, A.I. Kondyr and V.Ya. Prochorenko, "Fusion Technology 1990", North Holland Publ., Amsterdam 1991, p. 944.
10. A. Atanov, A. Chepovski, V. Gromov, G. Kalinin, V. Markov, V. Rybin, E. Saunin, G. Shatalov, A. Sidorov, Yu. Strebkov, V. Bondarenko, I. Chasovnikov, V. Vinokurov, V. Zemlyankin, V. Prohorenko, A. Kondar and A. Ukhlinov, Fusion Eng. Des. 14 (1991) 213.
11. A. Overs, L. Aufret and H. Houbas, Proc. 4th Intern. Conf. on Liquid Metal Engineering and Technology, Avignon, Oct. 17-23, 1988, Vol. 3, p. 710-1

ELECTROCHEMICAL STUDIES IN THE Na - Hg SYSTEM

H.U. Borgstedt and Z. Peric

Kernforschungszentrum Karlsruhe
Institute of Materials Science
P.O. Box 3640, D-76021 Karlsruhe, Germany

1. INTRODUCTION

Many metallic main group (s and p block) elements form intermetallic compounds with alkali metals or are significantly soluble in molten alkali metals [1]. The metals of the groups 2b, 3a, 4a and 5a are those which are soluble and form stable compounds with sodium. Their oxides and halides are less stable than the corresponding compounds of sodium, the metals can be isolated by means of reduction with liquid sodium. The combination of the reducing and dissolving properties and the wide range of the liquid state of sodium and solutions of metals in the molten alkali metal are the basis for the possibility to use molten sodium for extractive metallurgical processes. The solubility of metals in liquid sodium at 200 °C is shown in Figure 1 [2].

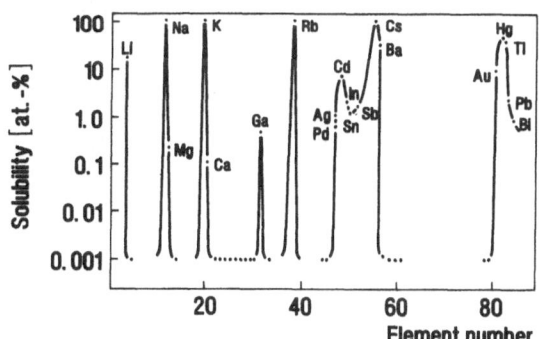

Fig. 1. Solubility of elements in liquid sodium at 200 °C (after C.C. Addision [2])

Such molten alloys containing sodium as solvent and one of the above mentioned metals as solutes have been studied in the past using electrochemical cells. Solid materials as for instance β"-alumina [3] or Na⁺ ions conducting glass [4] were acting as electrolytes. Such methods can be applied to determine solubilities as well as thermodynamic data.

Mercury is one of the metals which are soluble in liquid sodium and completely miscible in the liquid state above 353 °C, the mercury sodium phase diagram, taken from a recently published critical evaluation [5], is shown in Figure 2. Mercury is also widely used as process medium in zinc - mercuric oxide batteries [6]. The anode of this battery contains zinc amalgam, the cathode mercuric oxide, the cell reaction is represented by the equation

$$Zn + HgO \leftrightarrow ZnO + Hg.$$

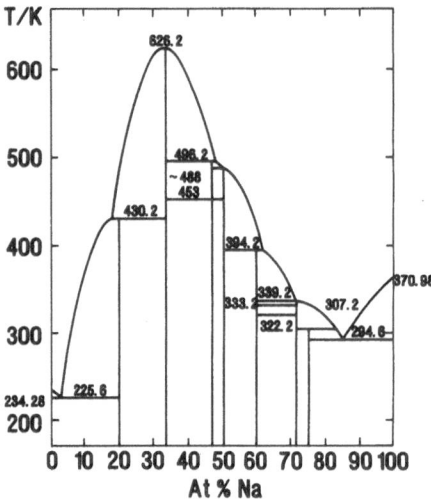

Fig. 2. The Hg - Na phase diagram

The used batteries contain metallic mercury containing small amounts of mercuric oxide and zinc oxide. Thus, mercury can be extracted from wasted batteries of this type, if they might be shredded in the first step of the process. The solution of mercury in sodium can easily be separated from the solid parts of the waste by filtration. A part of the sodium is consumed due to the formation of sodium oxide according to reactions as:

$$HgO + 2\ Na \leftrightarrow Hg + Na_2O$$

$$ZnO + 2\ Na \leftrightarrow Zn + Na_2O$$

The solution of mercury in sodium which is the product of the extractive step of the process has to be separated. The separation of the two metals of this molten alloy can easily be performed by means of an electrolysis through β"-alumina as solid electrolyte [7]. Electrochemical studies were performed in order to establish the data for the separation of the metals. The electrolysis might be more effective than the vacuum distillation, in which the energy consumption is high, and than the reaction with water, by which the sodium is totally consumed. The electrolysis recycles a large part of the process fluid sodium.

2. EXPERIMENTAL METHODS

The experimental set-up was placed inside a dry argon glove box in which the partial pressures of oxygen and moisture were below 0.1 Pa. The electrochemical cell (see Figure 3) consists of a metallic container which is equipped with an electrical heater. A hemispherically closed tube of β"-alumina of 25 mm diameter and 150 mm length is centrally mounted into the container. The remaining volume in the container is filled with sodium metal, the inner side of the β"-alumina tube is filled with mercury - sodium liquid alloy. The two metallic phases are connected to the electrodes. The voltage is supplied by a constant voltage generator.

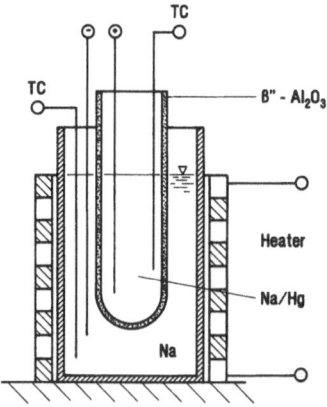

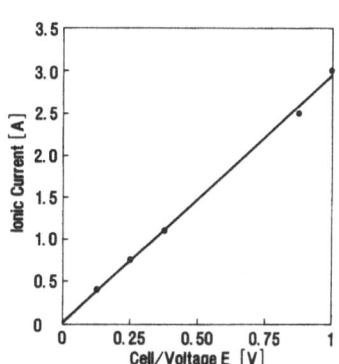

Fig. 3. Schematic of the electrochemical cell Hg-Na / β"-alumina /Na used within the dry Ar glove box

Fig. 4. The cell current as a function of the applied voltage (measured at 250°C)

The set-up could also be used to measure the cell voltage of the cell

Na amalgam / β"-alumina / Na

which is related to the chemical activity of sodium in the amalgam and, thus, to its composition. The chemical activity of Na in the amalgam, a_{Na}, is related to the cell voltage by the Nernst equation

$$\ln a_{Na} = - F E / R T$$

For those measurements, the electrodes were disconnected from the voltage supply and connected to a high impedance digital voltmeter. The temperature in the amalgam was measured using a calibrated Ni/Ni-Cr thermocouple in stainless steel jacket. Thus, the voltage of the cell could be measured as a function of temperature.

3. RESULTS OF EXPERIMENTAL STUDIES

The ionic current passing the solid electrolyte was measured as a function of the applied voltage at 250 °C in a chain with an amalgam containing more than 90 % Na. As is shown in Fig. 4, the current is linearly dependent on the voltage.

The electrolysis tests at 250 °C and a electrolysis voltage of 1.0 V are listed in Table 1. A series of eight experiments was performed in one amalgam, the composition of which was changed during the tests due to the transport of Na^+ ions through the electrolyte. The concentration of Na in the amalgam was calculated from its initial composition using the product of time and current in A·s. Some compositions were additionally checked by means of the determination of the melting point. The effect of the electrolyses on the composition of the amalgam is shown in Fig. 5.

Table 1. Electrolysis tests of Na amalgam with an initial Na content of > 98 at %

No.	I (A)	Time (min.)	Total time (min.)	EMF of the cell (mV)	a_{Na}	conc. of Na (at %)
1	0.52 *	145	145	0.7	0.98458	98.15
2	0.59 *	240	385	0.9	0.98022	97,94
3	1.34	240	625	1.1	0.97588	97.21
4	1.56	300	925	3.7	0.92115	94.37
5	1.06	240	1165	497	$1.62 \cdot 10^{-5}$	74.17
6	0.35	10	1175	491	$1.84 \cdot 10^{-5}$	73.10
7**	0.47	30	1205	555	$4.46 \cdot 10^{-6}$	70.18
8**	0.12	15	1220	746	$6.44 \cdot 10^{-8}$	69.54

* U = 0.5V   ** higher temperature of test, 320 °C in 7, 330 °C in 8

Figure 6 shows the relations between the composition of the amalgam and the cell voltage for two different sets of experiments with different amalgams. The second series was performed up to much lower concentrations of Na in the amalgam. The volume of the amalgam was drastically reduced by the leaching of Na. Thus, the electrode area was smaller due to the fact that the amalgam filled only a small part of the electrolyte tube. This may be the reason for the different relations in this case.

The chemical activity of Na in the amalgam is significantly reduced if the content of Na was lower than about 85 at %. This is shown in Fig. 7 which was obtained from the original results by means of conversion of the cell voltage into the chemical activity of Na, and calculating the contents from the initial composition of the amalgam and the amount of electro transport during the single steps of electrolysis.

4. DISCUSSION AND CONCLUSION

Electrochemical studies of sodium amalgams with the solid ionic conducting ceramic material β"-alumina have clearly shown, that the method has the capability to separate the two constituents by means of electrolysis. The electrolytic cells offer the possibility to determine the chemical activity of Na in the amalgam through the cell voltage. The coulometric titration allows the determination of concentrations of Na as well. The high chemical activity of Na in the amalgams containing more than 92 at % Na indicates only low interaction of the components in these liquid alloys. The significant decrease of the chemical activity of Na at concentrations below 90 at % should be due to interactions which are characterised by the formation of intermetallic phases. Such intermetallics which are the solid phases in equilibrium with the solution reduce the chemical activity of Na in the amalgam even in the liquid state. The findings are in agreement with the phase diagram of the Hg - Na system (see Fig. 2).

The combination of coulometric and thermal analytical methods was used to define some saturation concentrations of Hg in liquid Na. The logarithmic values which were obtained in this study are shown in Fig. 8 as a function of the reciprocal temperature. The regression line of these values results in the equation, valid for temperatures from the eutectic point at 294.6 K up to ~ 400 K:

$$\log (c_{sat} \, Hg \, /at \, \%) = 3.0115 - 551.6 \, (T \, /K)^{-1}$$

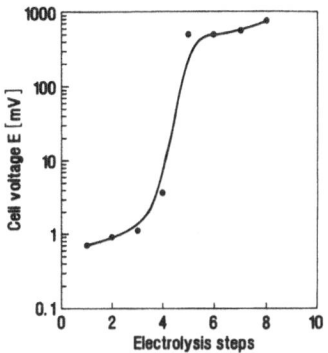

Fig. 5. Changes of the Na content of an amalgam sample due to the steps of electrolysis

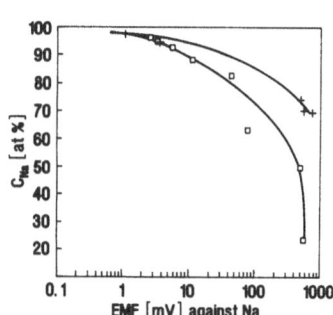

Fig. 6. Relation of cell voltages and compositions of two different initial amalgams

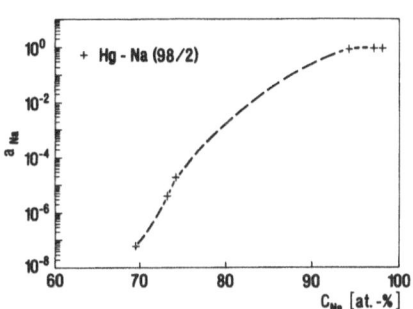

Fig. 7. Chemical activity of Na in amalgams at 250 °C in relation to the Na contents of the amalgams

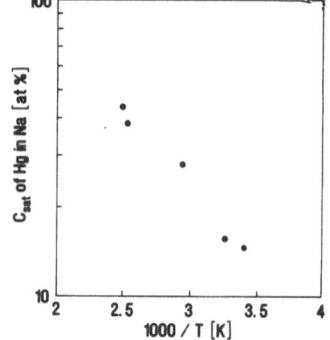

Fig. 8. Solubility of Hg in liquid sodium

Considerable amounts of mercury are soluble in liquid sodium at moderately high temperatures. Thus, the extraction of mercury from solid, dry waste is favourable. The dissolution is an exothermic process which should lead to a complete extraction. The adsorption of small amounts of Na on surfaces of the solid remainder of the batteries causes, however, the contamination of it with a part of the dissolved mercury. The extraction has, therefore, to be optimised to a process step which completely dissolves and removes mercury from the waste. This is necessary since the remaining waste should be free of traces of mercury to be deposed within ordinary deposits.

The technology of separation of mercury from the extraction medium, liquid sodium, can easily be based on the electrochemical process using the β"-alumina as solid electrolyte. Though the complete recovery of sodium from the amalgam needs electrical energy to overcome increasing cell voltages, the electrolysis seems to be a reliable and economic way to get separated the two components of the amalgam. The capability of mercury to dissolve several metals may result in a raw product which contains not only small amounts of sodium, but also silver, zinc, cadmium and other metals which are used in the miniature battery technique.

Acknowledgements

We gratefully acknowledge the chemical analyses contributed by Mrs. Dr. Ch. Adelhelm and the metallographic examinations made by Mrs. B. Bennek-Kammerichs.

References

1. H.U. Borgstedt, C.K. Mathews, "Applied Chemistry of the Alkali Metals", Plenum Press, New York 1987, p. 161.
2. C.C. Addison, B.M. Davies, R.J. Pulham, D.P. Wallace, "The Alkali Metals", The Chem. Soc., London 1967, p.290-308.
3. L. Hsueh, D.N. Bennion, J. Electrochem. Soc. 118 (1971) 1128-1130.
4. M.B. Dergacheva, G.R. Khobdaberganova, Z. Metallk. 79 (1988) 629-632.
5. C. Hirayama, Z. Galus, C. Guminski, "Metals in Mercury", Solubility Data Series, vol. 25, Pergamon Press, Oxford 1986.
6. Ullmann's Encyclopedia of Industrial Chemistry, 5th Edition, Vol. A3, VHC, Weinheim 1985, 363.
7. H.U. Borgstedt, Z. Peric, "Verfahren zur Rückgewinnung von Quecksilber", German Patent no. P 40 34 137, 5 June 1991.

BEHAVIOUR OF BETA ALUMINA SOLID ELECTROLYTES IN AN "AMTEC" ENVIRONMENT

M. Steinbrück, V. Heinzel, F. Huber,
W. Peppler, M. Voss, H. Will

Kernforschungszentrum Karlsruhe GmbH
Institut für Reaktorsicherheit
P.O. Box 3640, D-76021 Karlsruhe, Germany

1. INTRODUCTION

The AMTEC (Alkali Metal Thermo-Electric Converter) or SHE (Sodium Heat Engine) is a device for direct conversion of heat into electricity without the need of any moving parts [1-3]. The AMTEC has been considered for applications like remote terrestrial power generation, industrial co-generation, and space nuclear power generation [4, 5]. The beta-alumina solid electrolyte (BASE) is the crucial component in AMTEC systems. Today it is produced in large quantities for the sodium/sulphur cell, other similar energy storage devices, and sensors. These systems work at much lower temperatures compared with the AMTEC. The principle of AMTEC systems performance has been demonstrated many times; however the long-term reliability of the solid electrolyte and the other components has still to be proved.

2. AMTEC PRINCIPLE

The function of an Alkali Metal Thermo-Electric Converter is based on the isothermal expansion of sodium through an Na-β"-alumina electrolyte, which is permeable to sodium ions and is essentially impermeable to electronic charge carriers (Fig. 1).

The electrolyte separates a sodium high-pressure region (T = 900 ... 1300 K, p_{Na} = $5 \cdot 10^3$ ... $3 \cdot 10^5$ Pa) and a region with sodium vapour at low pressure (10^{-4}... 10^2 Pa) sustained by condensation between 400 and 700 K. By the activity gradient sodium ions pass the BASE from the high-pressure side to the low-pressure side leaving behind their valence electrons. These are conducted by an external electric circuit to the low-pressure side where they recombine with the sodium ions. Thus, electric energy is decoupled from the circuit. The sodium atoms evaporate to the condenser. From there the liquid sodium is pumped back to the high-pressure side where the heat is input.

The AMTEC is a typical low-voltage, high-current system with voltages around 1V and current densities about 0.5 A/cm^2, power densities of about 0.5 W/cm^2 are achievable.

Liquid Metal Systems, Edited by H.U. Borgstedt
and G. Frees, Plenum Press, New York, 1995

AMTEC systems can be used in any applications where heat at temperatures above 1000 K is available, such as solar heat, heat from solid, liquid or gaseous fuels and nuclear heat.

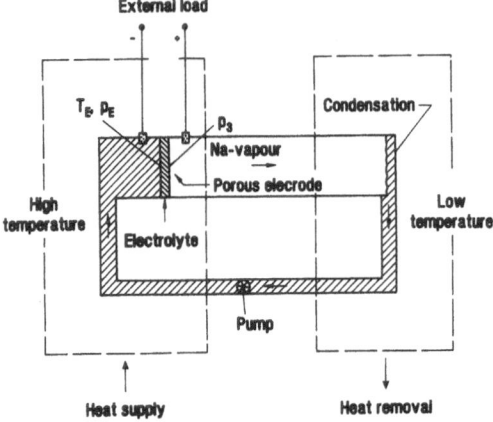

Fig. 1. Scheme of the thermodynamic AMTEC cycle

Due to its modularity AMTEC operates with an efficiency that is independent of size. Low maintenance requirements, silent and vibration-free operation and potentially little capital costs of fully developed AMTEC systems are further advantages which can be anticipated.

3. MATERIALS USED IN AMTEC CELLS

From the material scientist's point of view AMTEC is a very complex system. Ceramics as well as various metals are used for the various components of AMTEC systems. In Table 1 the main components and materials eligible for these components are compiled. A design of a single-tube AMTEC cell is schematically shown in Fig. 2.

Table 1. Main components and materials used for AMTEC systems

Component	Materials
Solid electrolyte	Na-β"-Al$_2$O$_3$ (MgO or Li$_2$O stabilized
Electrode	Mo, TiN$_2$, TiB$_2$, WPt, WRh ...
Current collector and lead	Ni, Cu, Mo
Feed-through	α-Al$_2$O$_3$, MgO
Housing, condenser, radiation shields	special steel, Nb1Zr
Brazing materials	alloys of Ni, Ti, Zr, Nb, Cu ...
Working fluid	Na (impurities)

All these materials are exposed to liquid or vapourised sodium at temperatures up to 1300 K. They have to withstand static and dynamic mechanical loads, e.g., due to differential pressure between the high- and the low-pressure chambers of up to $3 \cdot 10^5$ Pa. Moreover, the components have to survive thermal cycling between room temperature and operational temperature. The sodium ion current densities can be as high as 1 A/cm^2 for the solid electrolyte. Consequently, the electrode, the current collector and the leads have to be designed to have a very low electric resistivity. In addition, material combinations like the electrode on the low pressure side of the solid electrolyte or the joining between the solid electrolyte and the metallic housing have to be selected to be chemically and thermally compatible.

Tests of the metal to ceramic sealing and electrode/current collector design have been discussed elsewhere [6]. In this paper new results on the solid electrolyte are presented.

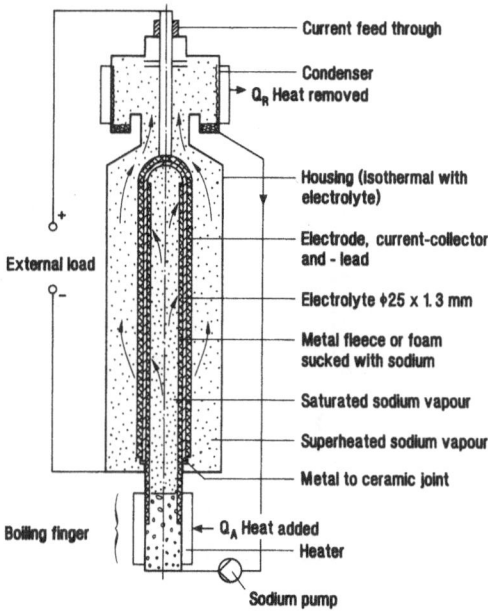

Fig. 2. Scheme of an AMTEC test cell with remote condenser, with the main components included.

4. EXPERIMENTAL RESULTS OF STUDIES ON BETA ALUMINA SOLID ELECTRIOLYTES (BASE)

Na-beta-aluminas are mixed oxides with the composition $Na_2O \cdot n\, Al_2O_3$ (n=5...11). The most important phases in this system are β- and β''-Al_2O_3, the latter showing a higher ionic conductivity. This conductivity of beta-alumina results from its crystal structure which consists of layers in which sodium ions are relatively mobile, separated by non-conductive so-called spinel layers of aluminium and oxygen ions. To stabilise the β-phase small amounts of Li_2O or MgO are added [7]. Today BASE is commercially produced on a large scale, above all for the Na/S-battery which works at temperatures of about 600 K. It has still to be proved that such electrolytes can be operated reliably over tenthousands of hours at high temperatures in AMTEC cells.

We have investigated various commercial BASE materials solely immersed in sodium at high temperatures, but also operated in AMTEC test cells.

4.1 Ionic Conductivity

The electrical resistivity of the BASE is an important property relating to the efficiency of AMTEC devices. Due to the lack of relevant data in the temperature range of interest, we have measured at high temperatures the ionic conductivity of BASE received from four manufacturers. An additional goal of the study was to compare various commercially produced materials. Figure 3 is a schematic of the set-up for these measurements in an Na/β-Al$_2$O$_3$/Na cell. Experimental details together with a discussion of results will be published shortly [8]. Therefore, in Fig. 4 only the summarised results are shown. No significant differences between the four electrolytes have been observed.

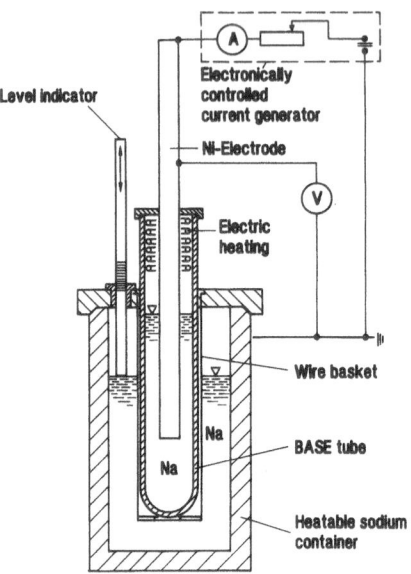

Fig.3. Schematic diagram of a set-up for ionic conductivity measurements.

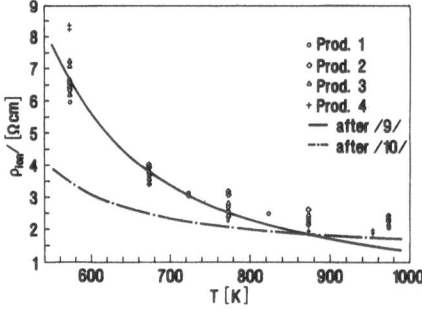

Fig.4. Ionic resistivity vs. temperature of commercial BASE tubes from various manufacturers compared with literature data.

In Fig. 4 two curves from the literature [9, 10] are plotted in addition; one shows better agreement with our results in the lower temperature range, the other in the higher temperature range.

The results indicate that BASE materials presently available obviously have been optimised with respect to the ionic conductivity.

4.2 Electronic Resistivity

A considerable electronic conductivity, σ_e, could cause a decrease in efficiency due to an electric bypass and may play a role in the degradation mechanisms of the solid electrolyte. It is well known that Na-β"-alumina changes colour and becomes grey or black when exposed to sodium. It is thought that blackening is associated with the electronic conductivity of the BASE due to reduction processes and the formation of electron-compensated oxygen vacancies [11]. To simplify the problem, the electronic conductivity is often neglected which has proved to be admissible for the Na/S-cell and in other low temperature applications. No adequate data are available on AMTEC operating temperatures. Only Weber published σ_e-values of BASE up to 1000 K [1].

We have measured the mean electronic conductivities of various BASE materials by the permeation method in an AMTEC related test set-up (Fig. 5).

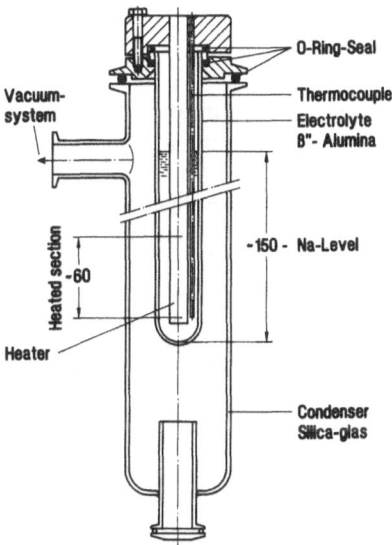

Fig. 5. Experimental set-up for permeation measurements of the electronic conductivity of BASE.

For that purpose, the amount of permeated sodium at constant temperature within a defined time under open-circuit conditions has been analysed and converted to σ_e- and ρ_e-values, respectively. In this context, the "mean electronic conductivity" is used because the permeation data were evaluated with the assumption made that the electronic transport number is constant over the sample thickness. A more detailed analysis of the data is in preparation [12].

In Fig. 6 the results are summarised in an Arrhenius plot. The mean electronic resistivities over the BASE wall thickness are in a range of $4 \cdot 10^4$ to $1 \cdot 10^5$ Ωcm at 1073 K and the activation energies amount to 1.1 eV for all samples. These values correspond to mean electronic transference numbers of te = $(2 \dots 5) \cdot 10^{-5}$.

The results indicate that the electronic conductivity is negligible in terms of efficiency losses, but it has to be considered in connection with the degradation of the electrolyte via the formation of neutral sodium in the polycrystalline material. Moreover, Fig. 6 shows a significantly higher electronic resistivity for Li_2O-stabilised electrolytes compared with MgO-stabilised materials.

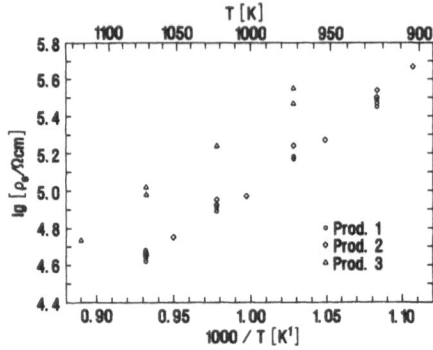

Fig. 6. Mean electronic resistivity of BASE. Prod. 1 and 2: MgO stabilized; Prod. 3: Li$_2$O-stabilized.

4.3 Bending Strength Tests

The bending strength of BASE has been measured at room temperature on a miniature version of a 4-point set-up, which allows a sufficient number of specimens to be measured for statistic evaluation, even from small tube sections.

In a first test series the influence was investigated of solely sodium immersion at high temperatures without electrochemical load on the solid electrolyte [13]. Figure 7 shows the results of tests on virgin samples (A), on samples after exposure to sodium for 1000 hours at 1274 K (B), and after subsequent heating at 770 K in vacuum (C). Compared to virgin BASE tubes, the ultimate bending strength of all samples has considerably decreased after exposure to sodium. However, after heating in vacuum the starting values of bending strength are reached again. It is supposed that the precipitation of neutral sodium within the polycrystalline material leads to internal stresses superimposing the test load. The behaviour observed on BASE samples after AMTEC tests was not the same. Figure 8 shows that there was no significant decrease in bending strength after AMTEC tests. Obviously, the very low sodium pressure on the outside of the electrolyte largely prevents neutral sodium from being formed.

It has to be mentioned that the sodium exposure tests were carried out at 1273 K, whereas the AMTEC tests were performed at only 1073 K.

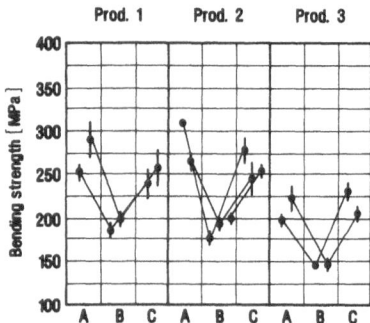

Fig. 7. Bending strength of virgin and sodium exposed BASE samples from several manufacturers. The vertical bars indicate the practical limits of error.

4.4 XRD-and ^{23}Na-NMR Studies

First results of x-ray diffraction and nuclear magnetic resonance spectroscopy studies confirm the precipitation of metallic sodium. X-ray diffractograms of virgin and exposed samples are nearly identical, only little line broadening was observed. This may point to internal stresses due to sodium precipitation. Metallic sodium in the ceramic was detected by ^{23}Na-MAS-NMR. Figure 9 shows two spectra of virgin electrolyte material and of BASE, respectively, after exposure to sodium for 1000 hours at 1273 K. The spectra are similar, except for one peak in the spectrum of the exposed sample with a chemical shift of 1125 ppm which can be attributed to metallic sodium.

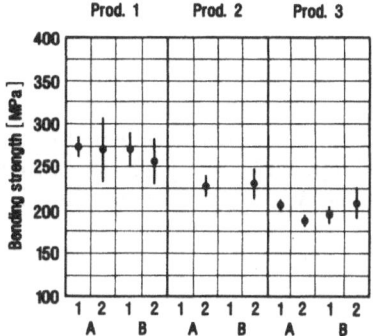

Fig. 8. Bending strength of BASE after AMTEC tests. High- and low-temperature ranges according to the axial temperature gradient.

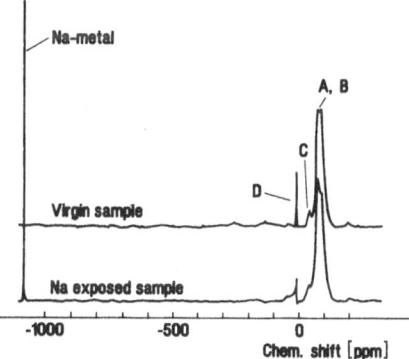

Fig.9. ^{23}Na-MAS-NMR spectra of virgin and sodium-immersed (1000 h, 1273 K) BASEs.

5. CONCLUSIONS

Commercially produced beta-alumina solid electrolytes seem to be optimised in terms of ionic conductivity. Within the range of errors there were no differences in resistivity values of materials received from several manufacturers. The ionic conductivity is high enough to build high-efficiency AMTEC systems.

The electronic transport numbers of the BASE materials have been measured to be in a range where they are negligible in terms of efficiency losses. But the electronic conductivity may play a role in degradation mechanisms via the precipitation of neutral sodium in the ceramic. The electronic conductivity of MgO-stabilised BASE is significantly higher than that of Li_2O-stabilised BASE.

The results of bending strength tests suggest a higher degree of sodium precipitation in the electrolyte material after exposure to liquid sodium compared with specimens examined after AMTEC tests. This may be of interest in stand-by operation of AMTEC systems. First NMR studies confirm the precipitation of neutral sodium in electrolytes exposed to sodium. Further investigations based among others on NMR and also on ESR spectroscopy should provide more information about the concentration and localisation of sodium in the BASE ceramics.

Even though up to now no results have been known which definitely rule out the application of $Na-\beta''$-alumina in AMTEC cells, the reliability of the BASE tube must be better understood before production of commercially applicable AMTEC systems is being envisaged.

Acknowledgement

The authors would like to thank Dr. Hartmann (Friedrich-Schiller-University Jena) for the NMR measurements, Dr. Skokan (KfK) for X-ray diffraction studies, and Mrs. Peric and Dr. Borgstedt (both KfK) for assistance in carrying out the sodium immersion experiments.

REFERENCES

1. N. Weber Energy Conversion 14 (1974) 1.
2. T. Cole, Science 221 (1983) 915.
3. F. Huber, V. Heinzel, W. Peppler, H. Will, KfK-Nachrichten 22 (1990) 152.
4. E. Sasakawa, M. Kanzaka, A. Yamada, H. Tsukuda, "Performance of the Terrestrial Power Generation Plant Using the Alkali Metal Thermo-Electric Conversion (AMTEC)", Proc. 27th Intersociety Energy Conversion Engineering Conference, Vol. 3, p. 143 (1992).
5. R. K. Sievers, C. P. Bankston," Radioisotope Powered Alkali Metal Thermoelectric Converter Design for Space Systems", Proc. 23rd Intersociety Energy Conversion Engineering Conference, Vol. 3, p. 159 (1988).
6. V. Heinzel, F. Huber, W. Peppler, H. Will, Key Engineering Materials 59-60 (1991) 381.
7. R. Stevens, J. G. P. Binner, J. Mater. Sci. 19 (1984) 695.
8. M. Steinbrück "Ionic Conductivity of Commercial Beta-Alumina Solid Electrolytes", Solid State Ionics, in preparation.
9. R. Knödler, Asea Brown Boveri, private communication.
10. T. Cole, N. Weber, T. K. Hunt, "Electrical Resistivity of Beta-Alumina Solid Electrolytes from 200 °C to 1000 °C", International Conference on Fast Ion Transport in Solids, Lake Geneva, Wisconsin, 1979, p. 277.
11. M. Barsoum "Degradation of Ceramics in Alkali-Metal Environments", in: Science and Technology of Fast Ionic Conductors, ed. by H. Tuller, M. Balkanski, Plenum Press, New York 1989, p. 241.
12. M. Steinbrück, H. Näfe "Electronic Conductivity of Polycrystalline Na^--Alumina at High Temperatures. II. Permeation Measurements", Solid State Ionics, in preparation.
13. M. Voss, Thesis, Kernforschungszentrum Karlsruhe, 1992.

EXPERIENCE IN OPERATING HEAVY LIQUID METAL MHD TWO-PHASE FLOW SYSTEMS

H. Branover and S. Lesin

Center for MHD Studies
Ben-Gurion University of the Negev, Beer-Sheva, P.O. Box 653
Beer-Sheva, 84105
Israel

1. INTRODUCTION

The physical principle underlying the generation of electricity in conventional generators (or alternators) is the Faraday effect- the creation of a voltage gradient in a copper coil installed in the rotor of the generator (that is forced perpendicular to a magnetic field created in the generator stator). The prime-mover of the generator rotor is usually a turbine of some kind or a reciprocating engine.

Rather than rotating a copper coil through a magnetic field, it is possible to realize the Faraday effect by forcing a conducting fluid (liquid metal) through a magnetic field. Fig.1 illustrates, schematically, one embodiment of this approach. Shown is a direct-current (DC) Faraday type MHD generator in which an electrically conducting fluid flows through a channel perpendicular to a magnetic field, thus creating an induced electric field across the two electrodes (located in two opposing faces of the channel). The resulting MHD generator is characterized by having no moving parts and by the production of DC (rather than alternating current (AC)) electricity.

LMMHD power systems may be divided into two major classes, dependent on how the electrodynamic fluid is driven through the system: those where the expanding gas performs work against the gravitational force field (gravitational systems), and those where the expansion forces act against a resisting Lorentz force in a magnetic field. Most of the present work at Solmecs concentrates on gravitational, natural circulation power plants using a single phase DC Faraday generator and called ETGAR [1,2].

The LMMHD energy conversion principle can be embodied in a large variety of power conversion system concepts representing different thermodynamic cycles, different combinations of working fluid, as well as different system architecture. Fig.2 illustrates the principles of the ETGAR system. The system consists of a vertical loop of LM, a mixer, a separator and an MHD generator. Thermodynamic working fluid (gas or vapor) is injected at a relatively high pressure into the mixer which is at the low elevation of the riser pipe(the left pipe in the scheme). The bubbles flow upwards, driven by buoyancy forces, and drag the liquid metal along.

Liquid Metal Systems, Edited by H.U. Borgstedt
and G. Frees, Plenum Press, New York, 1995

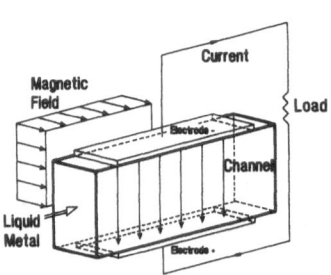

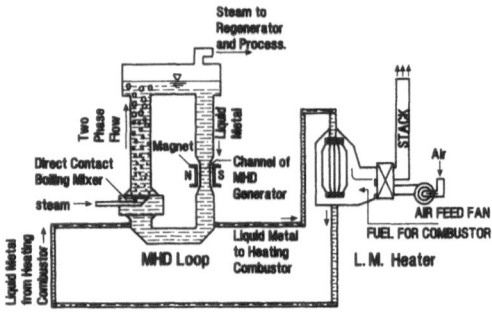

Fig.1. Schematic of the MHD Generator of DC-Faraday type

Fig.2. Schematic of the ETGAR type LMMHD Energy Conversion System

The gas expands nearly isothermally to a low pressure at the separator which is at the high elevation of the riser pipe. In the separator, suitably shaped as a cyclone, gas is driven out of the two-phase mixture. The liquid, now at a high elevation, with a portion of the two-phase flow kinetic energy preserved, enters the downcomer. Here, the potential energy is converted into mechanical energy. At the bottom of the downcomer pipe, the accumulated mechanical energy is converted directly into DC electrical power in the DC Faraday generator. The liquid is then returned to the mixer, where it enters with essentially the same pressure as the gas.

It is clear that for high power extraction, the LM density should be as high as possible, since the pressure differential that exists between the riser and the downcomer (due to the density difference) causes the LM to circulate in the system. From a techno-economical point of view, the most suitable LM for use in the ETGAR system is lead, or lead alloys.

Several advantages of the described system stem from the simplicity of its design and ease of control. The installation costs are much lower than with conventional turbine systems and the system components can be manufactured by any reasonably equipped, modern, mechanical workshop.

The major advantage of the ETGAR concept is related to the fact that the LMMHD energy conversion system (ECS) performs a very special type of thermodynamic cycle. The LMMHD cycles differ from the conventional turbine cycle mainly by the expansion process: in the LMMHD system, the expansion of the thermodynamic working fluid is nearly isothermal. This unique feature of LMMHD power conversion is due to the intimate contact between the thermodynamic working fluid (e.g., steam) and the LM. This contact results in a continuous heat transfer throughout the expansion process, from the LM (which has a very high heat capacity relative to the heat capacity of the vapor), to the vapor or gas. The direct contact heat transfer in the MHD expansion process leads to two important results: (1) it increases the average temperature of energy delivery from the heat source to the cycle, thus increasing the efficiency of the cycle, and (2) the isothermal expansion can be considered to be a process with an infinite number of reheat stages, and the additional heat delivery occurs without the need for extra reheaters.

The LMMHD EC technology appears to be especially attractive for cogeneration applications [3] (production of electricity and heat simultaneously), particularly for industrial cogeneration applications associated with heat sources which can deliver a large fraction of their energy at or above the high temperature of the cycle (constant temperature heat sources, such as solar collectors, fluidized bed combustors, boiling water, nuclear fission reactors, etc.).

The development of the ETGAR technology, which began in 1980 at the Center for MHD studies in the Ben-Gurion University, includes the following R&D activities:

(1) Investigation of physical phenomena;
(2) Development of a universal numerical code for parametric studies, optimization and design of the system;
(3) Development of component engineering;
(4) Building and testing of integrated, experimental, ETGAR type systems;
(5) Material studies;
(6) Economic evaluation of the system and comparison with conventional technologies;
(7) Development of a moderate scale industrial demonstration plant.

Items (1)-(6) have been accomplished while item (7) is in the beginning of progress.

2. MAIN R&D ACTIVITIES OF THE ETGAR PROGRAM

The ability to predict the characteristics of the two-phase liquid metal-steam (or gas) flow in the vertical upcomer pipe of an ETGAR system is of the most critical importance to the design of such systems and particularly to the assessment of its efficiency. Therefore, much attention and resources have been dedicated to the investigation of this phenomena. Numerous empirical correlations suggested by a number of researchers give dramatically different results for the same flows (the characteristics calculated using different correlations differ sometimes by a factor of 10 and even more). In view of that, a very thorough, wide ranging, experimental investigation was one of the first tasks of the ETGAR program. Two characteristics of the two-phase flow in the upcomer pipe are most important, namely, the void fraction and frictional pressure losses. When the void fraction is known, it is easy to calculate the slip ratio (the ratio between the velocities of the gas and the liquid).

Five different facilities have been specially constructed for the investigation of the two-phase flows [4]. The fluid combinations are mercury and steam, mercury and nitrogen, mercury and freon, mercury and steam, lead/bismuth and steam. The pipe diameters vary from 2.7 cm to 20.3 cm. Since the experimental investigation of such flows is a very difficult task and all the known measurement methods are incomplete and not fully reliable, a variety of experimental approaches have been implemented. In most experiments, instantaneous pressure distributions along the height of the upcomer were measured and the average void fraction was calculated numerically using the one-dimensional differential equation for the two-phase flow (assuming frictional pressure losses, according to one of the existing empirical correlations which has been experimentally verified in a separate study and incorporated into the ETGAR program).

The research carried out at the CMHDS led to significant improvements in the characterization of the two-phase flow phenomena expected in the riser of ETGAR systems. One of the most important outcomes is the development of a new empirical correlation which enables the reliable prediction of the velocity ratio between the LM and the steam (slip), the friction factor, as well as of the steam void fraction distribution along the riser [5].

In view of the importance of reducing the size of the bubbles for maintaining a small slip ratio, a number of studies have been undertaken, investigating a means to influence the liquid-gas interface phenomena (surface tension, stability, etc.). These studies have been carried out jointly with the Casali Institute of the Hebrew University, Jerusalem. Two means of influencing the interface have been checked: lowering of surface tension by use of surface active additives and absorption of small solid particles on the interface. The studies convincingly demonstrate the possibility to substantially influence the interface and thus lower the size of the bubbles.

The three key components, the mixer, separator and MHD generator, have been intensively tested regarding their performance characteristics, prediction and optimization of design. Mixers of different design were tested with a variety of working fluids. Two main questions have been addressed: (1) the possibility to influence the size of the bubbles along the height of the upcomer, and (2) minimizing pressure losses in the mixer. It turned out that no substantial differences in bubble size occurred when completely different mixer designs were used (mixers with injection of gas through porous elements, jet mixers, injection of gas in the direction of flow and injection in the opposite direction, etc.). It is possible to say that the two-phase flow has a very short "memory" about the conditions of its inception. The resulting conclusion was that preference should be given to a mixer design which minimizes the possibility of plugging by impurities. The jet mixer (Fig.3) best fits this condition. It should also be noted that the pressure losses in the jet mixer are the lowest, providing that there is no substantial velocity difference between the mixing streams.

The separator of an ETGAR system has to provide a complete separation between the two phases (the carry-under of gas into the downcomer reduces the system efficiency), while simultaneously preserving a maximal possible fraction of kinetic energy. Different types of separators were tested and analyzed. These include designs based either on gravitational forces or centrifugal forces (in which the flow rotates in a stationary structure). New separator designs that offer attractive performance were developed. One design example is shown in Fig.4.

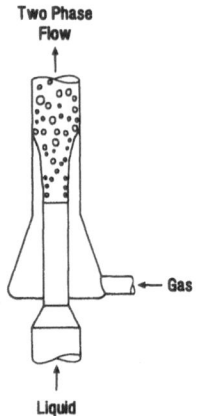

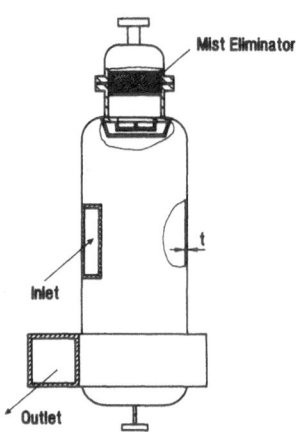

Fig.3. Schematic of the Jet Mixer   Fig.4. ETGAR Centrifugal Separator

A significant portion of the R&D effort was devoted to an experimental investigation of the performance of MHD generators under flow conditions expected in the downcomer of the ETGAR loop, as well as to the development of an analytic model for the prediction of the performance of the MHD generator.

DC Faraday type generators have been tested with mercury and lead/bismuth flows at temperatures up to 170 °C. Power outputs up to 10 kW have been achieved. For the prediction of the performance characteristics of generators with the means for reducing end effects (insulating partitions, high ratio between the length of the generator and the electrode spacing, etc.), the theoretical solution of Sutton et al. [6] has been used, as well as the results of numerical studies by Gershon and Lykoudis [7].

The MHD channel was built from four welded stainless steel plates. The inner surface is coated with enamel to give an electrical insulation. Copper electrodes are mounted to the channel by screws and a Viton O-ring to prevent leakage. The external electrical circuit of the generator (load) is represented by a piece of water-cooled copper pipe. The load resistance can be adjusted by changing the effective length and temperature of the copper pipe.

Fig.5. Photograph of the "ETGAR-3" Generator channel

The MHD channel used in the lead/bismuth system (ETGAR-3) is shown in Fig.5. The comparison of the experimental and theoretical results of MHD generators working within systems with different liquid metals shows very fair agreement [8].

In parallel with the above described R&D work, a computer code was developed [9] to simulate the overall performance of the ETGAR energy conversion system. The computer code system, referred to as ECSAN (Energy Conversion System Analyzer), was designed to be extremely versatile. Each of the systems' components are represented by a detailed analytical (or semi-analytical) model. Every model incorporates all the know-how accumulated from the R&D program.

The data base generated from component R&D programs was thoroughly tested, in an integral way, in the ER-4 (mercury-steam) and ETGAR-3 (lead/bismuth-steam) experimental systems. A very satisfactory agreement was found between the ECSAN code predictions and the experimental measurements of the overall energy conversion efficiency of ER-4 and ETGAR-3.

3. BUILDING AND TESTING OF INTEGRATED EXPERIMENTAL ETGAR SYSTEMS

In addition to a number of small scale, LM experimental facilities, two integrated systems have been designed, built and tested to date. The smaller system, ER-4, works with mercury and steam and generates less than 1 kW of electricity. The larger system, ETGAR-3, works with a lead/bismuth alloy and steam and generates up to 10 kW of electrical power. Both systems operate at temperatures up to 170 °C, and have steam condensation at a relatively high temperature (about 65 °C) to avoid deeper vacuum in the condenser. The design and construction of these facilities involved very extensive searches for optimal engineering solutions and especially for proper choice of materials. The systems were built to accomplish several tasks, including the following:

(1) Verification of performance characteristics predicted by calculations using the ECSAN computer code;
(2) Confirming the ability of an ETGAR facility to work continuously without failures;
(3) Behavior of confinement materials, electrodes, and working fluids under continuous operation conditions;
(4) Investigation of performance stability;
(5) Accumulation of experience in operating and controlling ETGAR systems.

3.1 ER-4 System

The ER-4 system served as an intermediate step in up scaling from "table-top" facilities to the ETGAR-3 pilot plant. This system also provides very convenient conditions for detailed experimental studies on a number of physical phenomena related to liquid metal MHD power facilities, in general, and natural circulation concepts, in particular. The ER-4 system is a relatively small facility with an extremely simple structure. The manufacture of several versions of each component is relatively inexpensive at this size level, and disassembly and assembly require just a few hours. This enabled economical testing and optimization of the design of a number of key components, such as MHD generators, separators, etc. The final design of these components for the ETGAR-3 facility was established on the basis of a critical analysis of ER-4 component tests. The operation of the ER-4 facility can be understood from the piping and instrumentation diagram presented in Fig.6.

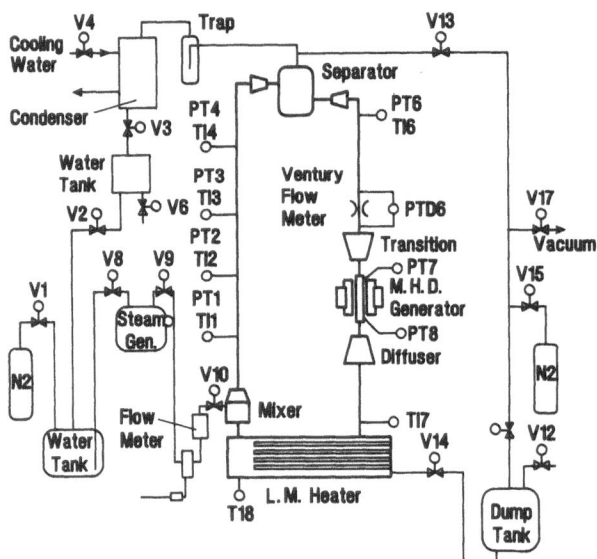

Fig.6. Schematic Diagram of the "ER-4" Experimental System

The ER-4 system consists of two vertical pipes, connected at the bottom and sharing a separator tank at the top. The left pipe is the riser and the right pipe is the downcomer (both are three inches). The system is filled with hot mercury (160 °C) and circulation of this occurs by introducing water steam as the thermodynamic fluid, through the mixer, into the riser. The two-phase fluid flows up the riser, driven by the pressure difference between the two columns of fluid. It is separated at the top and flows down the downcomer as single-phase LM. Electric power is generated in the single-phase LMMHD generator which is mounted in the downcomer. The operation parameters of the ER-4 system are given in Table 1.

The whole system was constructed using regular carbon steel, while the MHD generator channel was made of stainless steel plates. The LM heater (the lower part of the system) consists of four electrical heating elements (2 kW each). The LM flow rate was measured by a ventury flow meter. The pressure distribution along the main LM loop was measured using a regular pressure transducer (adaptable to room temperature) which was mounted remote from the hot mercury in the loop using mercury filled tubes.

Two different separators have been tested. One is gravitational, constructed of a cylindrical, horizontal vessel with an inlet and outlet located at its bottom. The second is a centrifugal unit with a tangential inlet and outlet, and a gas outlet located at the center of its top. The second separator showed much better performance. The facility has a semi-computerized control system (some of the valves are manual).

3.2 ETGAR-3 System

ETGAR-3 has been designed as an integrated power system, to demonstrate the system's unique features and performance potential and to function as a flexible test facility. The system demonstrated the basic engineering information needed to confidently proceed with an up scale industrial demonstration research project. The piping and instrumentation diagram is presented in Fig.7, and the isometric view is given in Fig.8. The main components are as follows: (1) riser, (2) downcomer, (3) separator, (4) LM heater, (5) mixer, (6) MHD generator, and (7) steam boiler. The working fluid (e.g., the liquid metal) is a eutectic lead/bismuth alloy (at 170 °C) and the thermodynamic fluid is water steam.

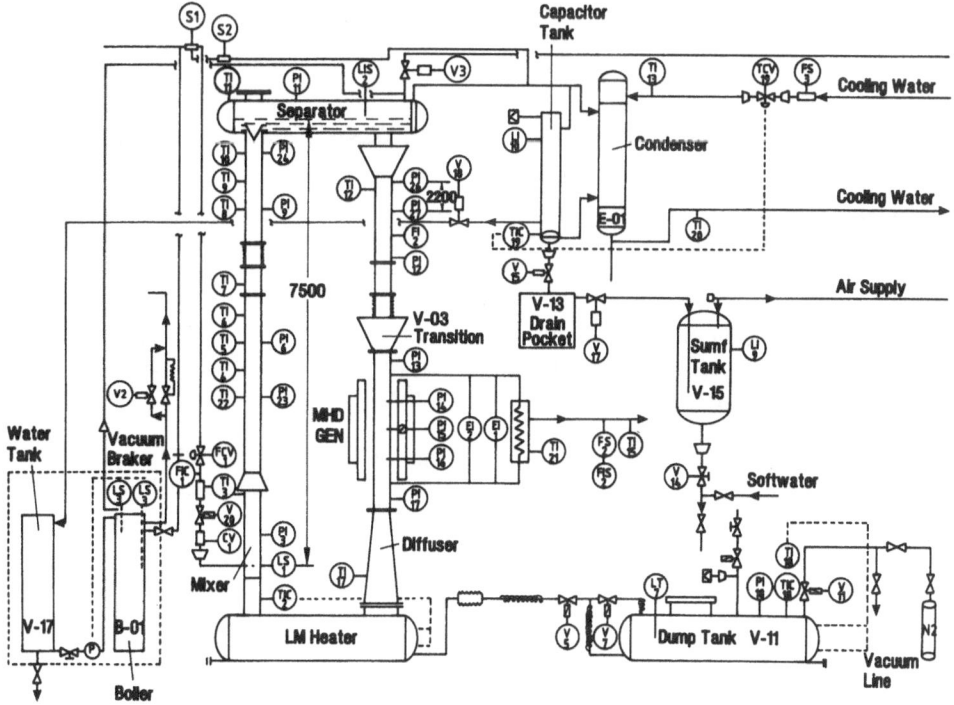

Fig.7. Schematic Diagram of the "ETGAR-3" System

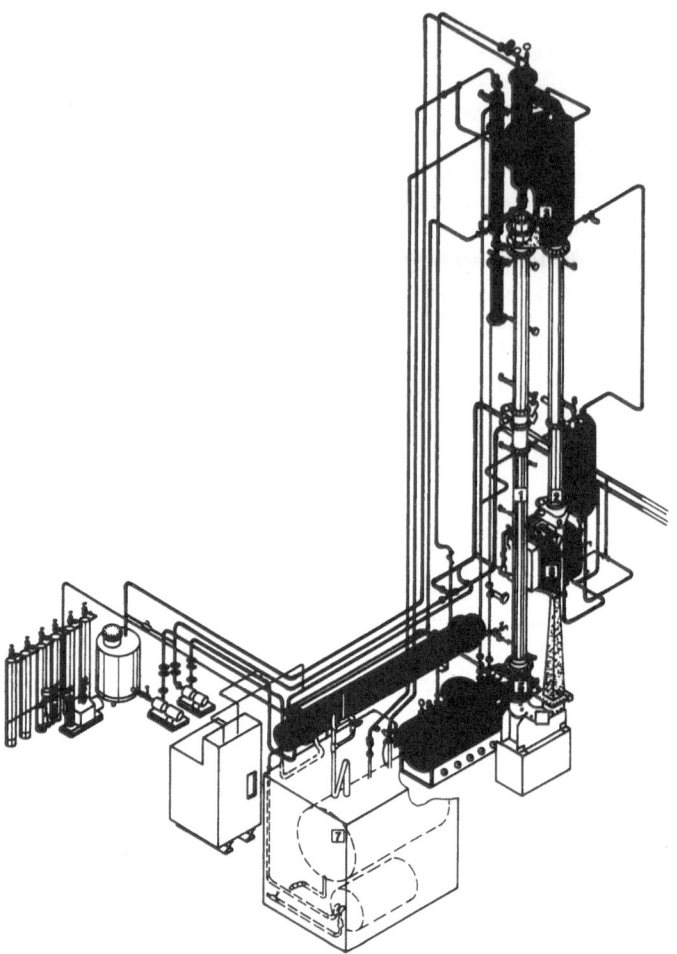

Fig.8. Schematic Isometric View of the "ETGAR-3" System

The diameters of the riser and downcomer pipes, also constructed using regular carbon steel, are both 8 inches. The parameters of the ETGAR-3 system are given in Table 1.The MHD channel was built similarly to the ER-4 MHD channel. ETGAR-3 has been heavily instrumented and fully computerized to facilitate the conduct of a comprehensive test program. The instrumentation was designed to produce data that can be used to evaluate the performance of every major system component, as well as the overall system and the variation of slip ratio in the upcomer as a function of the mixture quality. An overall view of the system is shown in Fig.9.

The ETGAR-3 system has been run for more than 3000 hours (accumulatively) and no major failures have occurred. There was no observable corrosion of the materials, aside from some corrosion of the copper electrodes in the MHD generator channel. The corrosion of the copper electrodes was due to the high solubility of copper in lead/bismuth.

It should be mentioned that the deterioration of the electrodes was not uniform. One electrode was damaged more than the other, and the damage was concentrated in the end zone. This can be seen in Fig.10 (which shows the photograph of one electrode after disassembly). The above are most probably attributable to erosion caused by the strong induced electrical currents. This result led to the conclusion that the electrodes in the real industrial system must be coated with a protective material which will remain stable when in contact with the hot LM. This material must be stable against both chemical corrosion and erosion by other processes.

Fig.9. Photograph of the "ETGAR-3" System

Fig.10. Photograph of one ETGAR-3 Electrode after Disassembly

The measured performance characteristic verified fairly the validity of the ECSAN code and its full relevancy for parametric studies and the design of ETGAR systems.

Table 1 Characteristic Parameters of the Experimental Systems

	ER-4 System Parameters	ETGAR-3 System Parameters
Liquid metal	Mercury	Lead Bismuth
Volatile liquid	Steam	Steam
High temperature in cycle	431.3 K	443 K
Low temperature in cycle	338.6 K	338 K
Mixer pressure	5.34 bar	4.9 bar
Thermal input	7.0 kW	97.5 kW
Average void fraction in upcomer	0.3	0.4
Effective height of the system	5.0 m	7.5 m
Upcomer diameter	0.078 m	0.203 m
Downcomer diameter	0.078 m	0.203 m
MHD generator channel width	0.02 m	0.06 m
MHD generator electrode spacing	0.10 m	0.15 m
MHD generator electrode length	0.35 m	0.525 m
Magnetic field	0 - 0.80 T	0 - 0.80 T

4. MATERIAL STUDIES

As previously mentioned, there were no problems of corrosion of the confinement materials which were observed during the operation of the experimental systems, except some problems regarding the copper electrodes. The latter problem will be more pronounced in future systems, since they are designed to operate at much more elevated temperatures (close to 500 °C). As a result, special studies have been performed in order to find the most efficient protection for the electrodes.

The necessity to provide the MHD generator with highly electroconductive electrodes wetted by liquid lead poses a material engineering problem of special difficulty. Copper appears to be the most promising material for generator electrodes, since copper has a high electroconductivity. However, the low stability of copper in molten lead (the solubility of copper in molten lead at 500 °C is 0.4% [10]) necessitates protecting the copper electrodes, perhaps with a coating formation.

The coating materials must have a high stability in molten lead and an acceptable level of electroconductivity. Tungsten satisfies these requirements. The specific electroresistivity of tungsten is only three times as much as that of copper ($7 \cdot 83 \cdot 10^{-8}$ and $2 \cdot 4 \cdot 10^{-8}$ ohm·m respectively [11]). As to the solubility of tungsten in a molten lead, the only information in the literature is about its negligible value [12].

Two methods of coating have been tested. The first one is a plasma spray tungsten coating on copper samples tested in molten lead at 500 °C. In this case, the coatings were deposited on a tungsten net welded to a copper substrate. The thickness of coatings is approximately 1 mm.

The coatings were tested using the rotating disk method (with a disk 60 mm diameter and 8 mm in height). The coated copper sample was rotated at 2400 rpm in molten lead by a steel shaft. The crucible for melting lead was manufactured from carbon steel. The melt's volume was approximately one litre. Nitrogen (purity of 99.7%) was introduced into the upper part of the crucible. The nitrogen flow rate was approximately 4 1/min. The melt temperature was measured by a chromel-alumel thermocouple which was placed in a pocket welded to the external surface of the crucible. The duration of the experiment was 100 hours.

Different rates of corrosion were observed on specific parts of the sample surfaces. Moreover, lack of reproducibility of the corrosion rates was observed on the various samples. A catastrophic local "corrosion" of copper was observed. The formation of yellow films on the sample surfaces and exfoliation of the coating nearby copper wetting also occurred. A photograph of one sample after 100 hours of testing is shown in Fig.11. Metallographic analysis shows that the tungsten layer thinned to 200-300 μm.

Fig.11. Photograph of a Plasma Spray Tungsten Coating on Copper Sample after 100 hours of testing

It is important to note that the grain boundaries of the tungsten coatings (the dimensions of the grains are approximately 50 μm) are etched and therefore the strength of cohesion between grains is weakened. These effects are characteristic of intercrystalline corrosion.

X-ray investigation of the sample surfaces following testing (X-ray diffractometer, Cu-Kα radiation) have shown that the next phases formed are: $Pb_{12}O_{29}$, $PbWO_4$ and Pb_2WO_5, where the first has the higher concentration. Also, it should be noted that some amount of powder of the lead oxides (PbO, Pb_3O_4 and $Pb_{12}O_{29}$) formed on container walls during testing.

Since the solubility of tungsten in molten lead at 500 °C is negligible, it follows from these experimental results that "corrosion" of tungsten coatings is caused by specific peculiarities in the plasma coating's structure and by hydrodynamic influence.

The second tested tungsten coating was formed on the copper disk using the CVD method. The CVD coating is performed by hydrogen reduction of hexafluorated tungsten. The thickness of the coatings was approximately 110 - 120 μm. Coatings obtained by this method at low temperatures (500 - 700 °C) have densities close to the theoretical values and a high adhesion strength with copper substrates [13].

The structure and the phase composition of the coatings after the test were investigated with optical and scanning electron microscopes, using EPMA method (microscope JEOL JSM-35CF, accelerative voltage 25kV, energy dispersive spectrometer). X-ray phase analysis (difractometer "Philips-14-10", CuKα radiation, graphite monochromator) was also performed.

The thickness and structure, including the grain size, and composition of the coating did not change during the test beyond values allowable within the limits of experimental error. In addition, no lead segregation on grain boundaries were found. A small quantity of lead oxide, PbO, was found on the sample's surface by X-ray analysis. Thus, we found no interaction between CVD tungsten coating with molten lead at 500 °C during the test period of 100 hours.

However, more prolonged tests are necessary in order to estimate the CVD tungsten coating's unsusceptibility to local corrosion damages under such conditions. Moreover, the formation of the tungsten oxides should result in a sharp increase in the electrode's electroresistivity and a considerable decrease in the generator efficiency. As a result, the interaction of the liquid lead and tungsten with the water vapor which is used as thermodynamic working fluid in a LMMHD generator will have to be considered.

Another issue which was studied is the intrinsic chemical stability of the molten lead/steam. Previous work with the ETGAR-3 system had established stability of lead/bismuth alloy with steam up to 200 °C (no dross formed).

The Liquid Metals Handbook [14] informs us that lead is corroded by steam in the presence of air and carbon dioxide, but that the metal is oxidized by water vapor alone only at white heat. This is consistent with the earlier British Treatise on Inorganic Chemistry [15] which classifies metals in terms of their reactivity with water, describing five experimentally observed sub-groups: (1) metals decomposing cold water, e.g. alkali metals such as sodium; (2) metals decomposing hot water, e.g. alkaline earth metals such as magnesium; (3) metals decomposing water at a red heat, e.g. zinc, tin, iron, nickel; (4) metals which decompose water only at a white heat, e.g. lead, copper; and (5) metals which do not decompose water at any temperature, e.g. mercury, silver and gold. In normal chemical parlance, red heat implies relatively low temperatures of 300-600 °C, while white heat, much higher temperatures, around 1000 °C. Consequently, no reaction of steam with lead is to be expected at 480 °C, the upper temperature envisioned for the next planned ETGAR systems.

5. NEXT STAGE OF THE PROJECT

The first moderate commercial-scale demonstration plant to be built in the framework of the ETGAR commercialization program will be the ETGAR-5 plant, designed to produce approximately 1.2 kW of electrical power output. It will be working with lead as the electrodynamic fluid and steam as the thermodynamic fluid.

Negotiations regarding a joint project with the Johns Hopkins University, MA, USA for the construction of the ETGAR-5 system are in advanced stages. The cooperative program includes extensive studies on several problems related improving construction materials and up scaling the ETGAR-type facilities.

The ETGAR-5 system has been designed to verify the performance predictions for an up scaled plant and to demonstrate the applicability of the presently developed LMMHD technology to continuous operation conditions. The continuous operation of the system will also provide the opportunity to acquire all necessary know-how in running and controlling these types of plants.

In the meantime, research and development activities are continuing in order to further optimize technical and commercial figures. One of the studies is related to the direct contact boiling process.

The steam in the ETGAR system will initially be generated by a conventional boiler and then injected through the mixer to the hot lead.

In order to improve construction and operation costs of the LMMHD systems, it is proposed to inject the thermodynamic fluid in its liquid phase (instead of steam) into the hot liquid metal in the "riser" branch of the loop. The boiling of the volatile liquid occurs under direct contact with the liquid metal, and thus avoids the need for an expensive external steam generator. It is also anticipated that the boiling process will lead to extensive mixing of the two-phase flow and hence a decrease of the slip between vapor bubbles and liquid metal. As a consequence, the loop efficiency is expected to increase.

The mixing of a cold volatile fluid and a hot nonvolatile liquid can lead to significant superheating. This is due to the absence of preferred nucleation sites, whereby the volatile fluid remains in a liquid phase far beyond its normal saturation temperature. In this situation, the phase transition can be an explosive type of boiling. Although the explosive mechanical energy released by the vapor explosion process is several orders of magnitude less than that which is typical of chemical explosions, the destructive potential is still high and may result in damaging effects. The evaporative fluxes, fluid acceleration, and departures from thermodynamic equilibrium in vapor explosions are substantially greater than those involved in the classical boiling approach.

In this respect, it is of fundamental importance to be able to predict the pressure shocks which can be generated and plan for means of prevention. Using a tabletop experimental system at the first stage of this study, it was found that the surrounding pressure can significantly change the explosive nature of the direct contact boiling process and even suppress it completely [16].

Thus, we can conclude that, from an engineering point of view, since the liquid water in the ETGAR system will be injected at the higher cycle pressure, the damage from the vapor explosion can be avoided.

The applicability of the current experimental results to a larger scale system using the "real" fluids (e.g., lead and water) has yet to be verified in detail. This will be done in the second phase of the research program - the OFRA system - which is under construction.

The OFRA facility will simulate a real LMMHD loop and is designed to operate with lead and eutectic lead/bismuth, at temperatures up to 480 °C, with the following research targets:

(1) Studies on the direct contact boiling phenomena of a multi-droplet bed of water in lead;

(2) Influence of different additives (surfactants, foaming materials, etc.), dissolved in the lead, on the two-phase flow characteristics, namely, void fraction distribution and phase velocity ratio, flow configuration and overall system performance;

(3) Studies of water reaction with molten lead over a wide range of temperatures, pressures and flow rates;

(4) Testing of the lead reaction with the construction materials in the presence of liquid/vapor water.

The OFRA system consists of two main vertical pipes as shown in Fig. 12. The left pipe (the riser) and the right pipe (the downcomer) share a separator at the top and a liquid metal (LM) heater at the bottom. The system is filled with a hot liquid metal which circulates by introducing the thermodynamic fluid through a mixer into the bottom part of the riser.

There are two general options: injection of steam water generated in an external steam boiler, or injection of preheated liquid water into the mixer. By using the second option, the steam will be created by direct contact with the liquid metal.

The bubbles which are created (directly in the case of steam injection, and by direct contact after boiling in the case of liquid injection) flow upwards, driven by buoyant forces, and in the process drag the liquid metal along.

Although the results on the tabletop system show that the explosive nature of the boiling process can be suppressed by ambient pressure, the experiments in the OFRA system are planned to begin at low working temperatures (150 °C) in order to reduce any potential risks. For this reason, an eutectic lead/bismuth will be used at the first stage. The working temperature will be increased gradually up to approximately 370 °C. At this higher temperature, if the boiling process is regular and no damage is inflicted on the system, the working fluid will be replaced by pure lead. By using lead, the working temperature of the last stage will be increased up to 480 °C, the designed working temperature for the real MHD power system. The influence of lead with special additives on the system's performance will be investigated at the final experimental step.

The metals will be stored in three different tanks, shown in the scheme as dump tanks A, B and C. Following the melting process, the appropriate liquid metal is pushed to the main loop by nitrogen gas.

The construction material for the system's components and pipes is 316L stainless steel. We decided to use high nickel content steel, even though it is not the most suitable material when working with lead at such an elevated temperature (480 °C). We feel that this decision will not have negative consequences since the planned working duration term at 480 °C is

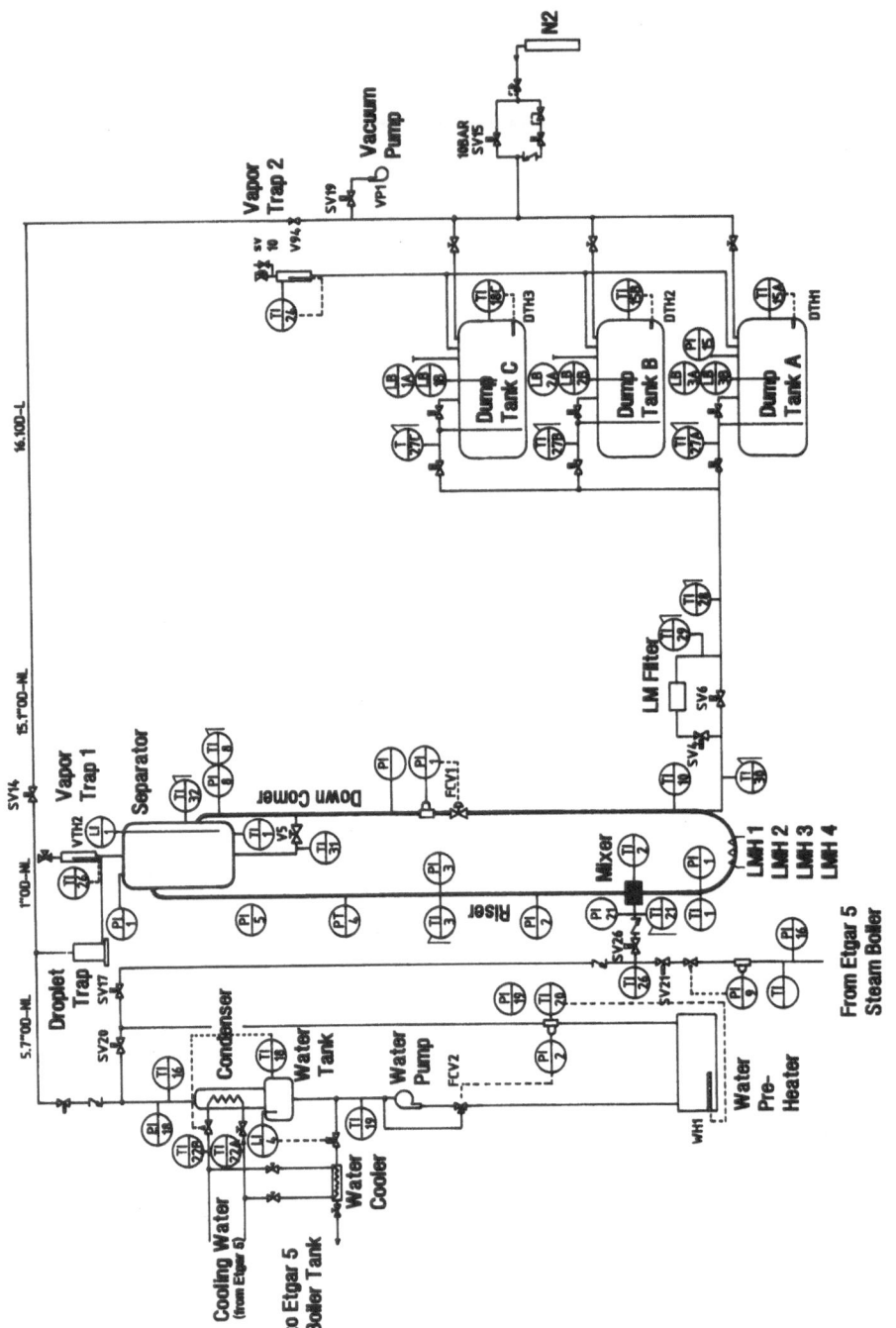

Fig.12. Schematic Diagram of the "OFRA" System

Fig.13. Photograph of the Lower Part of the "OFRA" System

relatively short (a few hundred hours in total). In addition, the ferritic T91 and the martensic HT9, which are the most suitable materials, are difficult to acquire in the small quantities which we need for the OFRA experimental system.

The system is designed to be fully computerized and operated completely automatically. The system is currently in the final stage of construction (Fig.13) and experiments are expected to commence during 1994.

REFERENCES

1. M. Petrick and H. Branover, "Liquid Metal MHD Power Generation - Its Evolution and Status," Proc. Single-and Multiphase Flow in an Electromagnetic Field, Energy Metallurgical and Solar Applications Seminar, Ben Gurion University, p.371, (1984).

2. Branover, H., El-Boher, A., & Lesin, S., "Liquid Metal Magnetohydrodynamics: a new Mode of Power Generation," Proc. 10th International Lead Conference, pp.383-399, Nice, France (1990).

3. Arthur D. Little, Inc. "The United States Cogeneration Market Relative to Solmecs LMMHD Technology," Prepared for Solmecs (Israel) Ltd. by E.J. Cook and D.E. Kleinschmidt, April 1989.

4. Y. Unger, A. El-Boher, S. Lesin and H. Branover, "Two-Phase Liquid Metal-Gas Flows in Vertical Pipes," Proc. of the 9th International Conference on Magnetohydrodynamic Electrical Power Generation, 2, pp. 743-752, Tsukuba, November 1986.

5. A. El-Boher, S. Lesin, Y. Unger and H. Branover, "Experimental Studies of Liquid Metal Two-Phase Flows in Vertical Pipes", Proc. of the 1st World Conference on Experimental Heat Transfer, Dubrovnik, Yugoslavia, Sept. 1988.

6. G.W. Sutton, H. Hurwitz and H. Poritsky, Jr., "Electrical and Pressure Losses in a MHD Channel Due to End Current Loops," Trans. AIEE, Comm. & Electronics, 80, (1962) 687-695.

7. Gershon, P. & Lykoudis, P., "Analytical Study of End Effects in Liquid Metal MHD Generators," Final Report prepared for ANL, Purdue University, Indiana, 1979.

8. Branover, H., Deckel, U., El-Boher, A. & Lesin, S., "Research and Development of Single Phase Liquid Metal MHD generators," Proc. of 27th Symposium Engineering Aspects of Magnetohydrodynamics, Reno, Nevada, pp.9.2.1-9.2.7, 1989.

9. S. Sukoriansky and G. Talmage, "A Computer Package for Analysis of Liquid Metal MHD Power Conversion Systems," Proc. of International Specialist Meeting on Mathematical Modeling of MHD Power Stations, Eindhoven University of Technology, Eindhoven, the Netherlands, April 28-29, 1986.

10. O.J. Cleppa and J.A. Weil, Amer. Chem. Soc., 83 (1951) 4848.

11. Handbook of Chemistry and Physics, 70th ed. CRS Press Inc., 1990.

12. J.R. Week, "Liquidus curves of nineteen dilute binary alloys of bismuth" ASM Trans. Quart., 58 (1965) 302.

13. A.I. Krasovski, R.K. Chuzhko, V.R. Tregulov,and O.A. Balakchovski, *Fluorine method of tungsten preparing*, Nauka, Moscow, 1981.

14. Liquid Metals Handbook, Second Edition Revised, USAEC/Navy, p.133, 1954.

15. J.W. Mellor, Comprehensive Treatise on Inorganic and Theoretical Chemistry, 1, Publ. Longmans, p.493, 1922.

16. Lesin, S., "Direct contact boiling of volatile liquid droplets in two-phase flow natural circulation systems", Ph.D. Thesis (in Hebrew), Ben-Gurion University, Chem. Eng. Dep., 1992.

LARGE SCALE FAUNA EXPERIMENTS ON THE INTERACTION OF SODIUM, CONCRETE, AND STEEL

W. Cherdron and W. Schütz,

Kernforschungszentrum Karlsruhe GmbH
Laboratorium für Aerosolphysik und Filtertechnik I
D-76021 Karlsruhe, Germany

1. INTRODUCTION

With respect to pipe ruptures and leakages in liquid metal fast breeder reactors, it can be assumed that relatively large amounts of liquid sodium will be poured or sprayed into an oxygen-containing atmosphere. Under reactor conditions, the sodium will burn immediately, leading to temperature and pressure rises in the containment, and the strong aerosol release may influence ventilation and filter systems. In addition to these consequences, which are well known, it must be taken into account that the burning sodium pool will also attack mechanical structures like steel and concrete. It is necessary to investigate the consequences of such events. Intense research on sodium fires has been carried out in the past. An overview on the consequences of pool fires has been given in ref. [1]. First results from KfK experiments on the behaviour of a burning pool on a concrete floor were reported in ref. [2]. In this case, we have a sodium-water reaction and possibly a sodium-silicon reaction in addition to the sodium oxygen reaction, and possibly an enrichment of hydrogen in the atmosphere. Since the sodium pool is contained in a steel collar during these experiments, there is also a mechanical, thermal, and chemical interaction of the reaction products with steel. In this paper, we present results from two large-scale experiments (600 litres of sodium on a 1 m^2 concrete area) on the interaction of a burning sodium pool with concrete and steel.

2. EXPERIMENTS

The experiments were performed in the KfK FAUNA vessel which is shown in Fig. 1. The main features of this vessel which has been designed to perform large fire experiments are listed in Table 1.

The principle of the experimental set-up is shown in Fig. 2. The liquid sodium was poured from a storage vessel onto a cylindrical concrete block of 1 m^2 surface area. The concrete was surrounded by a steel cylinder. Another steel cylinder on top of the concrete served as a collar to contain the sodium pool. Since the atmosphere in the vessel was normal air, a pool fire started immediately after the procedure of sodium filling. Temperatures were recorded in the sodium pool, in the concrete block, and in the vessel atmosphere.

Liquid Metal Systems, Edited by H.U. Borgstedt
and G. Frees, Plenum Press, New York, 1995

Table 1. Design features of the reaction vessel

Volume	200 m^3
Overpressure	10 bar
Wall thickness	30 mm
External water cooling system	60 m^3/h
Evacuable	
Automatic gas supply (O_2, CO_2, N_2)	
Energetic reaction inside up to (e.g. sodium fire on 12 m^2)	1 MW
Sodium storage capacity	3000 kg
at	550°C, 3 bar
External sodium supply	1500 kg
at	550°C, 10 bar

Fig. 1. The sodium fire facility FAUNA

The principle of the experimental set-up is shown in Fig. 2. The liquid sodium was poured from a storage vessel onto a cylindrical concrete block of 1 m^2 surface area. The concrete was surrounded by a steel cylinder. Another steel cylinder on top of the concrete served as a collar to contain the sodium pool. Since the atmosphere in the vessel was normal air, a pool fire started immediately after the procedure of sodium filling. Temperatures were recorded in the sodium pool, in the concrete block, and in the vessel atmosphere.

The consumed oxygen was periodically replaces throughout the test. The hydrogen content of the atmosphere was monitored as well as the aerosol concentration and composition.

Two different types of concrete were used, namely shielding concrete (containing small iron spheres), and ordinary construction concrete of SNR-300. In previous tests, we had already observed rather mild reactions with the first type, but violent reactions with the second type. The composition of the two types of concrete is shown in Table 2.

For both experiments the initial conditions were the same:
- 0.6 m^3 of concrete at room temperature
- 500 kg of sodium at 500°C
- normal atmosphere
- vessel gas-tight closed
- shielding concrete (Exp. FB 6)
- construction concrete (Exp. FB 7)

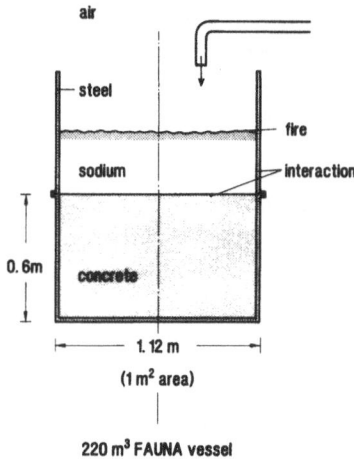

air

steel

fire

sodium

interaction

0.6m

concrete

1.12 m

(1 m² area)

220 m³ FAUNA vessel

Fig. 2. The principle of the experimental set-up inside the FAUNA vessel

3. RESULTS

In the experiment with the shielding concrete, and a ferritic steel collar for the sodium pool (FB 6), we found only a weak reaction which is shown in Fig. 3.

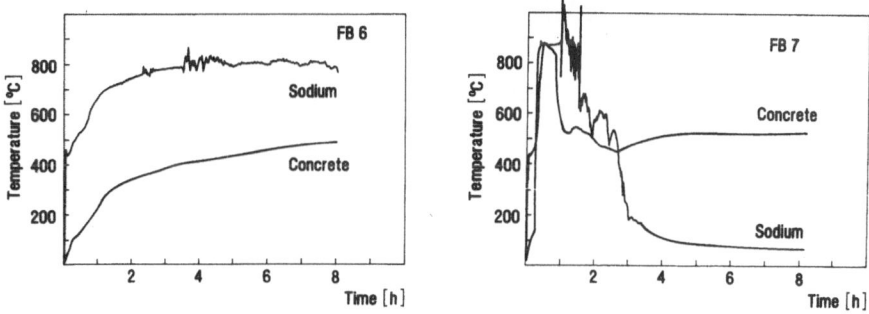

Fig. 3. The course of the temperature inside the sodium and the concrete during the experiments FB6 and FB7

The sodium pool temperature increased slowly from 500°C to 800°C after 4 hours and then remained almost constant. Some temperature fluctuations were observed, probably due to minor silicon interactions. The concrete temperature increased slowly to 500°C. However, the steel collar failed after 8 hours due to interaction with the pool resulting in a sudden spill of sodium (containing reaction products including hydride) into the vessel and a strong increase of the sodium fire. Fig. 4 shows photos of a section of the failed ferritic steel collar and a very interesting rising of the grain size.

In the case of the construction concrete (FB 7), and a stainless steel collar, we found a quite different result as shown in Fig. 3. A violent reaction of sodium with construction concrete was observed, leading to sodium boiling. However, we saw no failure or significant traces of reaction on the stainless steel collar.

Tab.2. The Composition of the two Types of Concrete

CONSTRUCTION CONCRETE	
Blast Furnace Cement	280 kg/m^3
Electrostatic Filter Filler	100 kg/m^3
Sand 0 / 2 mm	726 kg/m^3
Stone Clips (Basaltic) 8 / 11 mm	1014 kg/m^3
Pebble Stones 16 7 32	181 kg/m^3
Water	165 kg/m^3
Density	2450 kg/m^3
Total Water	160 l/m^3
IRON-SERPENTIN CONCRETE (SHIELDING CONCRETE)	
Blast Furnace	300 kg/m^3
Fe-Granulates 0 / 0.85 mm	475 kg/m^3
1 /2 mm	1000 kg/m^3
Serpentine 3 / 5 mm	169 kg/m^3
8 / 12 mm	240 kg/m^3
12 / 20 mm	686 kg/m^3
Magnetide 0 / 8 mm	525 kg/m^3
Water	160 kg/m^3
Density	3550 kg/m^3
Total Water	292 l/m^3

It is recommended that sodium-concrete reactions must be avoided. If there is no steel liner, three different ways of mitigation are suggested:

- Use of a concrete type which is resistant against sodium. However, this type is not yet available; only protective layers based on a concrete composition exist.

- Use of a concrete type which reacts with sodium in such a way that the reaction zone forms, together with reaction products, a layer which stops further penetration of sodium into the concrete.

- Install sloped concrete floors, which direct the liquid sodium into flow channels and catch pans. In that case, a concrete type which shows a mild reaction would be sufficient.

On the other side, continuing research may further improve the mild reaction qualities of a shielding concrete type composition, and finally lead to an almost non-reactive type of concrete.

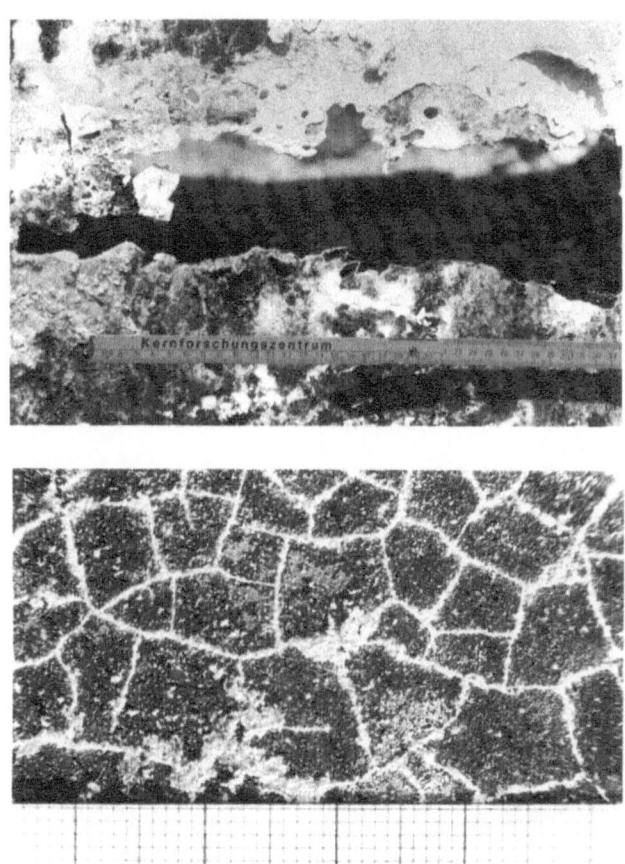

Fig. 4. Corrosion of the ferritic steel and grain size enlargement
during the experiment FB 6

References

1. W. Cherdron, Jahrestagung Kerntechnik 1992, p. 187.
2. W. Cherdron, Jahrestagung Kerntechnik 1989, p. 231.

NACOWA EXPERIMENTS ON RELEASE AND TRANSFER OF CESIUM, ZINC, AND IODINE FROM A SODIUM POOL UNDER EFR CONDITIONS

J. Minges and W. Schütz

Kernforschungszentrum Karlsruhe GmbH.
Laboratorium für Aerosolphysik und Filtertechnik I
D-76021 Karlsruhe, Germany

1. INTRODUCTION

EFR is a common European project for a 1500 MW el fast breeder reactor [1]. According to the first consistent design of 1991, the primary sodium tank will contain a sodium pool of 17 m diameter and 545 °C surface temperature, an argon cover gas of 0.85 m height, and an air-cooled upper closure with the 'low' temperature of 120 °C (simplified sketch see Fig. 1).

Thus, a strong temperature gradient will exist across the cover gas, causing convection, aerosol formation, and related phenomena. More recently, the design has been changed to an upper closure temperature above 300°C by installing a thermal shield. For a better understanding of the conditions during normal reactor operation, and in case of accidental release through a leak in the upper closure, it is necessary to study the aerosol system including heat transfer across the cover gas, mass deposition on cover plate and other structures, and enrichment of volatile fission products, and other radioactive materials in the aerosol. In addition, it is interesting to compare the sodium aerosol behaviour in helium cover gas with the argon case, since fundamental differences occur. The main intent of this paper is to present experimental enrichment factors for cesium, zinc, and iodine for the cover gas system (vapour and aerosol), and for the sodium deposits in the cover plate area.

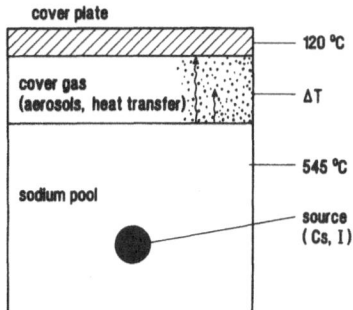

Fig. 1. Simplified sketch to illustrate the main conditions of the NACOWA program, in accordance to the first consistent design of EFR

Liquid Metal Systems, Edited by H.U. Borgstedt
and G. Frees, Plenum Press, New York, 1995

2. EXPERIMENTAL FACILITY

We used the 300 litre NACOWA sodium test vessel (0.6 m diameter, 1.1 m height) which is equipped with an air-cooled cover plate and several installations to characterise the aerosol and to determine temperatures and radiative heat transfer (see Fig. 2). The cover plate has windows for observation of the sodium surface and of cover gas phenomena (Fig. 3). Experimental parameters are: Sodium pool temperature (250 - 550 °C), cover gas height (12 to 114 cm), cover plate temperature (120 - 300 °C), type and amount of admixtures (cesium, iodine, zinc), and type of cover gas (argon or helium).

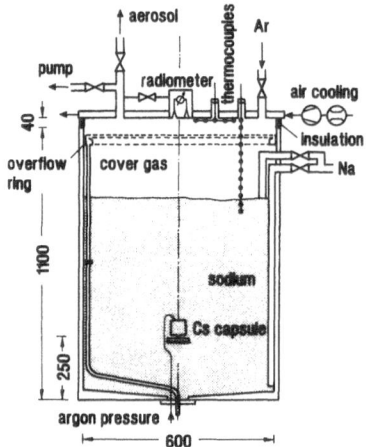

Fig. 2. NACOWA test vessel (dimensions in mm)

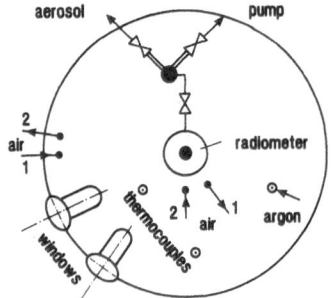

Fig. 3. Top view on the air-cooled cover plate

Cover gas samples are taken by a wash bubbler method as illustrated in Fig. 4. With this arrangement, it is also possible to take pool samples during a test. The collected amount of sodium is chemically analyzed on its content of the relevant species. Only stable isotopes and no tracer materials are used. A second method to collect cover gas samples, especially for particle size measurements, is the use of impactors.

The concentration is determined by chemical analysis of the amount of sodium which is trapped between A and B. The position of A is moveable between plate and pool surface.

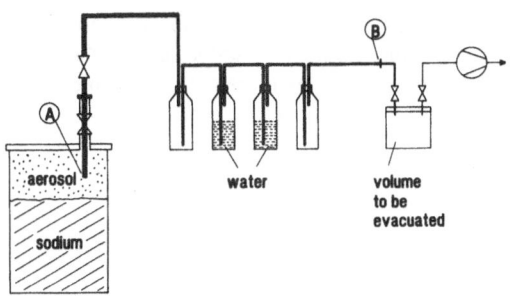

Fig. 4: Wash-bubbler method to determine sodium mass concentrations in the cover gas.

3. ENRICHMENT FACTORS

We assume that a contamination of the sodium pool has occurred, e.g. by a fuel pin failure. By evaporation or other processes, this contamination will be transferred to the cover gas aerosol (or vapour), and to the deposits in the plate area. The enrichment factor EF for species X is defined as:

$$EF = (X / Na)_{sample} : (X / Na)_{pool}$$

where X and Na refer to the relevant masses.

4. RESULTS ON AEROSOLS, HEAT TRANSFER, AND ARGON VERSUS HELIUM

The difference in density between hot argon gas directly above the pool surface and cooler gas below the cover plate causes a natural convective gas circulation between pool surface and cover plate. Sodium vapour, after evaporation from the pool, is transported by the convective gas stream from hot into colder areas, where it forms sodium aerosols due to supersaturation. Evaporation and convection increase strongly with rising pool temperature. Mass concentrations of airborne sodium (aerosol plus vapour) in the cover gas as a function of the pool temperature were determined from wash-bubbler and impactor measurements. The onset of aerosol formation is around 350°C, and the partition between aerosols and vapour appears to be very sensitive to the roof temperature. At 500°C and no roof insulation (120°C roof temperature), the pool surface is barely visible due to strong aerosol formation. With a thermally insulated roof (approx. 250°C), the visibility is much better, although the airborne amount of sodium is similar. In general, we observe a steep rise of airborne sodium with increasing pool temperature, reaching concentration values up to 36 g/m^3 at reactor conditions (545°C) but not yet the onset of a plateau (see Fig. 5).

Concerning aerosol particle sizes, we find 50% aerodynamic mass median diameters which are relatively large, approaching values near 8 μm (see Fig. 6).

Radiative heat transfer data were gained using a windowless thermo-electronic radio-meter. At 500°C, we have values of the order of 1 kW/m^2 at the cover plate. Sodium aerosol droplets act mainly as reflectors due to their low emissivity. An estimate of the aerosol effect on the radiative heat transfer suggests that the aerosol causes a small enhancement which is partly due to reflection of radiation off the side walls. Total heat transfer data were gained from temperature difference and flux of the cooling air. These values exceed the radiative part significantly. With helium as cover gas, we observe almost no aerosol formation. Obviously, we have mainly sodium vapour and stable layers. Since a mixture of helium and sodium vapour has a much higher molecular weight than pure helium, natural convection (and, thus, transport into cooler regions) is suppressed.

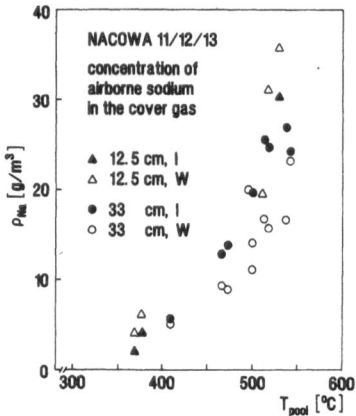

Fig. 5. Sodium mass concentrations, determined from wash-bubbler and impactor measurements during tests N11, N12, and N13.

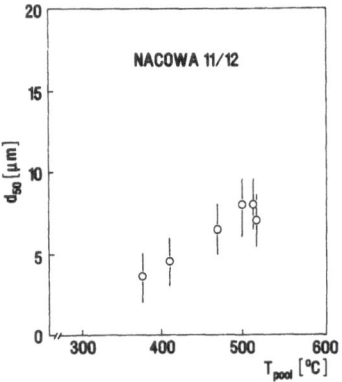

Fig. 6. Mean values (50% of mass distribution) of size spectra versus pool temperature from test N12.

5. RESULTS ON ENRICHMENT OF CESIUM AND ZINC

The vapour pressure of cesium, sodium, and zinc versus temperature is shown in Fig. 7, and the ratios Cs/Na, and Zn/Na in Fig. 8. According to the vapour pressure ratios, we expect enrichment factors of the order of 20 for cesium, and of the order of 0.3 for zinc at reactor conditions.

Several tests were performed with cesium sources (1 g or 5 g) inside the sodium pool to simulate cesium release after a fuel element failure. Enrichment factors of the order of 15 are found in the aerosol, in accordance to the Cs/Na vapour pressure ratio (see Fig. 9). However, much larger factors may be found for the cover plate deposits, especially on cold spots (Fig. 10). These results will be published in more detail in ref. [2].

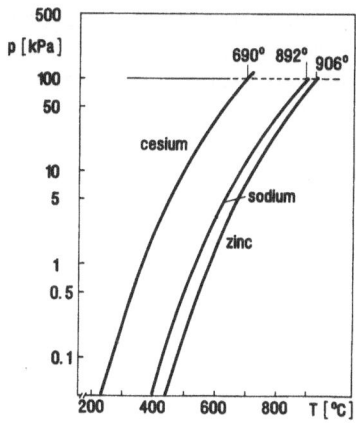

Fig. 7. Vapour pressure of cesium, sodium, and zinc.

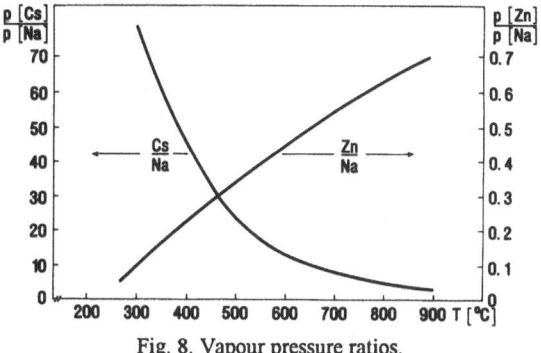

Fig. 8. Vapour pressure ratios.

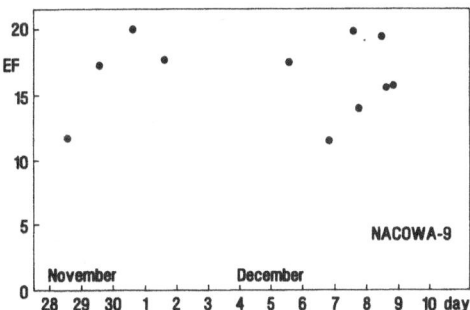

Fig. 9. Cesium enrichment factors from cover gas samples, taken during test N9.

Zinc is not a fission product but radioactive ^{65}Zn may play a role after interaction and subsequent neutron irradiation of sodium with alloy which has a Zn-containing protective oxide layer. Zinc levels have been detected, for example, in sodium samples of the KNK reactor [3]. We have performed one test with 50 g zinc powder mixed with 245 kg of sodium. A sodium pool sample which was taken from the pool surface region had a zinc concentration of $2.28 \cdot 10^{-4}$ which is 11% more than the theoretical value. So, we can assume an almost homogeneous mixture. We found zinc enrichment factors of the order of 20 in the aerosol, significantly above the vapour pressure (Tab. 1). Similar values are found for the deposits (Fig. 11). It is possible to explain these large numbers from thermodynamic considerations of the sodium-zinc-system. Such considerations were performed by Thorley et al. [4] for the conditions of the PFR reactor, and the authors report that enrichment factors of the order of 100 are possible.

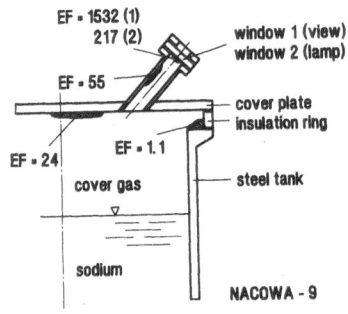

Fig. 10. Cesium enrichment factors from deposits on cover plate and upper rim of test vessel, test N9.

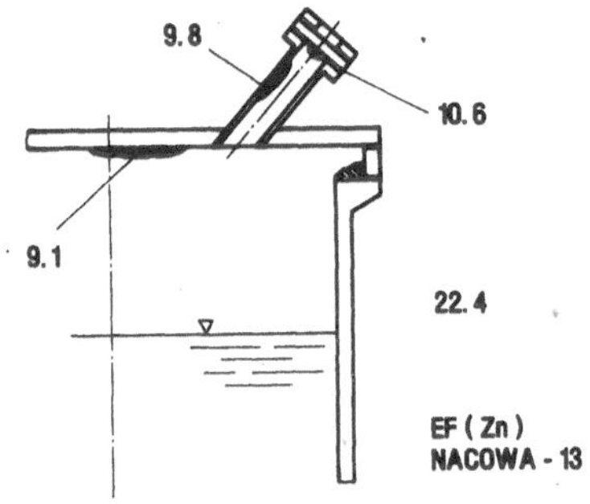

Fig. 11. Zinc enrichment factors from deposits on cover plate and upper rim of test vessel, test N13

6. RESULTS ON ENRICHMENT OF IODINE

In our most recent experiment, we have studied the enrichment of iodine, using an 0.59 g iodine source in a 245 kg sodium pool. We measured enrichment factors exceeding 700 in the cover gas but decreasing with time (see Fig. 12),

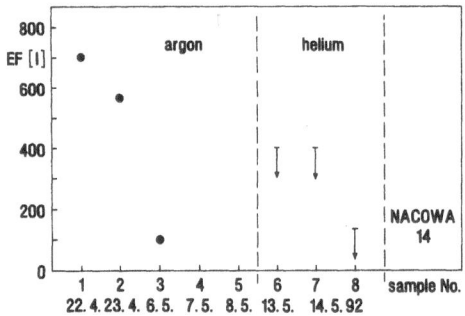

Fig. 12. Iodine enrichment factors from cover gas samples, taken during test N14.

and exceeding 800 in the deposits (see Fig. 13)
Almost 20% of the iodine inventory was found in the plate deposits at the end of the test. Iodine tests will continue in the near future.

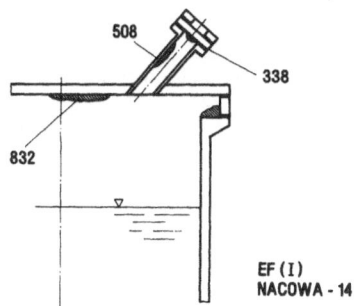

Fig. 13. Iodine enrichment factors from deposits on cover plate and upper rim of test vessel, test N 14

7. CONCLUSIONS FROM NACOWA EXPERIMENTS

Under conditions with low plate temperature, we find :
- sodium mass concentration in argon cover gas up to 36 g/m^3 average particle diameter near 8 µm.
- cesium enrichment factors near 15, possibility much higher in cover plate deposits
- iodine enrichment factors up to 800
- zinc enrichment factors near 20, much larger than vapour pressure ratio
- radiative heat transfer from pool to plate near 1 kW/m^2
- argon cover gas: dense aerosol
- helium cover gas: almost no aerosol but vapour
- temperature profiles indicate convection for argon and stable layers for helium.

Table 1. Enrichment for zinc from wash-bubbler samples during NACOWA - 13

sample	T_{pool} [°C]	T_{plate} [°C]	cover gas height [cm]	cover gas	sampling point: distance to roof [cm]	enrichment factor EF(Zn)
13/1	512	179	12,5	Ar	6	20,1
13/2	516	179	12,5	Ar	6	12,2
13/4	538	156	33	Ar	16	6,4
13/5	529	192	12,5	He	6	32,2
13/7	525	192	12,5	He	6	38,8
13/8	545	154	33	He	16	18,9
13/10	545	148	33	He	16	16,9
13/11	545	148	33	He	3	22,9
13/12	537	189	53	He	21	10,4
13/13	540	180	53	He	21	11,0
13/15	482	177	12,5	He	6	19,0
13/18	490	138	33	He	16	23,8
13/19	490	138	33	He	0	16,6
13/21	481	171	12,5	He	0	13,2

References

1. EFR Consistent Design-Justification of main design features. A000/0/168 A (Dec.1991).
2. J. Minges, W. Schütz; Nucl. Eng. Design 147 (1993) 17-22.
3. Ch. Adelhelm, H.U. Borgstedt; Nachuntersuchung natriumseitiger Dampferzeugerrohrabschnitte aus dem KNK II. (1984, unpublished).
4. A.W. Thorley, A. Blundell, and R. Lloyd; Chemical behaviour of zinc in cover gas environments. IAEA Specialists Meeting on LMFBR Cover Gas Purification, Richland/Washington, Sept. 1986, IWGFR / 61 p. 179-189.

THE DISASSEMBLY AND SAFE DISPOSAL OF
ALKALI-METAL SYSTEMS

Norbert Schwarz

Bereich Engineering, Hauptabteilung Energie- und Anlagentechnik
Österreichisches Forschungszentrum Seibersdorf Ges.m.b.H.
A-2444 Seibersdorf, Austria

1. INTRODUCTION

Starting in 1968 a continuous liquid metal research program was conducted at the Austrian Research Centre Seibersdorf for approximately 20 years. In the beginning fast reactor research with respect to material testing and safety requirements in sodium were the main topics. Later on emphasis shifted to the investigation of topping cycles with potassium as the working fluid. After the cancellation of international co-operation, research activities went on for some years, but in 1989 the decision was taken to end all liquid metal research to give room for new projects and activities. As a result of this decision the following test facilities had to be dismantled: sodium: 2 test facilities with a total inventory of 400 l sodium; potassium: 2 test facilities and 7 smaller experimental rigs with a total inventory of 800 l potassium.

The operation of all systems had been without severe problems because of a high motivation of the operation crew and a high standard of instrumentation. This safe operation record was accomplished by conducting regular safety experiments to simulate different accident conditions. These experiments were completely documented and the information was distributed in internal reports [1-3].

2. DISASSEMBLY OF SODIUM SYSTEMS

The first concentrated effort to remove a test facility was done with a sodium natural convection loop. At first the sodium was dumped into the dump tank and removed from the rest of the system. It was then decided to dissolve the sodium which was still expected to adhere to different internal parts of the components with methyl alcohol. This was done by pumping the alcohol through the loop and controlling the hydrogen emitted at the exit of the expansion tank. Although the system was circulated until no hydrogen could be detected, it turned out that there remained still considerable amounts of sodium in parts of the system which gave rise to unexpected explosions during the following cutting and water-cleaning process of different components. The second great sodium system was therefore drained and the whole system cut into small pieces, so that they could be thrown into a prepared water basin as long as the adhering sodium residues or plugs were not too great, or the weight of the steel components heavy enough. A water depth of 1 m proved to be an efficient explosion dampener and aerosol filter. Nevertheless, one has to be very careful not to throw components into the water which do not sink immediately. These components were cleaned with spray water from fire hoses applied from a considerable distance. For this procedure a place is needed where the aerosol formation and the explosions do not bother any persons. I am certain that these places become more and more scarce these days. The sodium itself remained in the storage tanks and will be used as a catalyst in an industrial process.

3. DISASSEMBLY OF POTASSIUM SYSTEMS

From different experiments we knew that, because of the high reaction rate of potassium with respect to sodium, it would not be possible to get rid of the potassium system in a way similar to the procedures used with the sodium systems. A literature survey did not prove to be very helpful because of ambiguous and contradictory statements, as for example: -Most of the explosions involving potassium have now been attributed to its storage with incomplete exclusion of air, if the layer of dioxide is driven into the underlying metal by dry-metal-cutting or a hammer blow, very violent explosions occur [4].
- Several shock sensitive experiments were carried out including impact tests (with metallic potassium and potassium superoxide), again no explosion was detected. However, explosions have been reported, the worst occurring in Czechoslovakia which resulted in two deaths. If the presence of organic reactants is suspected, then great care should be exercised and remote handling and disposal techniques should be used [5].
-In old samples of potassium metal it seems quite possible that access of moisture to the top of the sample could result in the formation of local pockets of potassium hydroxide (and carbonate) solution, which, when disturbed, entering into contact with the metal may cause an explosion [6].
That all these statements may be part of the truth does not make it easier to find the best way to cope with an obviously complicated and in some instances dangerous procedure. We therefore tried to get the additional counsel of colleagues who were experienced or had to cope with similar problems. After this discussion the following possibilities were taken into consideration:
-Recurrent filling and draining of the loop with a higher alcohol (ethyl carbitol). Controlling the dissolution process by measuring the hydrogen content in the cover gas.
-Purging the loop with inert gas under controlled addition of moisture. Control of hydrogen in the exit stream.
-Vacuum destillation and transfer of the potassium into the dump tank.
-Dissolution of the potassium with liquid ammonia under cryogenic conditions.
The first three proposals looked feasible. Not to make any mistakes at the start of our attempts (which could have had a negative influence on all the following undertakings), we decided to start with an experiment. A complicated but small system was tried to clean with alcohol. Since there were pockets of potassium which could not be purged and filled with fresh solution, the experiment failed: the alcohol reacted explosively with the oxide and reaction products in the internal parts and the flow of alcohol plugged completely. Disassembly of this partially filled loop became a dangerous activity because of burning alcohol and igniting potassium. Moreover, this experiment made clear that one could not be sure that tubes and components could be purged freely, although it was tried to drain the potassium completely out of all systems. Therefore we concluded that we would better not try to purge the system with moist inert gas because of the danger of pockets and incomplete

reaction zones. The problem would be when to decide to start disassembling the system. The same argument holds for vacuum destillation because it seems impossible, even with the best measuring techniques, to decide when the system will be completely cleaned. Since there was no clear recommendation what to do, we decided to start with peripheral parts of the system where no potassium should be expected, hoping to find individual solutions for smaller subsystems which could be valved or cut off and treated independently. This decision turned out to be the beginning of an essential experience:
"learning by doing"

4. FIRST STEPS

The largest system we had to dispose of was a high temperature loop, described in [7] and shown in a simplified flow scheme in Fig. 1.

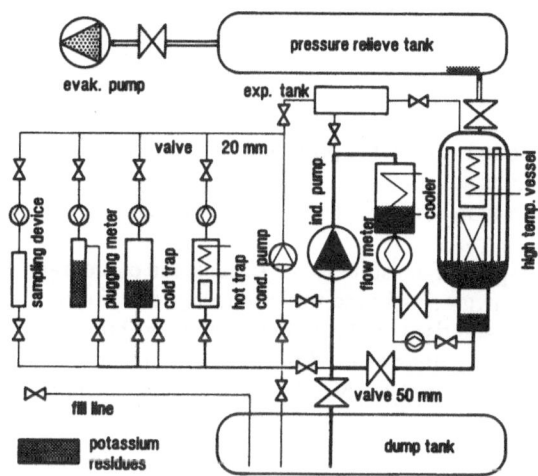

Fig. 1. Schematic flow scheme of potassium experimental facility

After the loop was drained and the potassium completely transferred to the transport barrels, we started to disconnect a pressure relief tank of 3 m³ volume and found that after a small amount of air had leaked into the tank its surface became hot. We purged with argon and could suppress further reactions. Under outflowing argon we inspected the disconnected tube and found yellow reaction products and appreciable aerosol films in the inside of the vessel. Carefully, small parts of the yellow crust were ripped off and thrown to the ground. After about half an hour only small patches of liquid remained of this super oxide; it had reacted and dissolved completely in air under ambient conditions. Therefore it was decided to rearrange the vessel in such a way that the generated liquid could flow freely out of the vessel and that the evolving hydrogen gas could be vented at the highest point. After 24 hours approximately 2 l of liquid were collected and the reaction seemed to have come to an end. Since this procedure worked so well with aerosols on the surface, with the possible contact of vacuum pump oil and the presence of yellow potassium super oxide, we were hopeful that a similar procedure would work with all the other components as well. On the other hand, it was to be feared that a lot of potassium would have to be expected in different parts of the system. In fact, 80 l of potassium were found during the disassembly and had to be disposed of. The decision now was clear to make a step by step solution, letting the potassium react with air under ambient conditions, keeping in mind that some very stringent precautions had to be undertaken. The simple model which was the basis of all further activities is given in Fig. 2.

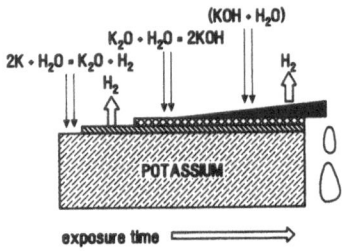

Fig. 2. Simple model of potassium-air-reaction

Fig. 2 shows the reaction of metallic potassium with moist air and the steps which occur until equilibrium conditions are reached:

Potassium reacts with moisture, builds up a surface film of the oxide and releases hydrogen

With the further transport of moisture to the surface, the oxide transforms to hydroxide

The hydroxide is very hygroscopic and forms a hydroxide solution

The equilibrium is characterised by the diffusion of water through the hydroxide layer, the rebuilding of the oxide reaction zone and the release of hydrogen gas through the solution. The kinetics of this process is strongly influenced by the ambient conditions humidity and temperature.

5. TOWARDS A ROUTINE PROCEDURE

Further inspection of the loop showed that we had to cope with the following problems:
- tubes filled with potassium .components with greater potassium residues
- gas tanks and vacuum lines with aerosol layers
- valves, pumps and other devices with inaccessible potassium filled cavities

A survey of where these potassium quantities were found is given in Fig. 1.

For these problem areas the following solutions were realised:

- Simple components like tubes were cut into small pieces and transferred to a catch pan where the sodium was melted out. Valves were heated and drained. Quantities in the order of 1 kg potassium metal were left to react with air in the catch pans under the following precautions, see Fig. 3.

- The catch pans were inclined so that the generated solution could flow away from the metal to a drain hole and into a separate vessel

- Blowers were arranged in such a way that the generated hydrogen was blown away efficiently. This assured that only minimal amounts of hydrogen could form an explosive mixture with ambient oxygen.

- The emptied tubes were left in an inclined position so that air could circulate and the solution could flow away. A further function of the blower is its cooling effect. The exothermic reaction of water and potassium could lead to the melting of the metal and to the disruption of the protective reaction layer.

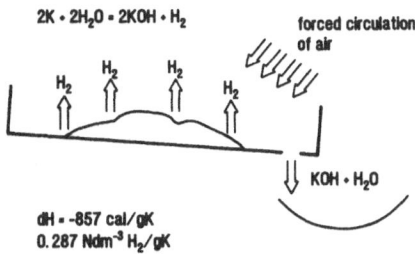

Fig. 3. Leaching of potassium in catch pans

- For some components it was necessary to drain potassium into transport barrels which could be done without greater problems although inert gas connections and drain lines had to be installed provisionally. Depending on geometry, each component then had to be treated individually for which the following measures were taken:
- position the components in such a way that the hydrogen solution may flow to the outside unrestricted
- assure a turbulent air flow in all parts of the interior with strong blowers that no dangerous hydrogen concentration can build up
- the highest point should have a vent hole so that in case of a blower failure the hydrogen can dissipate.
- Components which could not be drained (e.g. cold trap with Raschig-rings, residue container with very impure potassium...) were treated similarly as described above, although greater amounts of potassium had to be reacted. As an example, the actual situation of one of these components is described in Fig. 4.

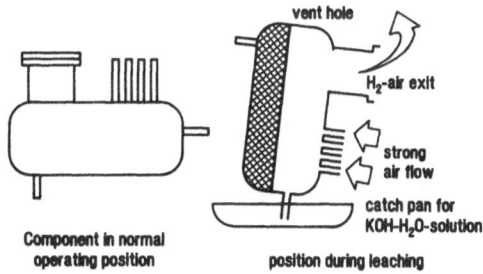

Fig. 4. Cleaning procedure for partially filled components

The cleaning operation for this unit, which contained 12 kg of impure potassium, took place in early spring with low temperatures and low absolute humidity. The procedure lasted for 40 days until reactions were complete. The generated hydroxide solution was used to neutralise the refuse water of the research centre. All the leaching procedures described above were performed under a protective roof in the open air.

- If necessary, greater vessels and complicated structures were cut apart with a plasma torch. This torch proved very effective even if potassium residues cought fire. Our experience shows that within approximately half an hour of initial air exposure of potassium metal, components can be handled without greater difficulties, so that it is possible to complete all necessary measures to ensure a safe leaching process if all is well prepared in advance.

- All components, when reactions had come to a stop, were treated with water. Smaller ones were thrown into the water basin (tubes, valves), the other ones were splashed with a water hose and left in the rain to get a final wash, so that everything, if not needed for other purposes, could finally be thrown to the scrap.

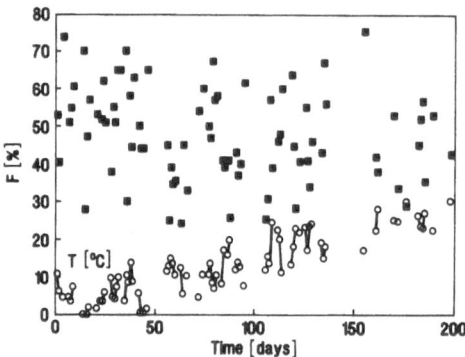

Fig. 5. Ambient climatic conditions during the "disassembly campaign"

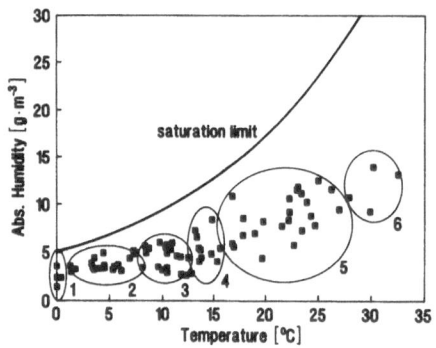

Fig. 6. Qualitative description of experiences gained

Region 1: Almost no generation of hydroxide and solution. White oxides slowly vanish.
Region 2: Small reaction rates
Region 3: Controlled reaction rates and good handling possibilities
Region 4: Hydroxide solution is generated rapidly
Region 5: Danger of solution-metal-explosions
Region 6: Potassium ignites before the hydroxide solution process starts

6. QUALITY DESCRIPTION OF REACTION RATES

During the "disassembly-campaign" ambient conditions were measured on site where the exposures took place. These data are given in Fig. 5. Since our activities lasted intermittently from January to July, a great spectrum of ambient conditions (relative humidity and temperature) was collected:

The data of Fig. 5 are replotted in Fig. 6 to show only the relationship between temperature, absolute humidity and the saturation limit. These data can be grouped together to give a qualitative description of observed reaction rates with respect to temperature and humidity conditions.

Although burning potassium does not give rise to a higher safety risk, the ultimate reaction of the formed oxides and carbonates to hydroxide is dispersed to regions where the aerosol settles down .This may give rise to environmental problems. From this description it follows that the ideal time for a disassembly process, as described above, will be when temperatures are in the order of 5 to 15 °C, with a corresponding relative humidity of approximately 50%. For lower temperatures, a higher humidity does not give rise to an unexpected behaviour; reaction rates remain low. The described method could be applied to sodium systems as well, the difference being that the moisture content of the air should be raised locally by a small steam generator.

7. SUMMARY AND CONCLUSIONS

A method was tested for the disassembly of potassium systems. This method uses moist air to react with potassium. The generated hydrogen is vented off with strong blowers. The generated potassium hydroxide, which is very hygroscopic, transforms into hydroxide solution. This solution has to be drained so that it cannot get into contact with metallic potassium. Since the melting point of potassium is only 63 °C the reaction heat of the oxide formation could melt the potassium at higher ambient temperatures, therefore the blowers have an additional cooling function. For simple components and complicated structures individual solutions had to be realised. This demonstrated that the described procedures work well for the disassembly and the safe disposal of potassium systems.

References

1. Schwarz, N: Sicherheitsexperimente und Brandversuche mit Alkalimetallen, ÖFZS Bericht No. 4208, März 1983.
2. Schwarz, N, Komurka, M: Die Handhabung von Kalium, ÖFZS Bericht No. 4207, März 1983.
3. Schwarz, N, et al.: Kalium/Wasser-Reaktionen in der Dampfphase, ÖFZS Bericht No. 4327, Juli 1985.
4. Bretherick, L: Relation of Lithium and Potassium Explosions, Chem. Brit., 14 (1978) 426.
5. Sloan, S. A.: Partially Oxydized Potassium - An Explosion Hazard?, Chem. Brit., 14 (1978) 597.
6. Davis, A. C.: Potassium Fires, Chem. Brit., 15 (1979) 179.
7. Schwarz, N.: Operating Experience with Potassium Systems, Proc. of the 3rd Int. Conf. on Liquid Metal Engineering and Technology, BNES London, 1984, Vol. 3, p 177-183.

INDEX